# Plants of Central Asia

## Volume 5

# Plants of Central Asia

*Plant Collections from China and Mongolia*

## Volume 5
## Verbenaceae—Scrophulariaceae

V.I. Grubov
L.I. Ivanina
O.V. Tscherneva

CRC Press
Taylor & Francis Group
Boca Raton  London  New York

CRC Press is an imprint of the
Taylor & Francis Group, an **informa** business

A SCIENCE PUBLISHERS BOOK

CRC Press
Taylor & Francis Group
6000 Broken Sound Parkway NW, Suite 300
Boca Raton, FL 33487-2742

© 2002 by Taylor & Francis Group, LLC
CRC Press is an imprint of Taylor & Francis Group, an Informa business

First issued in paperback 2019

No claim to original U.S. Government works

ISBN 13: 978-0-367-44712-0 (pbk)
ISBN 13: 978-1-57808-116-5 (hbk)

**Visit the Taylor & Francis Web site at**
**http://www.taylorandfrancis.com**

**and the CRC Press Web site at**
**http://www.crcpress.com**

# Notes

PLANTS OF CENTRAL ASIA. From the Materials of the V.L. Komarov Botanical Institute, Academy of Sciences of the USSR, Vol. 5. Verbenaceae-Scrophulariaceae. Compilers: V.I. Grubov, L.I. Ivanina and O.V. Tscherneva. 1970. Nauka, Leningrad Division, Leningrad.

This volume is the fifth of illustrated lists of Central Asian plants (within the People's Republics of China and Mongolia) published by the Komarov Botanical Institute, of the Academy of Sciences of the USSR, based on the Central Asian collections of famous Russian travellers and explorers (N.M. Przewalsky, G.N. Potanin and others) as well as of Soviet expeditions and preserved in the Herbarium of the Institute.

The present volume deals with the description of families Verbenaceae, Labiatae, Solanaceae and Scrophulariceae which contain several interesting endemic genera and species that are important for understanding the developmental history of Central Asian flora. Ill. 9 plates and 3 maps.

V.I. Grubov
*Editor-in-Chief*

# Contents

Note                                                              v

Introduction                                                      1

Taxonomy                                                          5

Special Abbreviations                                            5

Abbreviations of Names of Collectors                            5

Abbreviations of Names of Herbaria                              6

Family 104. Verbenaceae Jaume—V.I. Grubov                       7

Family 105. Labiatae Juss.—O.V. Tscherneva                      8

Family 106. Solanaceae Juss.—V.I. Grubov                      108

Family 107. Scrophulariaceae Juss.—(L.I. Ivanina             124

Addenda to Russian Original of vol. 5                         219

Index of Scientific Names of Plants                           232

Index of Distribution Ranges                                  242

# INTRODUCTION[1]

This is the fifth volume of the series Plants of Central Asia and covers the two very large families of order Tubiflorae—Labiatae and Scrophulariaceae—and two smaller Central Asian families Solanaceae and Verbenaceae.

Family Labiatae is represented in the studied territory by 41 genera and 157 species, of which only 1 genus and 22 species are endemic. The Central Asian regions of the former USSR (desert and desert-steppe Kazakhstan, Northern and Central Tien Shan and Eastern Pamir) would add to this list 7 more genera and about 60 species. Family Scrophulariaceae is represented in the studied territory by 18 genera and 152 species, of which 23 species are endemic. Two more genera and about 60 species would add to this list from the Central Asian part of the former USSR.

These two large families, however, are of little florogenetic interest. In Central Asia, they represent plants that have migrated relatively recently, mainly from the Mediterranean and, to a lesser extent, from Eastern Asia *Nepeta, Stachys, Elsholtzia, Phyllophyton, Pedicularis, Lagotis, Oreosolen*). Most species of these families are confined to the mountain borders of our region in the south-east, north-west and specially the west. They are predominantly high-mountain—alpine and hill-steppe—as well as steppe species. Only a few of the species of Labiatae and Scrophulariaceae have contributed to the desert flora of Central Asia, for example *Lagochilus ilicifolius, Schizonepeta annua, Eremostachys molucelloides, Dodartia orientalis*. The endemic species of these families are less indigenous and are essentially neoendemics. The recently established lone endemic Tien Shan monotypic genus *Metastachys*, very poorly differentiated from genus *Stachys*, is no exception. The more typical subendemic East Siberian-Central Asian genus *Cymbaria* with 2 steppe species is clearly of East Asian origin.

In Central Asia we found the so called terminal segments and links of taxonomic chains of Labiatae and Scrophulariaceae extending from outside. There are no indigenous local taxonomic groups—sections, cycles or series. This is also true even for such a large genus as *Pedicularis*.

---

[1]By V.I. Grubov

Distribution of the high-montane monotypic genus *Lancea* should be regarded as a very prominent, but not the only proof of direct floristic links in the past. The major part of the distribution range of *Lancea tibetica* falls in the Himalayas, Tibet and adjoining montane regions of West China (see Map 3) while a small isolated part has been found in the main Hangay mountain range in Northern Mongolia with a disjunction of a thousand kilometres, now falling in the desert expanse of Central Mongolia. Similar instances of distribution are known in other families too, for example among *Megadenia* from Cruciferae and *Mannagettaea* from Orobanchaceae detected by Popov (1954) in the Eastern Sayans and Rhododendron from Ericaceae. Such disjunct distribution ranges represent more a heritage of the cold Pleistocene Age when the alpine species of East Asia may have penetrated uninterruptedly along the high Mongolian plateaus to the north. Popov's view (1954) that *Mannagettaea* and *Megadenia* represented Paleogene sub-tropical relics, in our opinion, is not adequately substantiated. Neither their distribution ranges, confined to the high mountains in the eastern fringes of Tibet (Qinghai and Weitzan), nor the floristic affinities of host plants (*Salix, Caragana, Ribes*) on which all the known species of *Mannagettaea* parasitise favour such a conclusion. The relationships of these genera, if not extended theoretically, fall within the boreal forest and Mediterranean floras.

Family Solanaceae is far more interesting. This very ancient pantropic family is represented in Central Asia by only 9 genera and 21 species of these, 1 genus and 4 species are endemic and 3 more species subendemic. Endemic genus *Przewalskia* (see Map 2), surprisingly, is characteristic of the most younger Tibetan province whose flora may have formed only after the Glacial period (see Introduction, Vol. 1). This genus is undoubtedly close to genus *Scopolia* but is well distinguished by its unique characteristics and presence of major distinctive features. It was first established by C.I. Maximowicz as a monotypic genus but it is now clear that it comprises 2 distinct species. These are rather small rosetted plants with narrowly tubular axillary flowers and typical swollen fruit in which the accrescent inflated calyx forms an outer coarse membrane for the small pod, dehiscent with an operculum as in *Scopolia*. Such is the fruit structure in subendemic eastern Tibetan species *Scopolia tangutica* but the calyx accrescent in fruit, unlike in *Przewalskia*, is open, cup-shaped, its fringe closed at the top with folds only at the end, as the fruit ripens. The seeds of *Scopolia tangutica* and both species of *Przewalskia* are so similar that they could hardly be differentiated from external characteristics. *Scopolia tangutica* has a large branched stem but the early phase of growth of this species in spring resembles a rosette—a shaft with developed leaves emerges from the soil all at once. Among the eastern Tibetan species, *Przewalskia tangutica* has only a short underground stem and an ovate calyx nearly closed on top but a surface stem is sometimes seen in southern Tibetan *P. shebbearei* but the calyx in fruits is open at the top, with large teeth. The thought arises spontaneously that *Przewalskia* may have developed as a branch of the Tertiary genus *Scopolia* in the course of evolution adapted to the severe conditions of high mountains through neoteny. Since this genus is undoubtedly young, such a high tempo of evolution can only be explained in this manner. Moreover, in the case of *P. shebbearei*, this process is not complete—the surface stem is not yet fully reduced and the accrescent in calyx in fruit is not completely closed above, resembling its form in *Scopolia tangutica* (see Plates IV and V). The latter is a forest and high-montane species of Qinghai and the south-eastern border of Tibet; on the contrary, compared to its closely related Chinese forest species *S. sinensis* Hemsl. it represents somewhat the first step towards such a process of evolution.

Another interesting endemic of the same type was the new species of genus *Mandragora* described here from eastern Tibet—Weitzan. According to established concepts, this ancient Mediterranean genus was known in East Asia from the single high-montane species *M. caulescens* distributed in Kam, Eastern Himalayas and in eastern Southern Tibet. Unlike the Mediterranean large-leaved rosetted species, *M. caulescens* has a surface stem, small leaves and broadly campanulate nutant flowers. Our new, proper Tibetan species *M. tibetica* undoubtedly represents the closest relative of Kam-Himalayan species (it has in particular the same form of leaves and scales with the same characteristic ciliate pubescence as in *M. caulescens*) but grows in even more severe dry and cold high uplands of eastern Tibet and is a small densely rosetted plant. Thus, even in this case, evolution evidently proceeded along the line of inherited juvenile stage of the vegetative phase.

The large (with more than 100 species) extratropical genus *Lycium* in the Old World is confined mainly to southern Africa and the Mediterranean and is represented in our territory only by 6 species, of which 1 is endemic and 2 subendemic. The endemic Junggar-Tien Shan *L. flexicaule* and subendemic Mongolian *L. potaninii*, according to A.I. Pojarkova (1950), belong to a series of eastern Asian species while the subendemic Central Asian *L. truncatum* belongs to a series of eastern Mediterranen species. Thus we find here rare instances of 2 branches of an old genus arriving from opposite directions taking root in the Central Asian flora, although it must be conceded that all of these species are very closely related. Unlike the above erythrocarpous species common in Central Asia, melanocarpous *L. ruthenicum* is an extensively distributed Mediterranean desert species.

In our area, there are only 2 species from the large pantropic family Verbenaceae; they provide very striking proof of the direct penetration into Central Asia of the typical eastern Asian warm-temperate forest genus *Caryopteris* (some ten species distributed in China, Korea and Japan). While

*C. tangutica* enters only from the east into the hilly forest Qinghai belt preserving the typical mesophytic features of the genus (tall shrub with broad dentate leaves), its closest relative *C. mongholica* enters the composition of desert-steppe and desert flora of Central Asia to a limited extent over a vast distribution range and is a subendemic. It acquired a totally xerophytic habit—low subshrub with densely pubescent entire, narrowly lanceolate leaves. Its affinity with *C. tangutica* is revealed not only by its flowers, but also by var. *serrata* with interrupted dentate leaves found from time to time along the southern fringe of its distribution range (see Map 1). This species evidently penetrated into Central Asia from the eastern boundary where it encountered no orographic barriers, as early as in the Tertiary period and became a total native long ago.

In this volume, the maps of distribution ranges of species of families Labiatae and Scrophulariaceae were drawn by their respective authors and of families Verbenaceae and Solanaceae by O.I. Starikova, senior laboratory assistant. She also translated all Chinese texts from labels of herbarium specimens and from floristic literature. Further, she rendered much assistance to the authors in drawing up distribution ranges and references to the geographic distribution of species for which the authors express their sincere gratitude. Plates of plant drawings were prepared by artists G.M. Aduevska (plate IV) and T.N. Shishlova.

# TAXONOMY

## SPECIAL ABBREVIATIONS

### Abbreviations of Names of Collectors

| | | |
|---|---|---|
| A. Reg. | — | A. Regel |
| Bar. | — | V.I. Baranov |
| Chaff. | — | J. Chaffanjon |
| Chaney | — | R.W. Chaney |
| Ching | — | R.C. Ching |
| Chu | — | C.N. Chu |
| Czet. | — | S.S. Czetyrkin |
| Divn. | — | D.A. Divnogorskaya |
| Fet. | — | A.M. Fetisov |
| Glag. | — | S.A. Glagolev |
| Gr.-Grzh. | — | G.E. Grum-Grzhimailo |
| Grub. | — | V.I. Grubov |
| Gus. | — | V.A. Gusev |
| Ik.-Gal. | — | N.P. Ikonnikov-Galitzkij |
| Ivan. | — | A.F. Ivanov |
| Kal. | — | A.V. Kalinina |
| Kashk. | — | V.A. Kashkarov |
| Klem. | — | E.N. Klements |
| Kondr. | — | S.A. Kondrat'ev |
| Krasch. | — | I.M. Krascheninnikov |
| Kryl. | — | P.N. Krylov |
| Kuan | — | K.C. Kuan |
| Lad. | — | V.F. Ladygin |
| Ladyzh. | — | M.V. Ladyzhensky |
| Lee, Lee and Chu, Lee et al. | — | A.R. Lee |
| Lis. | — | V.I. Lisovsky |
| Li S.H. et al. | — | S.H. Li et al. |
| Litw. | — | D.I. Litwinow |
| Lom. | — | A.M. Lomonosov |
| Merzb. | — | G. Merzbacher |
| Mois. | — | V.S. Moiseenko |

| Nov. | — | V.F. Novitski |
|------|---|---------------|
| Pal. | — | I.V. Palibin |
| Pavl. | — | N.V. Pavlov |
| Petr. | — | M.P. Petrov |
| Pev. | — | M.V. Pevtsov |
| Pias. | — | P.Ya. Piassezki |
| Pob. | — | E.G. Pobedimova |
| Pop. | — | M.G. Popov |
| Pot. | — | G.N. Potanin |
| Przew. | — | N.M. Przewalsky |
| Rhins | — | J.L. Dutreuil de Rhins |
| Rob. | — | V.I. Roborowsky |
| Sap. | — | V.V. Sapozhnikov |
| Schischk. | — | B.K. Schischkin |
| Serp. | — | V.M. Serpukhov |
| Shum. | — | E.M. Shumakov |
| Sold. | — | V.V. Soldatov |
| Tug. | — | A.Ya. Tugarinov |
| Wang | — | K.C. Wang |
| Yun. | — | A.A. Yunatov |
| Zab. | — | D.K. Zabolotnyi |
| Zam. | — | B.M. Zamatkinov |

## Abbreviations of Names of Herbaria

| BM | — | British Museum of Natural History, Great Britain, London. |
|----|---|-----------|
| E | — | Royal Botanic Garden, Edinburgh, Scotland, Great Britain. |
| K | — | The Herbarium, Royal Botanic Gardens, Great Britain, Kew, Surrey. |
| KYO | — | Department of Botany, Faculty of Science, Kyoto University, Sakyo-ku, Kansai, Kyoto, Japan. |
| Linn. | — | The Linnean Society of London, Great Britain, London. |
| PE | — | Institute of Botany, Academia Sinica, Peking (Beijing), China. |
| TAK | — | Tashkent State University. |
| TI | — | Botanical Institute, Faculty of Science, University of Tokyo, Hongo, Tokyo, Japan. |
| TK | — | P.N. Krylov Herbarium, Tomsk State University. |
| TO | — | Herbarium of the Instituto Botanico dell' Universita, Torino, Italy. |

# Family 104. VERBENACEAE Jaume[1]

## 1. Caryopteris Bunge.

Pl. mongh.-sinens. (1835) 27; Maxim. in Bull. Ac. Sci. St.-Pétersb. 23 (1876) 389, 31 (1886) 87; Pei in Mem. Sci. Soc. China, 1, 3 (1932) 162.

1. Leaves linear lanceolate, entire, rarely with a few interrupted teeth (var. *serrata* Maxim.), greyish-green above, light-grey beneath. Subshrub up to 0.5 m tall ........................ 1. **C. mongholica** Bunge.
+ Leaves lanceolate to broadly elliptical, roughly serrate-dentate, sharply dichromatic, dark green above, ash-grey beneath. Subshrub up to 1.5 m tall ............................................. 2. **C. tangutica** Maxim.

1. **C. mongholica** Bunge, Pl. mongh.-sinens. (1835) 28; Maxim. in Bull. Ac. Sci. St.-Pétersb. 23 (1876) 389, 31 (1886) 87; Franch. Pl. David. 1 (1884) 231; Forbes and Hemsley, Index Fl. Sin. 2 (1890) 264; Pei Ch. in Mem. Sci. Soc. China, 1, 3 (1932) 165; Walker in Contribs U.S. Nat. Herb. 28 (1941) 655; Grubov Konsp. fl. MNR (1955) 233; Chen and Chou, Rast. pokrov r. Sulekhe (1957) 89. —Ic.: Curtis's Bot.-Mag. 158, tab. 2916; Rev. hortic. (1872) 451.

Described from East. Mongolia. Type in Paris. Isotype in Leningrad. Map 1.

Steppes and desert rocky and rubble slopes of hills and mountains, rocks, flanks and sandy-pebble floors of gorges, river shoals, fine sands.

IA. **Mongolia:** *Cent. Khalkha, East. Mong.* (in montosis lapidosis Mongolia chinensis [prope Chaschatu], 1831—Bunge, typus; far east. record: near Khalun arshan south of Gan'chzhur, 1899—Pal.). *Val. Lakes* (far west. record: left bank of Tuin-Gol river). *Gobi-Alt., East. Gobi, Alash. Gobi* (nor. and east.). *Ordos, Khesi.*

III A. **Qinghai:** *Nanshan* (alpine belt of west. Nanshan near Kuku-Usu river, July 22, 1879; Khagomi area, July 10, 1880—Przew.; Sanchuan, mountains west of Dzhamba river, March 14, 1885, Pot.—far southern find of species; Humboldt mountain range, southern slope, Magyn-Dybsyn area in Chan-sai ravine, 2400–2700 m, along brook, July 24, 1895, Rob.—far western record of species; Loukhushan' mountain range, southern slope, July 17, 1908—Czet.; 5 km south of Aksai settlement on high northern foothills of Altyntag, Aug. 2, 1958—Petr.).

General distribution: East Siberia (river Selenga valley near the state border)* Nor. Mong. (Hent. west., Hang. east., Mong.-Daur. west.; far northern find of species—lower Iro river valley), China (Nor.: Pohuashan hills west of Peking (Beijing)—far eastern record; Nor.-West.— far nor.-west. Shanxi).

2. **C. tangutica** Maxim. in Bull. Ac. Sci. St. Pétersb. 27 (1881) 525, 31 (1886) 87; Forbes and Hemsley, Index Fl. Sin. 2 (1890) 265; Danguy in Bull. Mus. nat. hist. natur. 17 (1911) 344; Pei Ch. in Mem. Sci. Soc. China, 1, 3 (1932) 172; Hao in Engler's Bot. Jahrb. 68 (1938) 633; Walker in Contribs U.S. Nat. Herb. 28 (1941) 655. —*C. tangutica* var. *brachyodonta* Hand.-Mazz. in Acta Horti Gotoburg. 13 (1939) 336.

---

[1]By V.I. Grubov.
*Addition to the original for 2001.—V. I. Grubov.

Described from Nanshan. Type in Leningrad.

Sunny open rocky slopes of mountains and on rocks, from 2000 to 2400 m altitude.

IIIA. Qinghai: *Nanshan* (South Tetungsk mountain range, bald slope in lower belt, Aug. 10, 1872; North Tetungsk mountain range, 2300 m, rocks, Aug. 11, 1884—Przew., typus!; Chortenton temple, sunny hillside, 2100 m, clayey-rocky soil, Sept. 7; same site, southern slope, 2400 m, Sept. 8—1901, Lad.; South Tetung mountain range, middle zone on southern slope, rock crevices, July 28, 1908—Czet.; "Kan-Tsao-Tien, alt. 2000 m, Aug. 3, 1908, Vaillant"—Danguy, l.c.; "Kokonor: Min-ho-hsien, um 1900 m"—Hao, l.c.; "Tien-Tang-Ssu, on an exposed, gravelly river bank, Ching."—Walker, l.c.).

General distribution: China (Nor.-West., Cent.—Hubei, South-West.—Sichuan).

## Family 105. LABIATAE Juss.[1]

1. Style attached above base of ovary lobes; nut with lateroventral attachment and large scar, often covering 1/2 height of nut; corolla uni- or bilabiate but invariably with poorly developed upper lip.. ................................................................................................ 2.

+ Style attached at base of ovary lobes; nut with basal, rarely with basodorsal or ventral attachment and minute scar; corolla bilabiate or subregular ........................................................................ 4.

2. Stamens 4, well developed; corolla unilabiate. ............................ 3.

+ Stamens 2, 2 upper (posterior) ones reduced; corolla bilabiate .... 3. **Amethystea** L. (*A. coerulea* L.).

3. Flowers large, 16–20 mm long; upper lip poorly developed, bipartite or 2-lobed; lower lip considerably longer, with highly developed midlobe ................................ 1. **Ajuga** L. (*A. lupulina* Maxim.).

+ Flowers small, 5–8 mm long; upper lip totally lacking, all 5 lobes of corolla recurved, exposing stamens and style ....2. **Teucrium** L. (*T. scordioides* Schreb.).

4. Calyx bilabiate with short entire lips; upper lip usually with orbicular appendage, less commonly without appendage but then with small umbo, shedding at time of seed maturation; lower lip usually persistent on plant; upper lip of corolla 3-lobed ................ ........................................................................ 4. **Scutellaria** L.

+ Upper lip of calyx invariably without appendage; upper lip of corolla not usually 3-lobed. ................................................................ 5.

5. Stamens enclosed in tube of corolla or exserted, ascending upward or directed forward. ........................................................................ 6.

+ Stamens exserted, pendent as though resting on slightly arched 1-lobed entire lower lip of corolla ........................................................ 40.

6. Stamens enclosed in corolla tube ................................................ 7.

---

[1]By O.V. Tscherneva.

+ All stamens, or at least 2 longer ones, exserted from corolla tube .. ................................................................................................................. 10.

7. Corolla tube with well-developed hairy ring inside; filaments villous; leaves sagittate at base............................................................ ........................ 27. **Metastachys** Knorr. [*M. sagittata* (Regel) Knorr.].

+ Corolla tube with uneven and poorly developed hairy ring inside or without it. Filaments glabrous; leaves not sagittate at base. ..... 8.

8. Calyx with 10 divergent subulate teeth; corolla tube with irregularly and poorly developed hairy ring inside; upper lip of corolla bifid at tip .................................... 5. **Marrubium** L. (*M. vulgare* L.).

+ Calyx with 5 teeth; corolla tube without hairy ring inside; upper lip of corolla entire or slightly emarginate ............................................. 9.

9. Perennial; upper lip of corolla entire; leaves palmatisect; stem and petiole densely villous. ..................................... 6. **Lagopsis** Bunge.

+ Annual; upper lip of corolla slightly emarginate; leaves undivided, entire or indistinctly serrate; stem with finely crispate pubescence, petiole ciliate ......... 33. **Antonina** Vved. [*A. debilis* (Bunge) Vved.].

10 (6). Anther lobes divaricate, orbicular, joined at tip; arranged on single plane as pollen scatter. ......................................39. **Elsholtzia** Willd.

+ Anther lobes divaricate, divergent or parallel, oblong or ovate, not joined or indistinctly joined at tip into single lobe; not arranged on single plane as pollen scatter. ........................................................ 11.

11. Connectives of 2 anterior (lower) stamens linear filiform, jointed with anther filaments freely or fixedly, with regularly developed lobe on postenor end and some modified (reduced) ones on anterior end ............................................................................ 30. **Salvia** L.

+ Stamens differently arranged. ........................................................... 12.

12. Corolla and calyx subregular.............................................................. 13.

+ Corolla invariably distinctly bilabiate; calyx either bilabiate or subregular ................................................................................................ 14.

13. Only 2 lower (anterior) stamens developed; upper (posterior) underdeveloped, transformed into filiform staminodes, without anthers; nuts trigonous, glandular at tip....................... 37. **Lycopus** L.

+ All stamens similar, developed; nuts ovate, obtuse, without glands, sometimes with hairs at tip. ......................................... 38. **Mentha** L.

14. 2 fertile stamens with 2 approximate parallel, oblong-linear, slightly curved lobes disposed on somewhat enlarged connective and overhanging from it. ...................................... 31. **Perovskia** Karel.

+ Stamens 4, all fertile; if 2, connective not enlarged and anther lobes not linear or overhanging ...............................................................15.

15. Corolla with lips of unequal length, upper lip usually concave or convex, pubescent outside, rarely nearly flat but then either leaves

end in spine or calyx with oblique or indistinctly bilabiate limb....
...................................................................................................... 16.

+ Lips of corolla poorly differentiated, upper lip invariably nearly flat, leaves invariably without spine, calyx either subregular or (in *Thymus* L.) bilabiate ........................................................................ 37.

16. Upper stamens longer than lower, latter sometimes not developed.
...................................................................................................... 17.

+ Upper stamens shorter than lower ..................................................... 24.

17. Flowers somewhat inverted due to contortion of corolla tube; upper lip of corolla under which, in its normal position, lie stamens and style, occupies place of lower lip and vice versa; calyx with oblique limb. .......................................... 8. **Lophanthus** Adans.

+ Flowers not 'inverted'; if inverted (some species of *Phyllophyton* Kudo), calyx distinctly bilabiate ............................................ 18.

18. Two pairs of stamens not parallel to each other. ........................... 19.

+ Two pairs of stamens parallel, arcuately ascending under upper lip. ...................................................................................................... 20.

19. Upper stamens turned downwards, lower ascending; lobes of disc barely perceptible; midlobe of lower lip of corolla without unguiform taper ....................................................................................
..7. **Agastache** Clayt. ex Gronov. [*A. rugosa* (Fisch. et Mey.) Kuntze].

+ Upper stamens ascending, lower directed forwards; lobes of disc well developed; midlobe of lower lip of corolla with unguiform taper toward base. ........................................... 10. **Schizonepeta** Briq.

20. Upper lip of corolla with exserted fold inside; pedicel flattened ....
.............. 14. **Lallemantia** Fisch. et Mey [*L. royleana* (Benth.) Benth.].

+ Upper lip of corolla glabrous inside; pedicel not flattened .........21.

21. Calyx more or less tubular, with 5 teeth, with oblique limb, more rarely bilabiate, but invariably without nodules in corners between teeth ........................................................................................................22.

+ Calyx distinctly bilabiate with nodules in corners between all or some teeth. .................................................... 13. **Dracocephalum** L.

22. Calyx with oblique limb; anther lobes diverging at 180°, leaves longer than broad; leafy bracts differing from cauline leaves. ........
...................................................................................... 11. **Nepeta** L.

+ Calyx bilabiate; anther lobes diverging at right angle or parallel; leaves more or less orbicular or reniform, broader than long; leafy bracts similar to cauline leaves ............................................................23.

23. Anthers with perpendicular lobes, approximate such that they are cruciate; calyx without hairy ring inside ...................................... 12.
Glechoma L. (*G. hederacea* L.).

+ Anthers with parallel lobes; calyx with hairy ring inside ...............
.................................................................................9. **Phyllophyton** Kudo.

24. (16). Calyx bilabiate; its lower lip appressed to upper concealing mouth of calyx after anthesis ........................ 15. **Prunella** L. (*P. vulgaris* L.).
 + Calyx not bilabiate .............................................................................. 25.
25. Upper lip of corolla concave or galeate, more rarely convex or flat but invariably highly pubescent outside. ..................................... 26.
 + Upper lip of corolla flat, glabrous or very weakly pubescent. ........
 ..................... 29. **Chamaesphacos** Schrenk (*Ch. ilicifolius* Schrenk).
26. Acaulous plant with highly rugose leaves forming rosette .............
 ............................ 17. **Lamiophlomis** Kudo [*L. rotata* (Benth.) Kudo].
 + Plant with well-developed stem, foliated throughout length. ... 27.
27. Lobes of style unequal; posterior lobe much shorter than anterior; less commonly, similar, but then filaments of all or only upper stamens with appendage at base .......................................................... 28.
 + Lobes of style identical or subidentical; filaments without appendage at base. .................................................................................... 29.
28. Appendages of upper filaments spur-like, glabrous, or lacking; calyx teeth repandous, with long cusp (awn) emerging from notches or orbicular-ovate, with subulate cusp at tip; leaves invariably undivided ........................................................................... 18. **Phlomis** L.
 + Appendages of upper filaments fimbriate; teeth of calyx deltoid, shortly subulate; leaves pinnatisect, more rarely undivided. ..........
 ........................................................... 16. **Eremostachys** Bunge.
29. Lower lip of corolla with 2 hollow cornet-like appendages anthers with transversely dehiscent lobes ...................................................
 ..................................... 21. **Galeopsis** L. (*G. bifida* Boenn.).
 + Lower lip without appendages; anthers with longitudinally opening lobes ...................................................................................... 30.
30. Nuts (less distinctly even ovary lobes) acutely trigonous, obtuse at tip. ........................................................................................... 31.
 + Nuts (and ovary lobes) ovate, more or less rounded at tip .............
 ........................................................................ 28. **Stachys** L.
31. Upper part of stem densely, imbricately foliated, villous-lanate, broadly rhomboid or orbicular leaves white-villous on both surfaces ............................ 19. **Eriophyton** Benth. (*E. wallichii* Benth.).
 + Plants of different habit ................................................................ 32.
32. Corolla more or less with elongated tube considerably exserted from calyx; calyx teeth quite soft, not subulate or prickly; lateral lobes of lower lip of corolla small, with 1 or more acute teeth along margin
 ........................................................................ 22. **Lamium** L.
 + Corolla with tube enclosed in calyx or slightly exserted; calyx teeth more or less subulate or prickly ..................................................... 33.

33. Subshrubs with spines (modified bracts) at base of whorls (sometimes spines present in leaf axils as well); upper lip of corolla bipartite or emarginate at tip .................................. 26. **Lagochilus** Bunge.

+ Herbs and subshrubs, without spines in whorls and leaf axils; upper lip of corolla undivided ................................................................ 34.

34. Anther lobes strongly divergent; leaves ovate, undivided ......... 35.

+ Anther lobes parallel; leaves of different form, deeply split. ...... 36.

35. Flowers small, 5–7 mm long; corolla not longer or insignificantly longer than calyx, tube without hairy ring inside; nut with short, erect hairs at tip ..............................................................
.................. 23. **Chaiturus** Willd. [*Ch. marrubiastrum* (L.) Spenn].

+ Flowers markedly large, 1–1.5 (2) cm long; corolla slightly longer than calyx, tube with hairy ring inside; nut glabrous ......................
.......................................... 20. **Stachyopsis** M. Pop. et Vved.

36. Corolla pink, tube enlarged upward, with hairy ring inside and upper lip faintly concave, more or less narrowed at base; calyx funnel-shaped, not longer than 9 mm, with 5 nerves. ....................................
............................................................................... 24. **Leonurus** L.

+ Corolla yellow, tube narrow, without hairy ring inside, upper lip galeate, not narrowed at base; calyx tubular-campanulate, very large, 13–18 mm long, with 10 nerves (5 of them less distinct .......
.............................................................. 25. **Panzeria** Moench.

37 (15). Only 2 lower stamens fertile, upper transformed into staminodes or lacking .................................................................... 32. **Ziziphora** L.

+ Stamens 4, normally developed ......................................... 38.

38. Calyx regular or subregular ................................................ 39.

+ Calyx bilabiate ................................................. 36. **Thymus** L.

39. Calyx with 15 nerves, glabrous in throat; nuts glabrous or pubescent at tip ................................................................ 34. **Hyssopus** L.

+ Calyx with 10–13 nerves, with hairy ring in throat; nuts finely glandular at tip. ..................................... 35. **Origanum** L. (*O. vulgare* L.).

40 (5). Annual; calyx hairy in throat, 5-toothed, upper tooth of calyx in fruit broader than rest, membranous, broadly ovate, decurrent on calyx tube, its fringes concealing 2 adjoining teeth; filaments of upper stamens often with appendages in the form of hairy tuft .........
...................................................... 41. **Ocimum** L. (*O. basilicum* L.).

+ Small shrub; calyx without hairs in throat, bilabiate, slightly inflated in fruit, with unmodified teeth; filaments without appendages .....
.................. 40. **Isodon** (Schrad.) Kudo [*I. pharicus* (Prain) Murata].

## 1. Ajuga L.
Sp. pl. (1753) 561; Maxim. in Bull. Ac. Sci. St.-Pétersb. 29 (1883) 180.

1. **A. lupulina** Maxim. in Bull. Ac. Sci. St.-Pétersb. 23 (1877) 391; Hance in J. Bot. (London) 16 (1878) 111; Forbes and Hemsley, Index Fl. Sin. 2 (1902)

315; Danguy in Bull. Mus. nat. hist. natur. 17 (1911) 346; Dunn in Notes Bot. Gard. Edinburgh, 6 (1915) 194; Paulsen in Hedin, S. Tibet, 6, 3 (1922) 45; Kudo in Mém. Fac. Sci. and Agr. Taihoku Univ. 2 (1929) 286; Rehder and Kobuski in J. Arn. Arb. 14 (1933) 30; Walker in Contribs U.S. Nat. Herb. 28 (1941) 655. —*A. lupulina* Maxim. f. *humilis* Sun in Acta Phytotax. Sin. 11, 1 (1966) 36. —Ic.: Maxim, l.c. tab. 3, figs. 10–15.

Described from Qinghai (Tetung river basin). Type in Leningrad.

Sandy banks of rivers, clayey-rocky soils mountain slopes as well as humus-rich soils in alpine belt of mountains.

IIIA. Qinghai: *Nanshan* (in mountains south of Tetung river, along mountain slopes and southern ravines, July 13, 1872, typus!; South Kukunor mountain range alpine zone, 3150–3500 m, June 7, 1880—Przew.; South Kukunor mountain range, both slopes, clayey-rocky soil, more rarely wet humus soil, 3600 m, Aug. 7, 1901—Lad.; Kukunor lake, Uiyu area, humus soil, Aug. 13, 1908—Czet.; Ganshiga river valley, left tributary of Peishikhe river [discharging in Tetung in area of a stud farm], 3350–3720 m, 1958—Petr.; "Jong-Ngam, alt. 3200 m, July 7, 1908, Vaillant"—Danguy, l.c.; "Ta P'an Shan, moist, grassy slopes, very common, Ching"—Walker, l.c.). *Amdo* (alpine zone of Mudzhik range, June 18, 1880—Przew.).

IIIB. Tibet: *Chang Tang* ("between Naktsong-tso and Selling-tso, 4636 m, Sept. 11, 1901, Hedin"—Hedin, l.c.), *Weitzan* (along Yantszytszyan river, 3900 m, along sandy banks of river, June 23, 1884—Przew.; Yantszytszyan basin, Chzhabuvrun area, 4200 m high, on wet humus and wet clay, July 10, 1900—Lad.; "Radja and Yellow River gorges, alt. 3050 m, No. 14129, June, 1926 [typus *A. lupulina* f. *humilis*]; Jupar Range, No. 14373, Rock"—Rehder and Kobuski, l.c.).

General distribution: China (North, North-West, South-West).

## 2. Teucrium L.
Sp. pl. (1753) 562.

1. T. scordioides Schreb. Fl. Vert. Unilab. (1774) 37; Juzepczuk in Fl. SSSR, 20 (1954) 50; Fl. Kirgiz. 9 (1960) 10; Fl. Kazakhst. 7 (1964) 297. —Ic.: Fl. Uzbek. 5, Plate 25, fig. 2; Fl. Kazakhst. 7, Plate 35, fig. 2.

Described from Crete island. Type in Geneva (?).

Wet, sometimes solonetz meadows, marshy sites, sandy and rocky banks of rivers, brooks, lakes and riverine floodplains.

IB. Kashgar: *Nor.* (Kurlya, along water front, No. 5885, July 14, 1958—A.R. Lee (1959)). *East.* (nor. boundary of Turfan valley, north-west of Toksun, Pacha-salgan picket, bog pool, Sept. 3, 1929—Pop.).

IIA. Junggar: *Jung. Gobi* (Dutai area, between Shikho and Chipeitsza, wet meadows, Aug. 3, 1947—Shum.; 3–4 km nor.-west of Kuitun settlement on old Shikho-Manas road, sasa zone, herbage-sedge swampy meadow, July 6, 1957—Yun. et al.; 2 km north of Kuitun settlement in cattle-breeding farm region, swampy meadow, No. 395, July 6, 1957—Kuan). *Tien Shan* (Savan dist., from Paotai to Shaomyn'tsz, along Manas river, on wet site, No. 1598, June 29, 1957—Kuan). *Dzhark.* (along Ili river near Kul'dzha, May 26, 1877—A. Reg.).

General distribution: Fore Balkh., Jung.-Tarb.; Europe, Mediterr., Balk.-Asia Minor, Fore Asia, Caucasus, Middle Asia, West. Siberia (south).

14

Note. In Chen and Chou's "Rast. pokrov r. Sulekhe" (1957), *Teucrium* sp. was cited for Khesi region. Without plant material, it is difficult to comprehend to which species of the genus this reference pertains. We have not seen any material of this genus from Khesi. region.

### 3. Amethystea L.
Sp. pl. (1753) 21.

1. **A. coerulea** L. Sp. pl. (1753) 21; Bunge in Ledeb. Fl. alt. 1 (1830) 19; Turcz. in Bull. Soc. natur. Moscou, 25 (1852) 416 (Fl. Baic.-dah. 2, 2); Franch. Pl. David. 1 (1884) 245; Diels, Fl. C. China (1901) 552; Forbes and Hemsley, Index Fl. Sin. 2 (1902) 310; Dunn in Notes Bot. Gard. Edinburgh, 6 (1915) 190; Kudo in Mém. Fac. Sci. and Agr. Taihoku Univ. 2 (1929) 302; Krylov, Fl. Zap. Sib. 9 (1937) 2294; Kitag. Lin. Fl. Mansh. (1939) 377; Volkova in Fl. SSSR, 20 (1954) 70; Grubov, Konsp. fl. MNR (1955) 233; Fl. Kazakhst. 7 (1964) 299; Dashnyam in Bot. zh. 50 (1965) 1641. —Ic.: Curtis's Bot. Mag. tab. 2448; Fl. Kazakhst. 7. Plate 35, fig. 4.

Described from Siberia. Type in London (Linn.).

On sandy-pebble floors of gorges, shoals and banks of rivers, rubble and rocky slopes of mountains talus and rocks.

IA. **Mongolia:** *Cen. Khalkha* (upper Kerulen, near foot of Bain-Erkhit, mountain range, in steppe, 1899—Pal.; near Ikhe-Tukhum-Nor lake, on way to Mishikgun, June 1923—Zam.; near foot of Bichikte, dry river-bed and close to it, loam and pebble, Aug. 31, 1925—Gus.; Ubur-Dzhirgalante river, between sources and Agit mountain, rubble bed of dry streams, Sept. 2, 1925—Krasch. and Zam.; Bichikte-Dulan-Khada ridge, dry bed, Aug. 28, 1926—Glag.; Utat, Aug. 8, 1927—Terekhovko). *East. Mong.* (lower Kerulen, in Mergen-Khamar gorge, 1899—Pal.; Shilin-Khoto town, steppe, 1954—Ivan.). *Val. Lakes* (rocky southern slopes of dry desert mountains above Tuin-Gol river, Sept. 1, 1924—Pavl.). *Gobi-Alt.* (Dundu-Saikhan mountain ranges, southern slope, humus soil in midbelt of mountains, July 13, 1909— Czet.; Bain-Tsagan hills, rubble places on south. slope, Aug. 4; same site, rubble slope, Sept. 16—1931, Ik.-Gal.). *East. Gobi* (rocky slope of Del'ger-Khangai mountain range, July 30, 1931—Ik.-Gal.).

General distribution: Jung.-Tarb.; Fore Asia, West. Sib. (Altay), East. Sib. (south), Far East (south), Nor. Mong. (Hent., Hang., Mong.-Daur.), China (Dunbei, North, North-West, South-West), Korean peninsula, Japan.

### 4. Scutellaria L.
Sp. p.l. (1753) 598.

1. Upper lip of calyx on back with arcuately concave appendage rounded at tip ................................................................. 2.
+ Upper lip of calyx with only small umbo on back. .............................
...................................................................... 6. **S. kingiana** Prain.
2. Flowers in axils of ordinary leaves, gradually decreasing in size or aggregated into unilateral sided racemose inflorescence with leafy bracts, quite similar to caulous leaves but somewhat smaller. ...... 3.

+ Inflorescence not unilateral leafy bracts sharply different from caulous leaves, more or less membranous ........................................ 7.

3. Flowers in racemose inflorescence, leafy bracts similar to caulous leaves but smaller ................................................................... 4.

+ Flowers in axils of ordinary leaves, gradually decreasing in size ... 5.

4. Corolla blue; leaves dense, subcoriaceous, glabrous or very insignificantly pubescent on upper surface, pitted-punctate beneath; nuts small, up to 1 mm long. .............................. 3. **S. baicalensis** Georgi.

+ Corolla yellow; leaves not coriaceous, pubescent and grainy yellow on both surfaces; nuts very large; about 2 mm long ................. ........................................................................ 13. **S. viscidula** Bunge.

5. Leaves pitted-punctate beneath .......................................................... 6.

+ Leaves not pitted-punctate beneath ................... 4. **S. galericulata** L.

6. Leaves small, 0.3–1.4 cm long; internodes elongated, leaves shorter than internodes; nuts very finely tuberculate ..................................... ....................................................................... 1. **S. alaschanica** Tschern.

+ Leaves very large, 1–3.5 cm long; internodes shortened, leaves longer than internodes; nuts large and papillate-tuberculate ......... ......................................................... 10. **S. scordiifolia** Fisch. ex Schrenk.

7. General colour of flowers pink-violet or blue, very rarely grainy yellow; leafy bracts ovate-lanceolate or lanceolate, 0.5–1 cm long and 1.5–4 mm broad, concave, carinate, narrowed at base, green or violet; nuts finely tuberculate, black ............ 5. **S. grandiflora** Sims.

+ General colour of flowers yellow, sometimes with greenish or purple-violet spots on upper or lower lip of corolla; leafy bracts ovate or ovate-lanceolate, usually rather flat, more often membranous, light green or purple; nuts finely tuberculate, densely covered with short stellate hairs, grey ................................................................... 8.

8. Leaves without tomentose pubescence, green on both surfaces .. 9.

+ Leaves (at least in juvenile stage) with more or less well-developed tomentose pubescence beneath (sometimes on upper surface as well) ................................................................................................................. 10.

9. Leaves with comparatively more teeth (4–7 on each side), usually with scattered or fairly abundant long thickened hairs above, densely glandular-pitted beneath and with short or long, squarrose hairs along views ...................................................... 12. **S. supina** L.

+ Leaves with few teeth (1–4 on each side), some leaves sometimes entire, with diffuse or fairly dense (especially beneath) pubescent thickened hairs and stalked glands on both surfaces ....................... ................................................................................... 8. **S. paulsenii** Briq.

10. Leaves pinnately incised-dentate, generally deeper than half breadth of half-blades ................................................. 9. **S. przewalskii** Juz.

+ Leaves comparatively less deeply crenate or dentate ................... 11.

11. Leaves green above, diffusely arachnoid, whitish beneath, densely appressed-tomentose; leafy bracts, calyces, pedicels and stem with long, rather thick lustrous hairs. ............................ 2. S. albertii Juz.

+ Leaves greenish on both surfaces, pubescence poorly manifest beneath; leafy bracts, calyces, pedicels and stems with long, rather thick hairs or hairs totally lacking ..................................................... 12.

12. Leafy bracts broadly ovate, shortly acuminate, with short appressed hairs all over surface and, additionally, with scattered fine stalked glands; stems mostly with intense anthocyanin coloration .............
..................................................................................... 7. S. krylovii Juz.

+ Leafy bracts ovate, rather thick squamose hairs in pubescence along with stalked glands; stems generally greenish or lilac in lower part
............................................................................... 11. S. sieversii Bunge.

1. S. alaschanica Tschern. in Novosti sist. vyssh. rast. (1965) 220.  —? S. rivularis auct. non Wall.: Walker in Contribs U.S. Nat. Herb. 28 (1941) 657.  —Ic.: Tschern. l.c.

Described from China (Alashan range). Type in Leningrad.

On open rocky-loamy slopes and precipices.

IA. Mongolia: *Alash. Gobi* (between Huang He river and mountains, 1871; Alashan mountain ranges slopes and precipices, July 9, 1873—Przew., typus!; Tszosto gorge, southern slope, May 18; Khote-Gol gorge, at all levels, rocky-humus soil, June 19—1908, Czet.; ? "Ho Lan Shan, exposed, dry, clayey cliffs, No. 1154, 1923, Ching"—Walker, l.c.).

General distribution: endemic.

Note. Highly characteristic species intermediate between *S. galericulata* L. and *S. scordiifolia* Fisch. but well differentiated from both. Known so far only from Alashan range where it has been collected time and again. The literature (Walker, l.c.) mentions the occurrence of *S. rivularis* Wall. in Alashan range but the study of *S. rivularis* described from Nepal shows that this species has extended into more humid regions of China: South, East and Taiwan. There are no further reports of *S. rivularis* from the Alashan range. The habitat conditions cited by Walker on the label correspond to those of our species and thus we assume that *S. rivularis* Walk. non Wall. does not differ from *S. alaschanica* Tschern. Unfortunately, it has not been possible to verify this assumption.

2. S. albertii Juz. in Bot. mater. Gerb. Bot. inst. AN SSSR, 14 (1951) 399; Juzepczuk in Fl. SSSR, 20 (1954) 147; Grubov in Bot. mater. Gerb. Bot. inst. AN SSSR, 19 (1959) 549; Fl. Kazakhst. 7 (1964) 311.  —Ic.: Fl. Kazakhst. Plate 36, fig. 5.

Described from Kazakhstan (around Dzharkent). Type in Leningrad.

Dry rubble slopes, dry pebble river-beds, exposed mottled rocks and wormwood steppes.

IIA. Junggar: *Jung. Alt.* (Urtak-Sary, West of Sairam lake, July 19, 1878—Fet.; Urtak-Sary, 1800 m, Aug. 4, 1878—A Reg.; Toli, on slope, No. 2480, Aug. 4; same site, No. 2762, Aug. 9; Ven'tsyuan', No. 1464, Aug. 14; north of Toli, 700 m, Aug. 16—1957, Kuan; along road from Borotala valley to Sairam-Nur lake, valley of Urtak-Sary river, wormwood-snakeweed steppe, Aug. 18, 1957—Yun. et al.) *Tien Shan* (Sairam, July 1877; Piluchi, north of Kul'dzha, 900–1200 m, July 22; Talki, July 1878—A. Reg.; Urumchi region, beyond Tasenku river, Biangou area, arid southern slope in spruce belt, Sept. 25, 1929—Pop.; high, about 200 m, right terrace of Manas river, near Chendokhoze river estuary, May 28, 1954—Mois.; Kuitun—Gobi, rocky ravine, No. 1089, June 27; Savan, on slope, No. 1294, July 8—1957, Kuan; Ulan-Usu river valley, worm-wood-chee grass steppe, July 16, 1957—Yun. et al.; along road from Urumchi to Karashar, on slope, 1570 m, No. 5942, July 21, 1958—Lee and Chu; B. Yuldus basin, residual mountain on right bank of Khaidyk river, 3–5 km from Bain-Bulak, Aug. 10, 1958—Yun.). *Dzhark.* (Aktyube, north of Kul'dzha, May 13; along Ili river, May 14—1877, A. Reg.; Ili, on slope, No. 3148, Aug. 7, 1957—Kuan).

General distribution: Jung.-Tarb. (southern slopes of Junggar Alatau).

3. **S. baicalensis** Georgi, Bemerk. Reise im Russ. Reich. 1 (1775) 223; Forbes and Hemsley, Index Fl. Sin. 2 (1902) 294; Kudo in Mém. Soc. Sci. and Agr. Taihoku Univ. 2 (1929) 269; Kitag. Lin. Fl. Mansh. (1939) 385; Walker in Contribs U.S. Nat. Herb. 28 (1941) 657; Juzepczuk in Fl. SSSR, 20 (1954) 103; Grubov, Konsp. fl. MNR (1955) 233; Dashnyam in Bot. zh. 50 (1965) 1641. —*S. macrantha* Fisch. ex Reichb. Ic. Bot. 5 (1827) 52; Franch. Pl. David. 1 (1884) 240; Dunn in Notes Bot. Gard. Edinburgh, 6 (1915) 177; Chen and Chou, Rast. pokrov r. Sulekhe (1957) 90. —Ic.: Reichb. l.c. tab. 488, fig. 681 (sub nom. *S. macrantha*); Fl. SSSR, Plate 5, fig. 5.

Described from Transbaikal. Type not known.

Rocks, rocky slopes, fine sand, in steppe zone.

IA. **Mongolia:** *Cis-Hing.* (Khaligakha area, arid sand steppe, July 23, 1899—Pot. and sold.), *East. Mong., Alash. Gobi* ("Nan Ssu Kou, No. 138, Ching"—Walker, l.c.) *Khesi* (Suchzhou, Sept. 11, 1890—Marten; "r. Sulekhe"—Chen and Chou, l.c.).

General distribution: East. Sib. (south), Far East (south), North Mongolia (Hent., Mong.-Daur.), China (Dunbei, North), Korean peninsula, Japan.

4. **S. galericulata** L. Sp. pl. (1753) 599; Forbes and Hemsley, Index Fl. Sin. 2 (1902) 294; Krylov, Fl. Zap. Sib. 9 (1937) 2296; Juzepczuk in Fl. SSSR, 20 (1954) 90; Grubov, Konsp. fl. MNR (1955) 234; Fl. Kirgiz. 9 (1960) 14; Fl. Kazakhst. 7 (1964) 302. —*S. galericulata* L. var. *genuina* Regel, Tent. Fl. Ussur. (1861) 118; Danguy in Bull. Mus. nat. hist. natur. 20 (1914) 84. —? *S. galericulata* var. *angustifolia* auct. non Regel: Danguy in Bull. Mus. nat. hist. natur. 20 (1914) 84. Ic.: Fl. Kazakhst. Plate 35, fig. 5.

Described from Europe. Type in London (Linn.).

Wet coastal and swampy meadows, coastal scrubs, wet banks of rivers, brooks and lakes, forest and forest-steppe belts.

IA. **Mongolia:** *Cis-Hing.* (Khalkhin-Gol river, Symbur area, Sept. 1, 1928—Tug.). *East. Mong.* (near Khailar town, wet meadow on river bank, April 7, 1951—S.H. Li et al. (1951); ? "Vallée du Kéroulen, June 1896, Chaff."—Danguy, l.c.). *Bas. Lakes* (Ulangom, under shadow of pea shrubs, along river, July 2, 1879—Pot.).

IIA. **Junggar:** *Jung. Gobi* (Guchen vicinity, July 1876—Pev.; Savan district, Mogukhu reservoir, near water, No. 1569, June 25; environs of Shikheitsz, No. 1172, July 2; north of Kuitun station, swampy meadow, No. 385, July 6; 3 km east of Kuitun station, in swamp, No. 431, July 7—1957, Kuan; north of Kuitun station, along Shikho-Manas road, grass-sedge swampy meadow, June 30, 1957—Yun. et al.). *Zaisan* (Ch. Irtysh river, left bank in Dzhelkaidar area, tugai (vegetation-covered bottomland), July 9; same site, right bank, in Burchum estuary, tugai, July 14—1914, Schischk.), *Dzhark.* (Kul'dzha, June 15; Suidun, July 16; Piluchi near Kul'dzha, July—1877, A. Reg.). *Balkh.-Alak.* (along Churchutsu river near Chuguchak, Aug. 10, 1840—Schrenk; around Chuguchak town, in swamp, No. 2817, Aug. 10; same site, in ditch, No. 1509, Aug. 12—1957, Kuan).

General distribution: Aralo-Casp., Fore Balkh., Jung.-Tarb., North and Cent. Tien Shan; Europe, Mediterranean, Balk. (north), Caucasus, Middle Asia, West. and East. Sib., Nor. Mong. (Hent.), China (north), Japan.

Note. *S. regeliana* Nakai, closely related to *S. galericulata*, is found in the Far East, China (Dunbei) and Korean peninsula, the main difference between them being the very large corolla in *S. regeliana*. In the territory studied by us, not a single report of this species was recorded but *S. galericulata* var. *angustifolia* Regel has been mentioned in literature (Danguy, l.c.) in lists of plants collected by Chaffanjon on Kerulen river. This name corresponds to *S. regeliana*. As we did not have occasion to verify the identification of this specimen, we treated it as *S. galericulata* with some uncertainty.

5. **S. grandiflora** Sims in Curtis's Bot. Mag. 17 (1803) tab. 635; Danguy in Bull. Mus. nat. hist. natur. 20 (1914) 85; Krylov, Fl. Zap. Sib. 9 (1937) 2299; Juzepczuk in Fl. SSSR, 20 (1954) 132; Grubov, Konsp. fl. MNR (1955) 234; Fl. Kazakhst. 7 (1964) 306; p. min. p. —*S. orientalis* L. var. *microphylla* Ledeb. Fl. Ross. 3 (1849) 395; Sapozhn. Mong. Alt. (1911) 382. —*S. tuvensis* Juz. in Bot. mater. Gerb. Bot. inst. AN SSSR, 14 (1951) 389; Juzepczuk in Fl. SSSR, 20 (1954) 133. —?*S. mongolica* Sobolevsk., in Bot. mater. Gerb. Bot. inst. AN SSSR, 14 (1951) 49; Juzepczuk in Fl. SSSR, 20 (1954) 185. —Ic.: Sims, l.c. tab. 635; Fl. SSSR, Plate 8, fig. 2.

Described from specimen grown from seeds received from Siberia. Type probably in London (K ?)

Rocky-rubble slopes, sandy-pebble banks of rivers, pebble beds and arid river-beds.

IA. Mongolia: *Khobd.* (Kharkhira hills, Kendulyun river, July 27, 1903—Gr.-Grzh.). *Mong. Alt.* (Yamadzhin mountain ranges, dry river-bed, July 11, 1877—Pot.; mountains between Urukta and Kobdo rivers, July 2; rocky-sandy bank of Kengurlen river, July 6; bank of Dzhirgalanta river flowing between Naryn and Shadzagai rivers, July 20—1898, Klem.; along washed pebble beds of hill streams in northern half of Tsagan-Gol valley, July 29; Tsagan-Bulun, Aug. 1; steppe, Khashatu river, Aug. 28—1899, Lad.; from Tal'-Nor lake to Delyun river, July 5, 1903—Gr.-Grzh.; Saksai river, dry ravine, July 31, 1909—Sap.; "Environs de Kobdo, sables, alt. 1500 m, Sept. 22, 1895, Chaff."—Danguy, l.c.). *Bas. Lakes* (steppe east of Ulan-Dab, compact rocky soil, June 23; Kharkhira river, along banks, July 9; Dzabkhyn river, July 18; Khara-Usa lake, on mountains along west. bank of lake, Aug. 19; Bukhon'-Shar mountain ranges, on dry bed in gorge, Aug. 30; Burgassutai river, dry pebble bed, Sept. 3; mountain ranges, dry river-bed, Sept. 4—1879, Pot.; bank of Bogden-Gol river, below Ulyasutai, July 6, 1894; steppe between Khara-Bur river and Dzelin collective farm, dry river-bed, Sept. 1; Ubsa-Nor, sandbank, Sept. 5—1895; near Kobdo, on hill, Aug. 22, 1896—Klem.; Khan-Khukhei range, Khangel'tsyk river, rocky steppe, June 24, 1924—Neiburg; pebble bed in Kharkhira river valley, Aug. 10; environs of Ubsanor basin near Kharkhira river discharge from mountains, rubble-pebble bed, Sept. 3—1931, Bar.). *Gobi-Alt.* (Baga-Bogdo, cottonwood terrace... 1925—Chaney; rocky northern slopes of Ikhe-Bogdo range branches, June 21, 1926—Kozlova; Artsa-Bogdo range, at interruption in river-bed, Aug. 6, 1926—Glag.; near foot of Khalga hill, Aug. 2; southern slope of Bain-Tsagan mountain ranges, rubble, Aug. 9; Dundu-Saikhan, rubble soil at foothill of Ulan-Khunde, Aug. 17; Dzun-Saikhan range, rubble near bank of brook at Yalo creek, Aug. 25; Bain-Tsagan range, Subulyur creek, rubble soil, Sept. 17—1931, Ik.-Gal.).

General distribution: West. Siberia (Altay), East. Siberia, Nor. Mongolia (Mong.-Daur).

Note. *S. grandiflora*, a Siberian-Mongolian plant, probably penetrates north-eastern Kazakhstan but only the Altay region. The Kazakhstan plants cited in Fl. Kazakhstana [Flora of Kazakhstan], 7, under the name *S. grandiflora*, however, represent a distinct species well distinguished from *S. grandiflora* Sims, in pubescence of the whole plant, size and form of leaves, size and pubescence of petiole, form and texture of leafy bracts, colour of corolla, and evidently should be called *S. turgaica* Juz. [in Bot. mater. Gerb. Bot. inst. AN SSSR, 14 (1951) 390. —*S. karkaralensis* Juz. in Bot. mater. Gerb. Bot. inst. AN SSSR 14 (1951) 391. —*S. grandiflora* auct. non Sims.: Fl. Kazakhst. (1964) 306, p. max. p.].

Within its distribution range, *S. grandiflora* is not uniform but we do not consider it possible to regard the race with much thicker stems, blue corolla and black nuts from Ulu Khem valley as an independent species *S. tuvensis* Juz. since not all the characteristics listed are permanent. The colour of the corolla varies from pink-violet to blue while glabrous as well as pubescent nuts are found. The first description of *S. grandiflora* makes no mention at all about nuts and later investigators reported for this species nuts with stellate pubescence. A study of a large number of specimens from the entire distribution range showed that in the valley of Ulu-Khem river, in the region of Katun' and Ini rivers, plants with glabrous as well as pubescent nuts are found. Apart from the pubescence of nuts, there is no difference at all between these plants. Within Mongolia, however, the race with pubescent nuts is extremely rare; nuts of all plants of this species here are finely tuberculate, black, without stellate pubescence.

### 6. S. kingiana Prain in J. As. Soc. Bengal, 59 (1890) 308.

Described from East. Himalayas. Type in London (K).

Rocks, at 4000–5000 m altitude.

IIIB. Tibet: *South* (Kang-ma, 60 miles north of Phari on the banks of Pe-na-mong Chu, Aug. 14, 1878—Dungboo [typus!K]; Dochen hills, Aug. 13, 1907—Stewart [K]; Everest, above Tinki, 4800 m, July 13, 1922—Morton [K]; same site, 1924—Kingston [K]; hill behind Gyantse, 4500 m, Aug. 13, 1936—Chapman [K]).

General distribution: Himalayas (east.).

Note. Very interesting high-montane species which, for some unknown reason, has been missed in several treatises on Tibet. Back of calyx without appendage and only with a small umbo—a characteristic of subgenus *Anaspis* (Rech. f.) Juz., whose members are extensively distributed in high mountain belts of Middle Asian.

### 7. S. krylovii Juz. in Sist. zam. Gerb. Tomsk. univ. 10, 8 (1936) 4; Krylov, Fl. Zap. Sib. 9 (1937) 2301; Juzepczuk in Fl. SSSR, 20 (1954) 149; Fl. Kazakhst. 7 (1964) 312, p.p.

Described from East. Kazakhstan (Zaisan, Blandy-Kum sand). Type in Leningrad.

Sand and sandy steppes.

IIA. Junggar: *Zaisan* (Alkabek, sand, June 6, 1908—B. Fedtschenko; Ch. Irtysh river, right bank below Burchum river, Sary-dzhasyk, Kiikpai well, hummocky sand, June 15;

between Burchum and Kaba rivers, Kiikpai well in Kara area, sand steppe, June 15—1914, Schischk.).

General distribution: Fore Balkh.; West. Sib.

Note. This species is closely related to *S. sieversii* Bunge but, unlike it, is essentially confined to sandy areas while *S. sieversii* grows on arid rubble-rocky slopes of foothills. These species are differentiated morphologically too: in *S. krylovii* the pubescence of leafy bracts and colour of stem differ.

8. **S. paulsenii** Briq. in Bot. tidsskr. 28 (1908) 233; Juzepczuk in Fl. SSSR, 20 (1954) 188; Fl. Kirgiz. 9 (1960) 28; Ikonnikov, Opred. rast. Pamira (1963) 210. —*S. oligodonta* Juz. in Bot. mater. Gerb. Bot. inst. AN SSSR, 14 (1951) 370; Juzepczuk in Fl. SSSR, 20 (1954) 187; Fl. Kirgiz. 9 (1960) 28; Fl. Kazakhst. 7 (1964) 317. —? *S. alpina* L. var. *cordifolia* auct. non Regel: Danguy in Bull. Mus. nat. hist. natur. 14·(1908) 132. —Ic.: Briq. l.c. fig. 1; Fl. SSSR, Plate 11, fig. 2 (sub nom. *S. oligodonta* Juz.).

Described from Middle Asia (Bordaba pass and Kara-Su river). Type in Copenhagen. Isosyntype (from Kara-Su river) in Leningrad.

Alpine meadows, rocky-rubble places, banks and dry beds of mountain rivers.

IA. Mongolia: *Mong. Alt.* (Angirty valley, July 26; in pass between Khartsiktei and Tsagan-Nur lake, July 27—1898, Klem.).

IB. Kashgar: *West.* (Kashgar mountains, July 1924—coll. ? [K]; west of Kashgar, Bostan-Terek, July 11, 1929—Pop.

II A. Junggar: *Cis-Alt.* (Kondagatai, on mountain slope, on gravel, Sept. 14, 1876—Pot.; Koktogon region, 2500 m, No. 1936, Aug. 17, 1956—Ching). *Jung. Alt.* (oberstes Dunde-Kelde Tal und oberstes Chustai Tal, July 5–6, 1908—Merzb.). *Tien Shan* (Muzart gorge, 1500 m, Aug. 16, 1877—A. Reg.; Danu region, Nos. 362, 483, July 21–22; Danu river, nor. slope, No. 2117, July 21—1957, Kuan; ulas of M. Yuldus, pebble bed in floodplain, 2500 m, No. 6339, Aug. 2, 1958—Lee and Chu; Manas river basin, above Danu-Gol, 6–7 km above peak to Danu pass, in talus, July 22, 1957—Yun. et al.).

IIIC. Pamir (Tarbashi, rocky-sandy banks of river, common, July 30, 1909—Divn.; "rochers, vallée de Tor-Bachi, alt. 3800 m, July 31, 1906, Lacoste"—Danguy, l.c.).

General distribution: Fore Balkh., Jung.-Tarb., Nor. and Cent. Tien Shan, East. Pam.

Note. High montane plant with fairly broad distribution range and relatively constant characteristics. A critical study of the material from the entire distribution range cast doubts on distinguishing the broader-leaved race from Junggar Alatau, Tien Shan and Chinese Junggar as an independent species, *S. oligodonta* Juz.

9. **S. przewalskii** Juz. in Bot. mater. Gerb. Bot. inst. AN SSSR, 14 (1951) 400; Juzepczuk in Fl. SSSR, 20 (1954) 148; Fl. Kirgiz. 9 (1960) 23; Fl. Kazakhst. 7 (1964) 311. —Ic.: Juzepczuk, l.c. (1951), fig. 3; Fl. Kazakhst. Plate 36, fig. 7.

Described from Issyk-Kul' lake. Type in Leningrad.

Dry exposed mountain slopes, sandy slopes of valleys and river-beds, grasslands, 600 to 3500 m.

IIA. Junggar: *Tien Shan* (along Kunges river, in forest, about 3500 m, on grassland, June 29, 1871—Przew.; Dzhagastai, Aug. 7; Dzhauku valley, 1800–2100 m, Sept. 1—1877, A. Reg.;

Burkhan-tau, June 5, 1878—Fet.; near Chinese picket, Aug. 20, 1878—Larionov; Sygashu, 600 m, May 4; Baga-Dzuslun—Bayanamyn, east. tributary of Dzhin river, 1500–1800 m, June 4; Borborogusun, 2700 m, June 5; Borgaty spring, 1500–1800 m, July 5—1879, A. Reg.; lake near Bogdo-Ula foothill, on mountain slopes, Aug. 29, 1898—Klem.; Kok-su valley, 1901—Littledale [K]; near Manas town, July 17, 1908—Merzbacher; Chzhausu region in intermontane basin, No. 3248, Aug. 11; Chzhausu to Shati, roadside, No. 843, July 12; 15 km west of Takes town, No. 3610, Aug. 17—1957, Kuan).

General distribution: Nor. and Cent. Tien Shan.

10. **S. scordiifolia** Fisch. ex Schrenk in Denkschr. Bot. Gesellsch. Regensb. 2 (1822) 55; Franch. Pl. David. 1 (1884) 240; Danguy in Bull. Mus. nat. hist. natur. 17 (1911) 15; Dunn in Notes Bot. Gard. Edinburgh, 6 (1915) 173; Kudo in Mém. Fac. Sci. and Agr. Taihoku Univ. 2 (1929) 263; Krylov, Fl. Zap. Sib. 9 (1937) 2298; Hao in Engler's Bot. Jahrb. 68 (1938) 634; Kitag. Lin. Fl. Mansh. (1939) 385; Juzepczuk in Fl. SSSR, 20 (1954) 99; Grubov, Konsp. fl. MNR (1955) 234; Fl. Kazakhst. 7 (1964) 303; Hanelt und Davažamc in Feddes repert. 70 (1965) 54; Dashnyam in Bot. zh. 50 (1965) 1641. —*S. galericulata* L. δ *scordifolia* Regel Tent. Fl. Ussur. (1861) 118; Danguy in Bull. Mus. nat. hist. natur. 20 (1914) 85. —**Ic.:** Fl. SSSR, Plate 5, fig. 3; Fl. Kazakhst. 7, Plate 35, fig. 8.

Described from Siberia. Type not known.

Steppe meadows, rocky meadow and steppe slopes of mountains river banks, coastal scrubs, forest fringes; adventitious weed.

IA. **Mongolia:** *Cis-Hing.* (Khaligakha area, dry sandy soil, Kulun-Buir-Norskaya plain, June 23; syrts (watershed uplands) along Abder river, June 25—1899, Pot. and Sold.; 5 km west of Tore-Gol river, herbage meadow, Aug. 7, 1949—Yun.; "Kailar, steppe sablonneuse, alt. 750 m, No. 1782, June 22, 1896, Chaff."—Danguy [1914] l.c.). *Cent. Khalkha, East. Mong.,* (near Dolon-Nora, 1870—Lom.; Lykse lake, Kulun-Buir-Norsk plain, July 5, 1899—Pot. and Sold.), *East. Gobi* (Inshan', June 25, 1871—Przew.).

IIIA. **Qinghai:** *Nanshan* (24 km south of Xining, mountain slopes in grass-forb steppe, 2650 m, Aug. 4, 1959—Petr.; "Kokonor: auf dem östlichen Nan-Schan, Aug. 3, 1930, Hao, No. 818"—Hao, l.c.; "Si-Ning-Fou, alt. 2400 m, July 12, 1908, Vaillant"—Danguy [1911].

General distribution: West. Sib., East. Sib., Far East (south), Nor. Mong., China (Dunbei, North, North-West, East), Korean peninsula, Japan.

11. **S. sieversii** Bunge in Ledeb. Ic. pl. fl. ross. 2 (1830) 10; id. in Fl. alt. 2 (1830) 394; Krylov, Fl. Zap. Sib. 9 (1937) 2300; Juzepczuk in Fl. SSSR, 20 (1954) 150; Fl. Kazakhst. 7 (1964) 314. —*S. soongorica* Juz. in Fl. SSSR, 20 (1954) 509, 148; Fl. Kazakhst. 7 (1964) 312. —*S. catharinae* Juz. in Bot. mater. Gerb. Bot. inst. AN SSSR, 14 (1951) 386; Juzepczuk in Fl. SSSR, 20 (1954) 145; Fl. Kazakhst. 7 (1964) 310. —*S. orientalis* var. *adscendens* Ledeb. Fl. Ross. 3 (1849) 395; Sapozhn. Mong. Alt. (1911) 382. —**Ic.:** Ledeb. Ic. pl. fl. ross. tab. 123; Fl. SSSR, Plate 10, fig. 2.

Described from East. Kazakhstan (Arkaul, Dolonkara and ChingizTau mountain ranges. Type in Leningrad.

Foothill plains, rubble-rocky slopes in desert steppes.

IIA. Junggar: *Jung. Alt.* (Karagaity area, near Yamat picket, mountain slopes of Dzhair, July 2, 1947—Shum.). *Jung. Gobi* (steppe along Kemerchek river, Aug. 15, 1906—Sap.; Shara-Sume region, 700 m, Sept. 9, 1956—Ching; pasture lands at Beidashanya foothill, 1000 m, Sept. 27, 1957—Kuan; 10 km north of Kosh-Tologoi settlement on Khobuk river, pass in small hillocky ridge, desert steppe, Aug. 4; border ridge 30 km east-north-east of Burchum, along road to Shara-Sume, steppe belt, on granite eluvium, July 5; submontane plain 30 km south of Shara-Sume, along road to Ch. Irtysh, on granite pillow lavas, July 7—1959, Yun.).

General distribution: Fore Balkh., Jung.-Tarb.; West. Sib. (south.), East. Sib. (south.).

Note. The species was first reported from 3 places: mountains of Arkaul, Dolonkara and Chingiz Tau; specimens described by A.A. Bunge do not indicate geographic locations and hence it is impossible to establish their origin.

*S. sieversii* is extensively distributed in the eastern regions of Kazakhstan, south-western regions of Siberia and north-western regions of China. Because of its comparatively large distribution range, this species is not quite uniform. We see no need, however, to distinguish individual races as independent species since, in our opinion, such races are the result of either the variability of the species or the result of hybridisation between *S. albertii* Juz. and *S. sieversii* Bunge but with greater inclination toward *S. sieversii*.

12. **S. supina** L. Sp. pl. (1753) 598; Krylov, Fl. Zap. Sib. 9 (1937) 2301; Juzepczuk in Fl. SSSR, 20 (1954) 183; Grubov, konsp. fl. MNR (1955) 234; Fl. Kazakhst. 7 (1964) 316. —*S. lupulina* L. Sp. pl. ed. 2 (1763) 835. —*S. oxyphylla* Juz. in Spisok rast. Gerb. fl. SSSR, 11 (1945) 149; id. in Fl. SSSR, 20 (1954) 182. —*S. alpina* auct. non L.: Danguy in Bull. Mus. nat. hist. natur. 20 (1914) 84. —*S. alpina* var. *lupulina* Benth. in DC. Prodr. 12 (1848) 412; Sapozhn. Mong. Alt. (1911) 382. —**Ic.:** Fl. SSSR, Plate 11, fig. 1 (sub nom. *S. oxyphylla* Juz.); Fl. Kazakhst. Plate 37, fig. 3.

Described from Siberia. Type in London (Linn.).

Southern rocky slopes and rocks in forest belt or, more often, in scrub-meadow steppe belt.

IA. Mongolia: *Mong. Alt.* (Upper Khobdos lake, left bank, July 19, 1909—Sap.; summer camp in Bulgan somon, along road to Kharagaitu-Khutul', on rim of ravine near lower forest boundary, July 27, 1947—Yun.; "Altai, alt. 2560 m, entre l'Irtich et Kobdo, Mongolie, Sept. 7, 1895, Chaff." —Danguy, l.c.).

IIA. Junggar:*Cis. Alt.* (Qinhe [Chingil] river gorge, on slope of gorge, No. 897, Aug. 2; Altay [Shara-Sume], No. 2456, Aug. 26—1956, Ching; 20 km north-west of Shara-Sume, Scrub-meadow steppe on southern slope, July 7; 25–30 km north of Koktogoi, right bank of Kairta river, Kuidyn valley, on southern rocky steppe slope, July 15—1959, Yun.). *Jung. Alt.* (Toli region, on slope, No. 980, Aug. 5; same site, Nos. 1235, 1270, Aug. 6; Toli dist., Dzhagistai, in forest, No. 125, Aug. 6; Albakzyn hills, No. 2646, Aug. 6—1957, Kuan).

General distribution: Aral-Casp., Fore Balkh., Jung.-Tarb.; Europe (European USSR), West. Sib. (south), East. Sib. (south).

13. **S. viscidula** Bunge, Enum. pl. China bor. (1832) 52; Forbes and Hemsley, Index Fl. Sin. 2 (1902) 298; Dunn in Notes Bot. Gard. Edinburgh, 6 (1915) 177; Kudo in Mém. Fac. Sci. and Agr. Taihoku Univ. 2 (1929) 262; Kitag. Lin. Fl. Mansh. (1939) 387.

Described from Nor. China (between Kalgan and Peking (Beijing)). Type in Paris. Isotype in Leningrad. Plate I, fig. 2.

Coastal sand and pebble beds in river valleys, arid loess slopes.

IA. Mongolia;*East. Mong.* (Boro-Gol, May 31, 1898—Zab.; Ourato, No. 2642, June 1866—David). *Ordos* (Obon-Shili, sandy soil, Aug. 19, 1884—Pot.). *Khesi* (along Yarlyn-Gol river, July 3, 1872—Przew.).

General distribution: China (Dunbei, North), Japan.

## 5. Marrubium L.
Sp. pl. (1753) 582.

1. **M. vulgare** L. Sp. pl. (1753) 583; Benth. Labiat. gen. et sp. (1834) 591; Hook. f. Fl. Brit. Ind. 4 (1885) 671; Knorr. in Fl. SSSR, 20 (1954) 235; Fl. Kazakhst. 7 (1964) 323. —**Ic.:** Fl. Kazakhst. Plate 38, fig. 3.

Described from northern Europe. Type in London (Linn.).

Along roadsides, fences, cultivated beds; as ruderal on river banks.

IIA. Junggar: *Tien Shan* (Dzhagastai, 900 m, Aug. [11], 1877; Khanakhai, 1200 m, June 24; Shara-Bugutal pass, 1800 m, Sept. 19—1878, A. Reg.) *Dzhark.* (Kul'dzha, 1876—Golike; Ili-Chapchal, near river, Aug. 5, 1957—Kuan).

General distribution: Aralo-Casp., Fore Balkh., Jung.-Tarb., Nor. Tien Shan; Europe, Mediterranean, Balk.-Asia Minor, Fore Asia, Caucasus, Middle Asia, Himalayas (Kashmir).

## 6. Lagopsis Bunge
Mém. Ac. Sci. St.-Pétersb. 2 (1835) 565; Ikonnikov-Galitzkij in Bot. mater. Gerb. Bot. inst. AN SSSR, 7 (1937) 39.

1. Corolla white or pale pink; plant with short appressed pubescence, dull green ................................................ 4. **L. supina** (Steph.) Ik.-Gal.
+ Corolla dark brown or yellow; plant with dense woolly pubescence all over or only in inflorescence .......................................................... 2.
2. Corolla yellow .................................................. 2. **L. flava** Kar. et Kir.
+ Corolla dark brown ........................................................................ 3.
3. Plant glaucous throughout due to dense woolly pubescence, leaves white-lanate on both surfaces; sessile glands scattered on leaf surface visible through pubescence on under surface ...........................
................................................... 3. **L. marrubiastrum** (Steph.) Ik.-Gal.
+ Plant less pubescent, leaves greenish, covered with glandular hairs; woolly pubescence almost lacking ...............................................
................................................... 1. **L. eriostachya** (Benth.) Ik.-Gal.

1. **L. eriostachya** (Benth.) Ik.-Gal. in Bot. mater. Gerb. Bot. inst. AN SSSR, 7 (1937) 42; Knorr. in Fl. SSSR, 20 (1954) 250; Grubov, Konsp. fl. MNR (1955) 234. —*L. viridis* Bunge in Mém. Ac. Sci. St.-Pétersb. 2 (1835) 566. —*Marrubium eriostachyum* Benth. Labiat. gen. et sp. (1834) 586. —*Molucella mongholica* Turcz. ex Ledeb. Fl. Ross. 3 (1849) 402.

Described from East. Siberia (East. Sayans). Type in Leningrad.

Sandy-pebble banks of rivers and lakes, on talus.

IA. **Mongolia:** *Mong. Alt.* (not far from Borogol-Daban, on rock debris, Aug. 18; Khara-Dzarga range, Sakhir-Sala river valley, in coastal pebble bed, Aug. 21—1930, Pob.).

General distribution: East. Sib. (Sayans), Nor. Mong. (Fore Hubs.).

Note. *L. eriostachya* was earlier reported only from around Hubsugul lake, i.e., from the region of Siberian mountain-taiga flora, and hence the report of this species in Mongolian Altay, Central Asia, is of special interest. It is possible that the specimens reported here are associated with residual patches of xerophilised taiga flora.

*L. eriostachya* is very similar in appearance to *L. marrubiastrum* (Steph.) Ik.-Gal., but differs from the latter in green leaves with pubescence of exclusively glandular hairs (lanate pubescence lacking) and very long bracts.

2. **L. flava** Kar. et Kir. in Bull. Soc. natur. Moscou, 15 (1842) 425; Ik.-Gal. in Bot. mater. Gerb. Bot. inst. AN SSSR, 7 (1937) 43; Knorr. in Fl. SSSR, 20 (1954) 249; Fl. Kazakhst. 7 (1964) 325. —*Marrubium flavum* Walp. Repertorium, 3 (1837) 856; Ledeb. Fl. Ross. 3 (1849) 403. —Ic.: Fl. SSSR, Plate 16, fig. 2; Fl. Kazakhst. 7, Plate 38, fig. 5.

Described from Junggar Alatau. Type in Leningrad.

Pebble beds and talus, rubble slopes in alpine belt.

IIA. **Junggar:** *Jung. Alt.* (toward Chubaty pass, 2700–3000 m, Aug. [2]; Kasan pass, 2700–3300 m, Aug. 10—1878, A. Reg.; Borotala river basin, below Koketau pass, July 21, 1909—Lipsky; Toli district, on slope, Aug. 6, 1957—Kuan). *Tien Shan* (in Kokkamyr mountains, 1800 m, July 28, 1878; Kum-Daban, northern slopes of Irenkhabirg mountains, 2700 [May 28]; Kumbel', 3000 m, June 3—1879, A. Reg.; Tien Shan slopes in gorges near foothills, pebble beds and clayey soil, June 5, 1879, Przew.).

General distribution: Jung.-Tarb.

Note. *L. flava* is closely related to *L. marrubiastrum* (Steph.) Ik.-Gal. and evidently represents its derivative. The difference between these 2 species lies mainly in the colour of the corolla (corolla in *L. flava* yellow, in *L. marrubiastrum* dark brown) and shape of lobes of its lower lip (in *L. flava*, midlobe of lower lip orbicular above, and undivided versus medially notched in *L. marrubiastrum*). These species differ ecologically as well: *L. flava* is found only in the upper mountain belt while *L. marrubiastrum* enjoys a much broader distribution.

3. **L. marrubiastrum** (Steph.) Ik.-Gal. in Bot. mater. Gerb. Bot. inst. AN SSSR, 7 (1937) 41; Krylov, Fl. Zap. Sib. 9 (1937) 2303; Knorr. in Fl. SSSR, 20 (1954) 248; Grubov, Konsp. fl. MNR (1955) 234; Fl. Kazakhst. 7 (1964) 325. — *L. incana* Bunge in Mém. Ac. Sci. St.-Pétersb. 2 (1835) 566. — *Molucella marrubiastrum* Steph. in Mém. Soc. natur. Moscou, 2 (1809) 8, excl. ic. —*Marrubium lanatum* Benth. in Labiat. gen. et sp. (1834) 587; Henderson and Hume, Lahore to Jarkand (1873) 331; Hook. f. Fl. Brit. Ind. 4 (1885) 671; Strachey, Catal. (1906) 144; Sapozhn. Mong. Alt. (1911) 382; Danguy in Bull. Mus. nat. hist. natur. 20 (1914) 85; Pampanini, Fl. Carac. (1930) 181. —Ic.: Ledeb. Ic. pl. fl. ross. 2, tab. 150.

Described from Altay. Type in Leningrad.

Talus, rocky slopes, rocks, rocky floors of gorges, sandy-pebble banks of rivers, moraine; from desert-steppe to alpine belt.

IA. Mongolia: *Khobd.* (Kharkhira river valley, pebble bed along bank, July 21, 1879—Pot.) *Mong. Alt.* (west.) *Gobi-Alt.* (Ikhe-Bogdo mountain range).

IIA. Junggar: *Tien Shan* (in east up to Bogdoshan mountain range).

General distribution: Jung.-Tarb. (Jung. Alt.); West. Sib. (Altay), Nor. Mong. (Hang.), Himalayas (west.).

Note: *L. marrubiastrum* has a relatively wide range but is distributed only in montane areas. In Khobd., Mong. Alt., Gobi-Alt., *L. marrubiastrum* is comparatively uniform, with insignificant variability in characteristics. Within East. Tien Shan and southern parts of Mong. Alt., however, plants are found that resemble more *L. flava* Kar. et Kir. A thorough analysis of all characteristics of these specimens compels us to treat them as *L. marrubiastrum* while concomitantly noting the presence in them of a lighter coloration of corolla and a somewhat different shape of lobes of the lower corolla lip. True, some transformation of characteristics has occurred here but total differentiation has not yet taken place.

The reference in "Fl. Kazakhstana" to *L. marrubiastrum* in Junggar Alatau is dubious since *L. flava* is widespread there; nonetheless it is highly probable that races transitional to *L. flava* may exist in Junggar Alatau, similar to their probability in East. Tien Shan noted above.

*L. marrubiastrum* from West. Himalayas is more pubescent with slightly more deeply divided leaves but in all the other characteristics, plants from West. Himalayas differ in no way from Altay plants.

4. **L. supina** (Steph.) Ik.-Gal. in Bot. mater. Gerb. Bot. inst. AN SSSR, 7 (1937) 45; Knorr. in Fl. SSSR, 20 (1954) 250; Grubov, Konsp. fl. MNR (1955) 234. —*Leonurus supinus* Steph. ex Willd. Sp. pl. 3 (1800) 116. —*Marrubium incisum* Benth. Labiat. gen. et sp. (1834) 586; Hance in J. Bot. (London) 20 (1882) 32; Franch. Pl. David. 1 (1884) 244; Kanitz, A növénytani (1891) 47; Forbes and Hemsley, Index Fl. Sin. 2 (1902) 299; Dunn in Notes Bot. Gard. Edinburgh, 6 (1915) 178; Kudo in Mém. Fac. Sci. and Agr. Taihoku Univ. 2 (1929) 247; Rehder and Kobuski in J. Arn. Arb. 14 (1933) 31; Kitag. Lin. Fl. Mansh. (1939) 381; Walker in Contribs U.S. Nat. Herb. 28 (1941) 656. — Ic.: Fl. SSSR, Plate 16, fig. 1.

Described from East. Siberia. Type in Berlin (?). Isotype in Leningrad. Plate II, fig. 1.

Meadow and steppe sloppe, coastal meadows as well as ruderal in ploughed fields and pasture corrals.

IA. Mongolia: *Alash. Gobi* ("Pei Ssu Kou, in large patches in dry or moist exposed places, common, No. 189, Ching"—Walker, l.c.), *Khesi* (Suanshantan valley, July 19, 1909—Czet.).

IIIA. Qinghai: *Nanshan* (midcourse of Huang He river, gentle rocky slopes, May; road to south and south-west up to Yarlyn-Gol river, July 3—1872; downstream of Huang He up to Balekun-Gomi area, June 4, 1880—Przew.; Lanchzha-lunva river, clayey and pebble substratum, May 14, 1885—Pot.; "Shang Hsin Chuang, along exposed roadsides, rare, No. 677, Ching"—Walker, l.c.; "Szi-ningfu, ad fin., June 1879, Nos. 48b, 50, Széchenyi"—Kanitz, l.c.). *Amdo* ("Radja and Yellow River Gorges, No. 14194, Rock"—Rehder and Kobuski, l.c.).

General distribution: East. Sib., Nor. Mong. (Mong.-Daur.), China (Dunbei, North, North-West, South-West), Korean peninsula, Japan.

## 7. Agastache Clayt. ex Gronov.

Fl. Virgin. (1762) 88; Kuntze, Revis. Gen. 2 (1891) 511.

1. **A. rugosa** (Fisch. et Mey.) Kuntze, Revis. Gen. 2 (1891) 511; Kudo in Mém. Fac. Sci. and Agr. Taihoku Univ. 2 (1929) 220; Kitag. Lin. Fl. Mansh. (1939) 374; Pojark. in Fl. SSSR, 20 (1954) 274. —*Lophanthus rugosus* Fisch. et Mey. Ind. Sem. hort. Petrop. (1835) 36; Franch. Pl. David. 1 (1884) 237; Forbes and Hemsley, Index Fl. Sin. 2 (1902) 288. —**Ic.:** Kom. and Alis. Opred. rast. Dal'nevost. kraya, 2, Plate 272; Fl. SSSR, Plate 14, fig. 2.

Described from China (?). Type in Leningrad.

Rocks, rocky and grassy slopes, often under shade of trees and shrubs.

**IA. Mongolia:** *East. Mong.* ("Sartchy, au bord d'un ruisseau dans les montagnes centrales, No. 2857, July 1866, David"—Franch, l.c.).

**General** distribution: Far East, China (Dunbei, North, Central, East, South-West, South), Korean peninsula, Japan.

**Note.** Fairly widely distributed plant in China. Only Franchet has cited this species from our territory.

## 8. Lophanthus Adans.

Fam. pl. 2 (1763) 194, 572, p.p.; em. Benth. in Bot. Reg. 15 (1829) post tab. 1282; em. Briq. in Engler-Prantl. Naturl. Pflanzenfam. IV, 3a (1897) 234; Levin in Tr. Bot. inst. AN SSSR, ser. 1, 5 (1941) 268.

1. Leaves cordate at base; calyx usually considerably enlarged upward, 8–11 mm long, slightly oblique in throat, its teeth largely broadly lanceolate; corolla 13–20 mm long .. 3. **L. schrenkii** Levin.
+ Leaves orbicular or truncated at base, rarely cordate; calyx 6–10 mm long, distinctly oblique or nearly bilabiate in threat its teeth deltoid; corolla 10–15 mm long .............................................. 2.
2. Calyx tubular, insignificantly enlarged upward, 6–8 mm long; central pedicels 3–7 mm long; corolla bluish-lilac ...................................
.................................................................. 2. **L. krylovii** Lipsky.
+ Calyx tubular-campanulate, usually considerably enlarged upward, 6–10 mm long; central pedicels 8–12 mm long; corolla blue ...........
.................................................. 1. **L. chinensis** (Rafin.) Benth.

1. **L. chinensis** (Rafin.) Benth. in Bot. Reg. 15 (1829) post tab. 1282; Kudo in Mém. Fac. Sci. and Agr. Taihoku Univ. 2 (1929) 221; Levin in Tr. Bot. inst. AN SSSR ser. 1, 5 (1941) 271; id. in Fl. SSSR, 20 (1954) 276; Grubov, Konsp. fl. MNR (1955) 234. —*Hyssopus lophanthus* L. Sp. pl. (1753) 569. —*Vleckia chinensis* Rafin. Fl. Tellur. 3 (1818) 89. —*Nepeta lophantha* Fisch. ex Benth. Labiat. gen. et sp. (1834) 464. —*Agastache lophanthus* Kuntze, Revis. gen. 2 (1891) 511. —**Ic.:** Tr. Bot. inst. AN SSSR, ser. 1, 5, 272, fig. 2.

Described from Nor. China. Type in London (Linn.).

Rocky and stony slopes, talus, granite exposures, floors of gorges, coastal pebble beds, sandy valleys; predominantly in steppe belt.

IA. Mongolia: *Khobd.* (along road to Kobdo, June 23, 1870—Kalning; environs of Ubsa lake, Kharkhira river, July 9, 1879—Pot.; Kharkhira mountain complex, Netsugun river—tributary of Namyur river, July 20; same site, Kendulyun river, July 27—1903, Gr. Grzh.). *Mong. Alt.* (road to Kobdo town along Botogon-Gol river, rocky soil, Sept. 1, 1899—Lad.; Taishiri-Ola range, Sakhtogai mountains, on rocks, Aug. 16; Urtu-Gol river valley, river edge, Aug. 17—1930, Pob.; nor. trails of Bus-Khairkhan range, ravine on upper 1/3 of trail, July 17; Tukhumyin-Khundei valley, on knolls and along ravines, Aug. 9—1947, Yun.; Bidzhiin-Gola gorge as road from Tamchi-Daba enters it, on talus on east. flank, Sept. 8; Bodkhon-Gola gorge, on rocks on its west. flank, Sept. 27—1948, Grub.) *Cent. Khalkha* (on bank of Shara-Bulyk, in sand, July 31; on bank of Ikhyn-Zaryn brook, in sand, Aug. 1—1893, Klem.; Del'ger-Khangai range, nor. creek valley, July 19, 1924—Pakhomov; along Urga—Ikhe-Tukhum-Nor lake road, upper valley of Sosyk-Khuduk, dry river-bed, July 8, 1925—Krasch. and Zam.; Del'ger-Khan massif, granite outcrops, rocky soil, Sept. 15, 1925—Gus.; environs of Ikhe-Tukhum-Nor lake, Ongon-Khairkhan mountain, June 1926—Zam.; on rocks at Choiren settlement, Aug. 1, 1926—Kondr.; Choiren-Ula, in crevices among rocks, July 7, 1941; Sorgol-Khairkhan town, along old road to Dalan-Dzadagad, in crevices among granite pillow lava structures, July 15, 1943—Yun.; old road to Dalan-Dzadagad, Bain-Ula town, on granite rocks, July 12, 1948—Grub.). *East. Mong.* (Nukhu-Daban, 1870—Lom.; between Dolon and Dzhirgalantu, July 29, 1898—Zab.; from Kerulen to Hinggan, 900–1200 m, Sept. 12, 1906—Nov.). *Gobi Alt.* (Ubten-Daban pass, Aug. 31, 1886—Pot.; Dundu-Saikhan mountains, river-bed, July 13, 1909—Czet.; Buur range, nor. creek, ravine slopes, July 1924—Pakhomov; Gurban Saikhan, at 2000 m, sand wash in broad deep canyon, fls. blue, 1925—Chaney; nor. slopes of Ikhe-Bogdo range offshoots, June 18, 1926—Kozlova; Ikhe-Bogdo range, Bityuten-Amo creek, creek slopes, Aug. 12, 1927—M. Simukova; Bain-Boro-Nuru, rocky slope, Sept. 10; Bain-Tsagan range, Khotun creek valley, dry river-bed, Aug. 11; Dundu-Saikhan mountains, under shade of rocks, Aug. 17; Dzun-Saikhan mountain, rocky mountain slope, Aug. 25—1931, Ik.-Gal.; Bain-Tsagan mountain range, hill slopes from trail to upper belt, July–Aug. 1933—Khurlat and M. Simukova; Dzun-Saikhan mountain, 6 km south of Dalan-Dzadagad, rocky steppe, June-July 1939—Surmazhab; Gurban-Saikhan range, east. fringe of Dundu-Saikhan range, rocky slope, talus with *Artemisia procera* shrubs, July 22, 1943; south. slope of Ikhe-Bogdo range, Narin-Khurimt, mountain slopes, 2300–2400 m, June 28, 1945—Yun.; same site, 2600–2800 m height, July 20, 1948—Grub.). *East. Gobi* (Khairkhan mountain at foot of rocks, July 23, 1909—Czet.; road from Ulan-Bator to Del'ger-Hangay, Khairkhan mountain, on rock, July 29; Del'ger-Hangay mountain, dry river-bed, July 30—1930, Ik.-Gal.). *Alash. Gobi* (along road from Bain-Barat area to Erdyni-Gol area, June 12, 1909—Czet.).

IIA. Junggar: *Tien Shan* (Nanshankou, humus soil near spring, June 6, 1877—Pot.; Koshety-Daban pass, June 5, 1879—Przew.).

General distribution: East. Sib., Nor. Mon. (Fore Hubs., Hang., Mong.-Daur.), China (North).

2. **L. krylovii** Lipsky in Acta Horti Petrop. 24 (1905) 122; Krylov, Fl. Zap. Sib. 9 (1937) 2305; Levin in Tr. Bot. inst. AN SSSR, ser. 1, 5 (1941) 274; id. in Fl. SSSR, 20 (1954) 277; Grubov, konsp. fl. MNR (1955) 235; Fl. Kazakhst. 7 (1964) 327. —**Ic.:** Tr. Bot. inst. AN SSSR, ser. 1, 5, 275, fig. 3.

Described from Altay. Type in Leningrad.

Alpine belt, dry rocky slopes and plateau, detritus and glacial moraines.

IA. Mongolia: *Mong. Alt.* (Bzau-Kul' lake, alp. tundra, July 11, 1906—Sap.).

General distribution: Jung.-Tarb.; West. Sib. (Altay and Naryn mountain range).

Note. Species of genus *Lophanthus* are very closely interrelated and difficult to distinguish. On analysing the entire material of this genus, E.G. Levin (1941) concluded that most of its species succeed geographically. These also include *L. chinensis* (Rafin.) Benth. and *L. krylovii* Lipsky. Differences between these species are insignificant although geographically nowhere contiguous. Moreover, these species are ecologically distinct as well; *L. krylovii* is an alpine species.

3. **L. schrenkii** Levin in Bot. mater. Gerb. Bot. inst. AN SSSR, 7 (1938) 218; id. in Tr. Bot. inst. AN SSSR, ser. 1, 5 (1941) 276; id. in Fl. SSSR, 20 (1954) 278; Fl. Kazakhst. 7 (1964) 328. —**Ic.:** Bot. mater. Gerb. Bot. inst. AN SSSR, 7, 10, 219; Fl. Kazakhst. 7, Plate 38, fig. 6.

Described from East. Kazakhstan (Bektau-Ata mountains hills). Type in Leningrad.

On rocks, among stones, in shaded gorges.

IB. Kashgar: *East.* (Tamirtyn-Gol, on mountain slopes, Aug. 23, 1895—Rob.).

IIA. Junggar: *Tien Shan* (Urumchi region, Tasenku river upper course, Biangou locality, on rocks in spruce belt, Sept. 24, 1929—Pop.; 30 km south-east of Urumchi, in hilly gorge, June 21, 1958—A.R. Lee (1959)).

General distribution: Fore Balkh.; Jung.-Tarb.; Mid. Asia.

## 9. Phyllophyton Kudo

in Mém. Fac. Sci. and Agr. Taihoku Univ. 2 (1929) 225. —*Pseudolophanthus* Levin in Tr. Bot. inst. AN SSSR, ser. 1, 5 (1941) 294.

1. Calyx with hairy ring inside ................................................................. 2.
+ Calyx without hairy ring inside .............................................................
.................................................. 5. **Ph. tibeticum** (Jacquem.) C.Y. Wu.
2. 2 stamens prominently exserted from corolla tube ........................ 3.
+ All stamens within corolla tube or 2 barely exserted .................... 4.
3. Plant green, all parts profusely glandular; corolla blue ..................
.................................................. 3. **Ph. nivale** (Jacquem.) C.Y. Wu.
+ Plant whitish, with silky pubescence; corolla lilac ...........................
.................................................. 1. **Ph. complanatum** (Dunn) Kudo.
4. Corolla with twisted tube, inverted, its upper lip falls somewhat below and lower above ................ 2. **Ph. decolorans** (Hemsl.) Kudo.
+ Corolla with straight tube, not inverted .........................................
.................................................. 4. **Ph. pharicum** (Prain) Kudo.

1. **Ph. complanatum** (Dunn) Kudo in Mém. Fac. Sci. and Agr. Taihoku Univ. 2 (1929) 225; C.Y. Wu in Acta Phytotax. Sin. 8, 1 (1959) 9. —*Nepeta complanata* Dunn in Notes Bot. Gard. Edinburgh, 6 (1915) 166. —*Glechoma complanata* Turrilla in Rep. Bot. Exch. Cl. Brit. Isles, 5 (1920) 659, in obs.; Marquand in J. Linn. Soc. London (Bot.) 48 (1929) 217. —*Dracocephalum rockii* Diels in Notizbl. Bot. Gart. Berlin, 9 (1926) 1030. —*Pseudolophanthus complanatus* Levin in Tr. Bot. inst. AN SSSR, ser. 1, 5 (1941) 296. —**Ic.:** Tr. Bot. inst. AN SSSR, ser. 1, 5, fig. 9, a.

Described from China (Yunnan). Type in London (K).
In rock debris in high mountain belt, from 4500 to 5200 m.

IIIB. Tibet:*Weitzan* (Yangtze river basin, Chamudug-La pass, 4700 m, on rock screes, covering wet clay, July 27, 1900—Lad.). *South.* ("On slate screes, rare. Atsa Pass, 4800–5200 m, Aug. 27, 1924, Ward"—Marquand, l.c.).
General distribution: China (South-West).

2. **Ph. decolorans** (Hemsl.) Kudo in Mém. Fac. Sci. and Agr. Taihoku Univ. 2 (1929) 225 (sub *Ph. decorans*); C.Y. Wu in Acta Phytotax, Sin. 8, 1 (1959) 10. —*Nepeta decolorans* Hemsl. in Hook. Ic. pl. 25 (1896) tab. 2470; id. in Kew Bull. (1896) 213; Hemsley, Fl. Tibet (1902) 194. —*Glechoma decolorans* Turrill in Rep. Bot. Exch. Cl. Brit. Isles, 5 (1920) 659, in obs. —*Pseudolophanthus decolorans* Levin in Tr. Bot. inst. AN SSSR, ser. 1, 5 (1941) 296. —**Ic.:** Hook. Ic. pl. tab. 2470; Tr. Bot. inst. AN SSSR, ser. 1, 5 fig. 14, $b_1$.
Described from South. Tibet. Type in London (K).
In high-mountains belt.

IIIB. Tibet: *South.* ("Gooring valley, 4950 m, July–Aug. 1895, Littledale, typus"—Hemsl. l.c.).
General distribution: endemic.
Note. *Ph. decolorans* has been described from a region where *Ph. pharicum* (Prain) Kudo is found. Only the type specimen is known. Hemsley's drawing depicts the corolla as oriented normally—a fact disputed by E.G. Levin (l.c.) who affirms for this species in 'inverted' corolla. We did not have occasion to study the type specimen of this species to verify its characteristics. However, if further investigations do not confirm 'inverted' corolla in *Ph. decolorans* on the basis of which this species is distinguished from *Ph. pharicum*, it should be treated as a synonym of the latter.

3. **Ph. nivale** (Jacquem.) C.Y. Wu in Acta Phytotax. Sin. 8, 1 (1959) 9. —*Glechoma nivalis* Jacquem. ex Benth. Labiat. gen. et sp. (1835) 737; Briq. in Engler-Prantl, Naturl. Pflanzenfam. IV, 3a (1897) 238. —*Nepeta nivalis* Benth. Labiat. gen. et sp. (1835) 737; Hook f. Fl. Brit. Ind. 4 (1885) 664. —*Pseudolophanthus nivalis* Levin in Tr. Bot. inst. AN SSSR, ser. 1, 5 (1941) 295. —**Ic.:** Tr. Bot. inst. AN SSSR, ser. 1, 5, Fig. 9, b.
Described from West. Himalayas. Type in Paris. Isotype in London (K).
Rubble slopes in high-mountain belt.

IIIB. Tibet: *South.* (Mt. Everest Exped., Arun Valley 3300 m, June 9; Dsakar Chu, 3750 m, July 6—1922, Morton [K]; Dsakar Chu, 4200 m, July 7, 1933—coll.? [K]; Bahmo Doptra La, 4200 m, June 29, 1938—Lloyd [K]).
General distribution: Himalayas (west.).

4. **Ph. pharicum** (Prain) Kudo in Mém. Fac. Sci. and Agr. Taihoku Univ. 2 (1929) 225; C.Y. Wu in Acta Phytotax. Sin. 8, 1 (1959) 9. —*Nepeta pharica* Prain in J. As. Soc. Bengal, 59 (1890) 306. —*Pseudolophanthus pharicus* Kuprian. in Bot. zh. 33 (1948) 235.
Described from East. Himalayas (Sikkim). Type in London (K).

IIIB. Tibet: *South*. (Khambajong, July 16, 1903—Younghusband [K]; Gyantse, 4500–4950 m, July 16, 1924 Ludlow [K]).

General distribution: Himalayas (east.).

Note. See note under *Ph. decolorans* (Hemsl.) Kudo.

5. **Ph. tibeticum** (Jacquem.) C.Y. Wu in Acta Phytotax. Sin. 8, 1 (1959) 10. —*Glechoma tibetica* Jacquem. ex Benth. Labiat. gen. et sp. (1835) 737, pro syn. —*Nepeta tibetica* Benth. Labiat. gen. et sp. (1835) 737; Hook. f. Fl. Brit. Ind. 4 (1885) 664; Hemsley, Fl. Tibet (1902) 194; Strachey, Catal. (1906) 142. —*Pseudolophanthus tibeticus* (Jacquem.) Kuprian. in Bot. zh. 33 (1948) 235.

Described from West. Himalayas. Type in Paris.

IIIB. Tibet: *South*. ("Near Rakas Tal, 4500–5100 m, Strachey et Winterbottom"—Hemsley, l.c.; "Lanjar, 4500–5100 m, Aug., Strachey et Winterbottom"—Strachey, l.c.).

General distribution: Himalayas (west.).

Note. *Ph. tibeticum* occupies a special position in the genus, representing somewhat a link between genera *Phyllophyton* and *Glechoma*. Possibly, it would be even more appropriate to treat it as a more xerophilised member of the latter. In *Ph. tibeticum*, the calyx lacks a hairy ring inside while arrangement of anther lobes is like that in species of *Glechoma*, i.e., the lobes diverge at a right angle. In all other characteristics, however, this is a typical representative of genus *Phyllophyton*.

## 10. **Schizonepeta** Briq.
in Engler-Prantl, Naturl. Pflanzenfam. IV, 3a (1897) 235.

1. Annual with leaves bipinnatisect into slender lobes; calyx teeth with awn-like cusp; corolla pale, whitish ... 1. **S. annua** (Pall.) Schischk.

+ Perennial with leaves incised into broad lobes, more rarely leaves almost undivided; calyx teeth without awn-like cusp; corolla bluish-violet ........................................................ 2. **S. multifida** (L.) Briq.

1. **S. annua** (Pall.) Schischk. in Spisok rast. Gerb. fl. SSSR, 10, 64 (1936) 72; Krylov, Fl. Zap. Sib. 9 (1937) 2314; Pojark. in Fl. SSSR, 20 (1954) 285; Grubov, konsp. fl. MNR (1955) 235; Fl. Kazakhst. 7 (1964) 330. —*S. botryoides* Briq. in Engler-Prantl, Naturl. Pflanzenfam. IV, 3a (1897) 237. —*Nepeta annua* Pall. in Acta Ac. Sci. Petrop. 1779, 2 (1783) 263. —*N. multifida* L. f. Suppl. (1781) 273 (non L. 1753). —*N. botryoides* Sol. in Ait. Hort. Kew ed. 1, 2 (1789) 287; Benth. Labiat. gen. et sp. (1834) 468; Hook. f. Fl. Brit. Ind. 4 (1885) 657; Sapozhn. Mong. Alt. (1911) 381; Simpson in J. Linn. Soc. London (Bot.) 41 (1913) 436; Danguy in Bull. Mus. nat. hist. natur. 20 (1914) 83. —**Ic.**: Pall. l.c. tab 12.

Described from Siberia. Type in Paris.

Rocky and rubble mountain slopes, trails, debris, sandy-pebble banks of rivers and floors of ravines.

IA. Mongolia: *Mong. Alt., Cen. Khalkha* (not far from Toli river bank, on cliff, June 18, 1895—Klem.; rock precipices and mountain slopes on Toli river, July 2, 1924—Pavl.; Del'ger-

Hangay mountains, rocky slope, July 30, 1931—Ik.-Gal.). *Bas. Lakes* (environs of Ubsa lake, Kholbo-Nor, pebble bed of dry brook, July 27; same site, Shibe river, dry river-beds, Sept. 1— 1879, Pot.; valley of Khobdo river near ferry point, along Khobdo-Ulangom road, meadow in Khobdo river valley, Aug. 23, 1944—Yun.). *Val. Lakes* (in dry creek, on rock debris on right bank of Tuin-Gol, July 9; on mountain on left bank of Tatsy-Gol, below Tatsy urton, July 18— 1893, Klem.; near Artso-Gol river estuary, rubble semi-arid land, July 25, 1922—Pisarev). *Gobi-Alt.*, East. Gobi (Khoir-Uldzeitu area, near Khutuk-Ula, desert steppe, June 25-Aug. 1930— Kuznetsov).

IB. **Kashgar:** *East.* (along road from Baiyankhe to Sansanko, in Turfan, on shaded slope of desert steppe, June 15, 1958 —A.R. Lee (1959)).

IIA. **Junggar:** *Tarb.* (Kotbukha, Aug. 10, 1876—Pot.). *Tien Shan* (Nanshankou, July 7, 1877—Pot.). *Jung. Gobi* (10 km north of Kosh-Tologoi settlement on Khobuk river, along road from Karamai to Altay, pass through gently hummocky ridge, desert steppe, July 4; 85 km south of Ertai [on Urungu], along road from Altay to Guchen, low-hillocky desert, in shale crevices, July 16—1959, Yun.).

**General distribution:** Fore Balkh.; West. Sib. (Altay), East. Sib. (Ang.-Sayan.), Himalayas (Kashmir).

2. **S. multifida** (L.) Briq. in Engler-Prantl, Naturl. Pflanzenfam. IV, 3a (1897) 235; Kudo in Mém. Fac. Sci. and Agr. Taihoku Univ. 2 (1929) 227; Krylov, Fl. Zap. Sib. 9 (1937) 2315; Kitag. Lin. Fl. Mansh. (1939) 384; Pojark. in Fl. SSSR, 20 (1954) 283; Grubov, konsp. fl. MNR (1955) 235; Fl. Kazakhst. 7 (1964) 331. —*Nepeta multifida* L. Sp. pl. (1753) 572, non L. f. 1781. —*N. lavandulacea* L. f. Suppl. (1781) 272; Franch. Pl. David. 1 (1884) 237; Forbes and Hemsley, Index Fl. Sin. 2 (1902) 290; Sapozhn. Mong. Alt. (1911) 381; Dunn in Notes Bot. Gard. Edinburgh, 6 (1915) 166. —**Ic.:** Reichb. Ic. bot. tab. DXXX, Fig. 726; Fl. SSSR, 20, Plate 14, fig. 3.

Described from Dauria. Type in London (Linn.).

Rubble and stony mountain slopes of rocks, dry ridges, debris, gully flanks, dry forest margins.

IA. **Mongolia:** *Cen. Khalkha* (east. slope of Bain-Khn mountain, 1899—Pal.; Sharkhai-Khuduk valley, river-bed, July 24, 1909—Czet.; Tsinkirin-Gol river valley, near Tsinkir-Dugang, south. rocky slope, steppe, July 23; east. rim of Bain-Ulan area, rocky slopes of mountains Sept. 8—1949, Yun.). *East. Mong.* (Muni-Ula mountains, July 1, 1871—Przew.; nor.-west. slopes of Malagaiten-Daban, July 23; syrt (watershed upland) between Kyrymty and Buin-Gol rivers, Aug. 1—1899, Pot. and Sold.; vicinity of Manchuria station, 1915—E. Nechaeva; Chaptsur, 1500 m, 1925—Chaney; 28–30 km south-east of Bain-Buridu, forb-grass steppe, Aug. 5; 30 km east-south-east of Bain-Tsagan somon, feather grass forb steppe, Aug. 6; 5 km west of Toge-Gol river, forb meadow, Aug. 7; 25 km south of Bain-Dung somon, near Irgai-Ula pass, grass-tansy steppe, Aug. 26—1949, Yun.). *Bas. Lakes* (on hill on left bank of Bogden-Gol river, July 18, 1895—Klem.) . *Gobi-Alt.* (Dzun-Saikhan, Yalo creek, willow groves on upper slope, Aug. 23, 1931—Ik.-Gal.).

IIA. **Junggar:** *Tien Shan* (along Dzhagastai river, Sept. 1, 1878—Larionov).

IIIA. **Qinghai:** *Nanshan* (along Tetung river, Aug. 9, 1880—Przew.).

**General distribution:** Tarb. (Saur); West. Sib., East. Sib., Far East, Nor. Mong., China (Dunbei, North), Korean peninsula.

## 11. Nepeta L.
Sp. pl. (1753) 570, p.p.

1. Annuals, with slender and short root.................................................2.
+ Perennials, with woody more or less thick root and often with distinct rhizome ..................................................................................4.
2. Calyx straight with erect nearly equal teeth; midlobe of lower lip of corolla erect, with protuberance at base; nuts glabrous, lustrous; involucre bracts more or less large, longer than cyme or false whorl ....................................................................................................3.
+ Calyx curved, bilabiate, lower teeth narrower and shorter than upper; midlobe of lower lip of corolla directed downward, glabrous, without protuberance; nuts pitted tuberculate; bracts of false whorl or cyme small, narrowly linear, shorter than calyx ........................... ................................................................ 10. **N. micrantha** Bunge.
3. Cauline leaves and leafy bracts reniform or semi-orbicular, bracts of false whorl like caulous leaves or spathulate, narrowed into long petiole ................................................................ **N. spathulifera** Benth.
+ Caulous leaves and leafy bracts ovate to narrowly lanceolate, rigidulous, involucre bracts of false whorl stiff, acuminate spinelike, with thick hard nerves, more or less longitudinally folded. ... ................................................................ 13. **N. pungens** (Bunge) Benth.
4. Flowers aggregated into large globose terminal head; involucre bracts of false whorls, comprising head, and bracts numerous, violet- large, longer than flowers, 1.5–2.3 cm long; low alpine plant with cuneate leaves. ............................... 9. **N. longibracteata** Benth.
+ Flowers aggregated into panicle or loose raceme or dense spicate inflorescence; involucre bracts of false whorls and bracts very short; plant with well-developed, fairly tall stem. ...............................5.
5. Plants usually unisexual; calyx straight, more less with nearly straight throat and erect subulate teeth.............................................6.
+ Flowers bisexual calyx with straight, oblique or bilabiate throat; teeth not subsulate. .........................................................................7.
6. Flowers in fork of cyme sessile; nuts with fairly crowded, very flat and small subacute tubercles, sometimes bulge more near tip bulging .............................................................. 16. **N. ucrainica** L.
+ Flowers in fork of cyme stalked; nuts glabrous or with sparse flat tubercles only in upper part and on back, at tip invariably with acute papilliform outgrowths and occasional hairs .......................... ................................................................ 11. **N. pannonica** L.
7. Stems foliated only in lower part, branched right from base; all false whorls on long peduncles, very rarely forming racemose or subpaniculate inflorescence ....................... 17. **N. yanthina** Franch.

+ Stems foliated almost throughout height, simple or branched; cyme or false whorl on short peduncle or sessile, gathered into loose raceme or dense spicate inflorescence, or in compound dense cymes, aggregated into dense racemes at ends of stems and axillary branches ................................................................................................. 8.

8. Flowers in compound dense cymes, aggregated into dense racemes at ends of stems and axillary branches; midlobe of lower lip of corolla concave, cyathiform, with large-toothed edges upcurved, without protuberance at base ............................................. 1. **N. cataria** L.

+ Flowers in cyme or false whorl aggregated into loose raceme or dense spicate inflorescence; midlobe of lower lip of corolla with protuberance at base, tapering toward its emarginated tip, with flat, nutant or subhorizontal procumbent lobes .................................... 9.

9. Flowers in terminal dense spicate inflorescence, 1–3 lower false whorls sometimes distant; calyx obconical .................................. 10.

+ Flowers aggregated into loose raceme; calyx tubular................. 18.

10. Calyx with straight throat and linearly subulate teeth, almost as long as tube; inflorescence narrowly cylindrical, spicate ........... 11.

+ Calyx with oblique throat or nearly bilabiate, teeth lanceolate or narrowly deltoid, 1/3–1/2 length of tube; inflorescence ovate, cylindrical or oblong-cylindrical ......................................................... 13.

11. Leaves lanceolate or oblong, small, up to 2.5 cm long and 0.8 cm broad, acuminate, more rarely (mainly lower ones) ovate, with 2–4 acute or subobtuse teeth on each side along margin. ......................
.................................................................. 12. **N. podostachys** Benth.

+ Leaves ovate, broadly ovate or ovate-cordate, very large or similar but small, invariably crenate-dentate or crenate along margin.. 12.

12. Stem straight, erect, sometimes poorly branched but lateral branches never growing like main stem; leaves large, 2–7 cm long and 1–3.5 cm broad. .................................. **N. laevigata** (D. Don) Hand.-Mazz.

+ Stem strongly branched, slightly ascending; leaves broadly ovate, small, 0.7–1.5 cm long and 0.6–1 cm broad. .....................................
.................................................................. 4. **N. discolor** Royle ex Benth.

13(10). Branched hairs present in prominent proportion of pubescence in plant.................................................................. **N. pamirensis** Franch.

+ All hairs in pubescence unbranched ............................................. 14.

14. Bracts shorter than calyx, usually linear or lanceolate ............... 15.

+ Bracts longer than calyx, foliaceous, oblong or orbicular .......... 17.

15. Axillary branches short, slender, invariably sterile, often underdeveloped .................................................... 3. **N. densiflora** Kar. et Kir.

+ Axillary branches elongate, foliaceous, all or most terminating in inflorescence .............................................................................. 16.

16. Caulous leaves green, with very short pubescence, lanceolate or oblong-ovate, with 3–4 spaced teeth along margin. ........................ ................................................................... 15. N. transiliensis Pojark.

+ Caulous leaves greyish on both surfaces due to dense pubescence, mostly rhomboid-ovate in outline, with approximate acute teeth along margin. ......................................................7. N. kokamirica Regel.

17. Stem stout, erect, profusely foliated; all bracts green; leaves oblong or oblong-ovate, crenate along margin. ............................................... .........................................................2. N. coerulescens Maxim.

+ Stems weak, ascending, moderately foliated; bract of upper flowers in inflorescence violet; leaves broadly ovate, argute along margin. ............................... 8. N. lamiopsis Benth. ex Hook. f.

18 (9). All leaves sessile, entire, oblong-lanceolate, tapered toward both ends, subobtuse ........................... 6. N. hemsleyana Oliv. ex Prain.

+ Only upper and middle caulous leaves sessile or all leaves on distinct petioles, ovate, ovate-oblong or ovate-lanceolate, serrate or dentate or crenate-dentate along margin, invariably acuminate ............. 19.

19. Leaves coriaceous, stiff, green on upper surface, greyish beneath, cyme or false whorl invariably on distinct peduncle, longer in lower false whorls .................................................. 5. N. erecta (Royle) Benth.

+ Leaves leptodermatous, not stiff, green on both surfaces; false whorls either subsessile or on barely distinct peduncles; only lower ones on fairly long peduncles ....................................................................... 20.

20. All leaves with distinct petioles, granular-glandular beneath; calyx and corolla with yellow glands ........................... 14. N. sibirica L.

+ Upper and middle caulous leaves sessile, cordate at base, lower caulous leaves petiolate, with sunken glands beneath; only small stalked glands in pubescence of calyx........ 18. N. wilsonii Duthie.

1. N. cataria L. Sp. pl. (1753) 570; Benth. Labiat. gen. et sp. (1834) 477; Hook. f. Fl. Brit. Ind. 4 (1885) 662; Hemsl. in J. Linn. Soc. London (Bot.) 26 (1890) 288; Diels in Engler's Bot. Jahrb. 29 (1900) 553; Forbes and Hemsley, Index Fl. Sin. 2 (1902) 288; Dunn in Notes Bot. Gard. Edinburgh, 6 (1915) 167; Kudo in Mém. Fac. Sci. and Agr. Taihoku Univ. 2 (1929) 229; Krylov, Fl. Zap. Sib. 9 (1937) 2309; Pojark. in Fl. SSSR, 20 (1954) 349; Fl. Kirgiz. 9 (1960) 51; Fl. Kazakhst. 7 (1964) 338.   —Ic.: Reichb. Ic. fl. Germ. XVIII, tab. 1242; Fl. Kazakhst. 7, Plate 39, fig. 4.

Described from Europe. Type in London (Linn.).

Meadows on mountain slopes, rubbish heaps, farms and kitchen gardens.

IIA. Junggar: *Tien Shan* (Dzhagastai, 600–900 m, Aug. 6, 1877; Sharabaguchi, 1200–1500 m, Sept. 20, 1878—A. Reg.).

General distribution: Aralo-Casp., Fore Balkh., Jung.-Tarb., Cent. Tien Shan; Europe, Mediterranean, Balk.-Asia Minor, Fore Asia, Caucasus, Mid. Asia, West. Sib., Far East, China (Central, South-West), Himalayas (Kashmir), Japan, India, North America (introduced).

2. **N. coerulescens** Maxim. in Bull. Ac. Sci. St.-Pétersb. 27 (1881) 529; Prain in J. As. Soc. Bengal, 59 (1890) 304; Forbes and Hemsley, Index Fl. Sin. 2 (1902) 289; Kudo in Mém. Fac. Sci. and Agr. Taihoku Univ. 2 (1929) 232; Marquand in J. Linn. Soc. London (Bot.) 48 (1929) 217; Rehder and Kobuski in J. Arn. Arb. 14 (1933) 31; Hao in Engler's Bot. Jahrb. 68 (1938) 634.  —*N. thomsoni* Benth. in Hook. f. Fl. Brit. Ind. 4 (1885) 658; Hemsley, Fl. Tibet (1902) 194.  —*Dracocephalum coerulescens* Dunn in Notes Bot. Gard. Edinburgh, 6 (1915) 171.  —*D. breviflorum* Turrill in Kew Bull. (1922) 154.

Described from Qinghai. Type in Leningrad. Plate II, fig. 3.

On clayey-rocky slopes, precipices and rocks at 3000–4900 m alt.

IIIA. Qinghai: *Nanshan* (South Kukunor range, Paidza-Gol river, rocks, 3300 m, Sept. 9, 1894—Rob.). *Amdo* (along Mudzhikkhe river, height 2700–2850 m, June 16, 1880—Przew., typus! "Radja and Yellow River gorges, Rock"—Rehder and Kobuski, l.c.; "auf dem Gebirge ja-he-mari, 4000 m; auf dem Ming-ge bei Tsi-gi-ganba, 3900 m, Hao"—Hao, l.c.).

IIIB. Tibet: *Weitzan* (on bank of Yantszytszyan, July 26, 1884—Przew.; Yantszytszyan basin, in Khichu river valley, on sunny side of precipice, clayey-rocky soil, 4050 m, July 14; Nruchu area, right bank, on conglomerate precipices of Yantszytszyan river, July 25—1900, Lad.). *South.* (Khambajong, Sept. 7, 1903—Young-husband [K]; Everest Exped., 1921—Wollaston [K]; above Kampa Dsong, 4500 m, July 17, 1922—coll. ? [K]; Everest Exped., Tinki Dsong, 4200 m, July 14, 1924—Kingston [K]; Sorogon, 15 miles from Gyantse, 4000 m, July 22, 1924; Gyantse, Kala, 4350 m, July 8, 1925—Ludlow [BM]; Dochen, 4200 m, Aug. 7, No. 856; Kala, 4200 m, Aug. 8, No. 879—1936, Chapman [K]; Everest Exped., Tenkye Dsong, 4200 m, June 29, 1938—Lloyd [K]; "Lanjar, 4900 m, Strachey et Winterbottom"—Hemsley, l.c.; "In sheltered gullies or under rocks in the alpine pastures, Atsa Tso, 4500 m, Aug. 26, 1924, Ward"—Marquand, l.c.).

General distribution: Himalayas (west, Kashmir).

3. **N. densiflora** Kar. et Kir. in Bull. Soc. natur. Moscou, 14 (1841) 725; Krylov, Fl. Zap. Sib. 9 (1937) 2313; Pojark. in Fl. SSSR, 20 (1954) 314; Grubov, Konsp. fl. MNR (1955) 235; Fl. Kazakhst. 7 (1964) 333.  —*N. saposhnikowii* Nick. et Plotn. in Bot. mater. Gerb. Gl. Bot. sada SSSR, 6 (1926) 20.  —*Nepeta* sp. Sapozhn. Mong. Alt. (1911) 381.

Described from Altay (Narym mountain range). Type in Leningrad. Talus and rocky slopes in alpine belt.

IA. Mongolia: *Mong. Alt.* (pass from M. Kairta to M. Ku-Irtysh, talus, July 17, 1908; upper course of M. Ku-Irtysh river—Sap.; Khargatiin-Daba, near summer camp of somon, alp. meadow, July 23; south. flank of Indertin-Gol river valley, near summer camp of somon, mountain steppe, July 24; along road to Kharagaitu-Khutul', alp. steppe, July 24—1947, Yun.).

General distribution: West. Sib. (Altay—Narym range).

Note. *N. densiflora* is reported only from Dzhaidak mountain (Narym range) where it has been collected time and again. In Mong.-Altay within Mongolia, however, plants gathered differ somewhat from *N. densiflora* in larger, ovate, obtusely dentate (more rarely almost entire) leaves. In our opinion, there is no justification yet to treat this broader-leaved race as an independent species, more so since high-mountain areas of Mongolian Altay have not been well studied to data. Evidently, the report from Dzhaidak mountains should be regarded as the westernmost and records in Mongolian Altay (A.A. Yunatov's collections) as easternmost boundary of the species. Possibly in future, the flora of intermediate regions will be studied more thoroughly and then it may be possible to correct the original description of the species. The shape of leaves evidently varies somewhat–lanceolate, oblong-ovate, ovate—but all of them have cuneate or orbicular-cuneate base.

4. **N. discolor** Royle ex Benth. in Hook. Bot. misc. 3 (1833) 378; Benth. Labiat. gen. et sp. (1834) 470; Hook. f. Fl. Brit. Ind. 4 (1885) 659; Hemsley, Fl. Tibet (1902) 194; Strachey, Catal. (1906) 142; Pampanini, Fl. Carac. (1930) 181; Persson in Bot. notiser (1938) 299. —*N. sabinei* Schmidt in J. Bot. (London) 6 (1868) 238. —Ic.: Schmidt, l.c. tab. 82, fig. 1–4 (sub nom. *N. sabinei*).

Described from Himalayas. Type in London (K).

High mountain areas, 3000–5000 m.

IIIB. Tibet: *South.* (Niti pass, 4500 m—Strachey and Winter-bottom [BM]).
General distribution: Himalayas (west., Kashmir), Afghanistan.

5. **N. erecta** (Royle) Benth. Labiat. gen. et sp. (1834) 482; Hook. f. Fl. Brit. Ind. 4 (1885) 663; Murata in Acta Phytotax. et Geobot. 16, 1 (1955) 15. —*Dracocephalum erectum* Royle ex Benth. in Hook. Bot. misc. 3 (1833) 380.

Described from India ("Kanaour"). Type in Liverpool.

IIIB. Tibet: *South.* ("circa Lhasa, Pulunka temple, No. 103292, Aug. 20; Panchogan, No. 103259, Aug. 23—1914,, Kawaguchi•—Murata, l.c.).
General distribution: Himalayas (west.).

6. **N. hemsleyana** Oliv. ex Prain, J. As. Soc. Bengal, 59 (1890) 305. — *Dracocephalum hemleyanum* (Oliv.) Prain ex Marq. in J. Linn. Soc. London (Bot.) 48 (1929) 218; Murata in Acta Phytotax. et Geobot. 16, 1 (1955) 13.

Described from Himalayas (not far from Fari town). Type in London (K).

Gravel terraces of rivers, 3000–4500 m.

IIIB. Tibet: *South.* (Gyangtse, July–Sept. 1904—Walton [K]; Gyamba, forms large clumps on open gravel terraces above the river, alt. 3300 m, Sept. 2, 1924—Kingdon Ward [K]; "circa Lhasa, Dha village, Aug. 23; Lhasa, Neesal temple, Sept. 25—1914, Kawaguchi"—Murata l.c.).
General distribution: Himalayas (east.).

7. **N. kokamirica** Regel in Acta Horti Petrop. 6 (1879) 358; Pojark. in Fl. SSSR, 20 (1954) 316; Fl. Kazakhst. 7 (1964) 334. —Ic.: Gartenflora, 29, tab. 1030.

Described from East. Tien Shan. Type in Leningrad.

Talus, rocky slopes from upper forest limit to alpine belt.

IIA. Junggar: *Tien Shan* (Kokkamyr mountains, 2100–2400 m, July 31, 1878, typus! Nilki near Kash river, 2100 m, June 8; Karagol, near pass to Nilki, 3000 m, June 17—1879; Beibeshan pass, 2100–2700 m, Aug. 31, 1880—A Reg.).
General distribution: Jung.-Tarb.

**N. laevigata** (D. Don) Hand.-Mazz. Symb. Sin. 7 (1936) 916; id. in Acta Horti Gotoburg, 13 (1939) 343; C.Y. Wu in Acta Phytotax. Sin. 8 (1959) 17.

—*N. spicata* Benth. in Wall. Cat. (1829) No. 2083; ej. Pl. Asia Rar. 1 (1830) 64;
ej. Labiat. gen. et sp. (1834) 470; Hook. f. Fl. Brit. Ind. 4 (1885) 659; Kudo in
Mém. Fac. Sci. and Agr. Taihoku Univ. 2 (1929) 228.   —*N. lamiopsis* auct.
non Benth.: Diels in Notes Bot. Gard. Edinburgh, 7 (1912) 151, 314 and
(1913) 378; Dunn, ibid. 6 (1915) 166.   —*Betonica laevigata* D. Don, Prodr. Fl.
Nepal. (1825) 110.

Described from West. Himalayas (Kumaon). Type in London (BM).

In coniferous and mixed forests, forest meadows, scrub, 2300–4000 m.

IIIB. Tibet: *South.* (occurrence possible).

General distribution: China (South-west), Himalayas (west., east., Kashmir), Afghanistan.

Note. Species with a fairly broad distribution range encompassing regions beyond the
territory under study. Within our flora, it may be found in South. Tibet. Plants differing somewhat from *N. laevigata* distributed in the western part of the distribution range are found in
Yunnan and partly in Sichuan. Many researchers have recorded the polymorphism of this
species.

Murata [Murata in Acta Phytotax. et Geobot. 16, 1 (1955) 15] cited *N. eriostachys* Benth.
from South. Tibet ("Lon-zonka village, July 31; circa Shigatse, Chusul station, Aug. 3; circa
Lhasa, mt. Panchogan, Aug. 23; mt. Sandok Petri, Sept. 25—1914, Kawaguchi"). *N. eriostachys*
Benth. and *N. discolor* Royle ex Benth. represent very closely related species and are found in
the high mountain areas of West. Himalayas but no one has reported them in the East. Himalayas regions. Having had no occasion to study the material cited by Murata, we restrain
from citing *N. eriostachys* for Cent. Asian flora. In our opinion, the entire material given under
this name pertains beyond doubt to one of the species of section Spicatae (Benth.) Pojark.,
more than to *N. laevigata* (D. Don) Hand.-Mazz.

8. **N. lamiopsis** Benth. ex Hook. f. Fl. Brit. Ind. 4 (1885) 659; Vautier,
Candollea, 17 (1959) 48.

Described from East. Himalayas (Sikkim). Type in London (K).

IIIB. Tibet: *South.* (Chaksam, Sangpo [Brahmaputra] Valley, July 4, 1905—Walton [K];
Lhasa, Sept. 21, 1905—Waddell [K]; Mt. Everest Exped., 1921—Wollaston [K]).

General distribution: Himalayas (east.).

9. **N. longibracteata** Benth. Labiat. gen. et sp. (1835) 737; Hook. f. Fl.
Brit. Ind. 4 (1885) 660; Hemsley in J. Linn. Soc. London (Bot.) 30 (1894) 118;
Deasy, In Tibet and Chin. Turk. (1901) 398; Hemsley, Fl. Tibet (1902) 194;
Strachey, Catal. (1906) 142; Pampanini, Fl. Carac. (1930) 182; Pojark. in Fl.
SSSR, 20 (1954) 303; Ikonnikov, Opred. rast. Pamira (1963) 210.   —**Ic.:**
Jacquem. Voy. Ind. Bot. tab. 137.

Described from West. Himalayas. Type in Paris.

Talus in high-mountain areas.

IIIB. Tibet: Chang Tang ("South end of Aru Tso, 4800 m, Aug. 4, 1896, Deasy and Pike"—
Hemsley, l.c.; "Dong Lung, 5050 m; e presso Chisil Gilgha, 5100 m; Campo Oltreremo 2°, 5100
m—Dainelli e Marinelli; Valle Shaksgam, 5025 m, Clifford"—Pampanini, l.c.). *South.* ("Balch
pass, 5100 m, Strachey et Winterbottom"—Hemsl. l.c.).

IIIC. Pamir ? (reported from adjoining regions).

General distribution: East. Pam.; Fore Asia, Himalayas (west., Kashmir).

10. **N. micrantha** Bunge in Ledeb. Fl. alt. 2 (1830) 401; Benth. Labiat. gen. et sp. (1834) 476; Danguy in Bull. Mus. nat. hist. natur. 20 (1914) 83; Krylov, Fl. Zap. Sib. 9 (1937) 2312; Pojark. in Fl. SSSR, 20 (1954) 383; Grubov, Bot. mater. Gerb. Bot. inst. AN SSSR, 19 (1959) 549; Fl. Kirgiz. 9 (1960) 52; Fl. Kazakhst.7 (1964) 339. —**Ic.:** Ledeb. Ic.pl. fl. ross. 5, tab. 412.

Described from East. Kazakhstan (Arkaul hills). Lectotype in Leningrad.

Sand knolls, dunes, more rarely on dry slopes in foothills and lower mountain belt.

IIA. Junggar: *Cis-Alt.* (south. slope of Chinese Altay, Ak-Su river, Burchum river system, June 10, 1903—Gr. Grzh.). *Tien Shan* (Karadzhal mountain range, right bank of Tutunkho river, high unflooded terrace, about 900 m high, June 7, 1954—Mois.). *Jung. Gobi* (Kholyt range, May 13, 1879—Przew.; Savan-Paotai, in sand, Nos. 770, 781, 851; north of San'daokhetsz, Sykeshu, No. 21, June 11; Manas river, west of Syaeda, sand, No. 902, June 14; near Shikho, on sand-dune, No. 218, June 25—1957, Kuan).

General distribution: Aralo-Casp., Fore Balkh., Jung.-Tarb., Nor. and Cent. Tien Shan; Mid. Asia.

**N. pamirensis** Franch. in Bull. Mus. nat. hist. natur. 2 (1896) 345; Pojark. in Fl. SSSR, 20 (1954) 333; Ikonnikov, Opred. rast. Pamira (1963) 211. —*N. pamiro-alaica* Lipsky in Tr. Bot. sada, 23 (1904) 230, p.p.

Described from Pamir (upper Vakhan-Dar'i). Type in Paris.

Rocky-rubble slopes, dry pebble beds, moraines, cobresia-sedge meadows, wormwood-bluegrass steppes in alpine belt.

IIIC. Pamir (occurrence possible since it is known from adjoining regions).

General distribution: East. Pam.; Mid. Asia (West. Pam.), Himalayas (west.).

11. **N. pannonica** L. Sp. pl. (1753) 570; Pojark. in Fl. SSSR, 20 (1954) 406; Fl. Kirgiz. 9 (1960) 53; Fl. Kazakhst. 7 (1964) 340. —*N. nuda* auct. non L.: Danguy in Bull. Mus. nat. hist. natur. 20 (1914) 83; Krylov, Fl. Zap. Sib. 9 (1937) 2309; Grubov, konsp. fl. MNR (1955) 235. —*N. turkestanica* Gandog. in Bull. Soc. Bot. France, 60 (1913) 26. —**Ic.:** Reichb. Ic. fl. Germ. tab. 1243, fig. III; Fl. Kazakhst. 7, Plate 39, fig. 5.

Described from East. Europe, Type in London (Linn.).

Forest borders, glades, wet meadows, meadoes in ravines, steppe slopes, 1200–3000 m.

IA. Mongolia: *Khobd.* (between boundary and Kobdo mountain, 1870—Kalning).

IIA. Junggar: *Cis-Alt.* (nor.-west. Shara-Sume, shrubby meadow steppe, July 7, 1959—Yun.), *Tarbag.* (Tarbagatai mountain range, north of Dachen town, on slope, Nos. 1525, 1534, 2900, Aug. 12, 1957—Kuan). *Jung. Alt.* (Toli district, Albakzin mountains, on slope, No. 2560, Aug. 6; same site, Nos. 1159, 1358, Aug. 7; 15 km north of Ulastai, Nos. 1236, 3859, Aug. 28—1957, Kuan). *Tien Shan* (nor. and east.).

General distribution: Aralo-Casp., Fore Balkh., Jung.-Tarb., North and Cent. Tien Shan; Europe (cent. and east.), Balk. (nor.), Caucasus, Mid. Asia, West. and East. Sib.

12. **N. podostachys** Benth. in DC. Prodr. 12 (1848) 372; Pojark. in Fl. SSSR, 20 (1954) 312; Fl. Kirgiz. 9 (1960) 45; Ikonnikov, Opred. rast. Pamira (1963) 211. —*N. maracandia* Bunge in Mém. Ac. Sci. St.-Pétersb. 7 (1851) 434. —*N. paulsenii* Briq. in Bot. tidsskr. 28 (1908) 235. —*N. discolor* auct. non Royle ex Benth.: Danguy in Bull. Mus. nat. hist. natur. 14 (1908) 132. — Ic.: Briq. l.c. fig. 2 (sub nom. *N. paulsenii* Briq.).

Described from Fore Asia (Kabula region). Type in London (K).

Rubble and rocky slopes of meadows and steppes, up to 4000–5000 m.

IIIC. Pamir ("Pentes est du Sasser-La, éboulis, alt. 5150 m, Sept. 7, 1906, Lacoste"— Danguy, l.c.).

General distribution: East. Pam.; Mid. Asia (Pam.-Al.), Fore Asia (east.), Himalayas (Kashmir).

13. **N. pungens** (Bunge) Benth. em. Stapf: Benth. Labiat. gen. et sp. (1834) 487, quoad nomen; Stapf, Bot. Ergebn. Polak. Exped. 1 (1885) 47; Krylov, Fl. Zap. Sib. 9 (1934) 2312; Pojark. in Fl. SSSR, 20 (1954) 429; Fl. Kirgiz. 9 (1960) 54; Fl. Kazakhst. 7 (1964) 342. —*N. pusilla* Benth. l.c. 488. —*N. fedtschenkoi* Pojark. in Fl. SSSR, 20 (1954) 524, 430. —*Ziziphora pungens* Bunge in Ledeb. Fl. alt. 1 (1829) 23. —Ic.: Ledeb. Ic. pl. fl. ross. 2, tab. 124; Fl. Kazakhst. 7, Plate 39, fig. 6.

Described from East. Kazakhstan (Arkaul hills). Type in Leningrad.

Rocky and rubble slopes of foothills to midbelt of mountains, on takyrs in deserts, in saxaul thickets and dry river-beds.

IIA. Junggar: *Jung. Gobi* (Shankhausyan, May 24, 1879—Przew.). *Zaisan* (Mai-kapchagai mountain, rocky slope, June 6, 1914—Schischk.). *Dzhark.* (Kul'dzha environs, 600–1200 m; Bayandai, 600–1200 m—May 6, 1878, A. Reg.).

General distribution: Aralo-Casp., Fore Balkh., Jung.-Tarb., Nor. Tien Shan; Mid. Asia.

Note. The lone species of series *Pungentes* Pojark. growing in Cent. Asian territory should be called *N. pungens*. *N. fedtschenkoi* Pojark. is simply a less glandulose race of *N. pungens* with blue corolla (glands are seen in pubescence of inflorescence of *N. fedtschenkoi* although the description of the species makes no mention of it). Throughout the distribution range it is difficult to establish any pattern in the variability of the various characteristics since all of them are highly variable. In a given region, more or less glandulose specimens can be found but almost invariably with stiff, spine-like accuminate involucre bracts of cymes. The typical form of *N. pungens* is found only in the central part of Kazakhstan.

In all the specimens studied, it was difficult to establish the real colour of the corolla since all the corollas in the herbarium were equally white (even in the type species of *N. fedtschenkoi*!). Special attention should therefore be paid to this characteristic when studying living plants. In our opinion, the blue coloration of the corolla of *N. fedtschenkoi* is also a variable feature.

In Turkmenia (Kopetdag, Bolshie Balkhany), Iran and Afghanistan, we found another race that is very closely related to this group. (*N. chenopodiifolia* Stapf = *N. microcephala* Pojark.), which differs insignificantly from *N. pungens*. Our material is inadequate to evaluate with total confidence the minute differences in the southern race as a characteristic of specific importance. It is highly possible that a study of substantial material will confirm the identity of the southern race with the true *N. pungens*.

14. **N. sibirica** L. Sp. pl. (1753) 572 (excl. syn. Buxb.); Pojark. in Fl. SSSR, 20 (1954) 342; Grubov, Konsp. fl. MNR (1955) 235; Fl. Kazakhst. 7 (1964) 337. —*N. macrantha* Fisch. Catal. Hort. Gorenk. ed. 2 (1822) 22, nom. nud.; Benth. Labiat. gen. et sp. (1834) 482, diagn.; Franch. Pl. David. 1 (1884) 238; Sapozhn. Mong. Atl. (1911) 381; Danguy in Bull. Mus. nat. hist. natur. 20 (1914) 83; Krylov, Fl. Zap. Sib. 9 (1937) 2307; Walker in Contribs U.S. Nat. Herb. 28 (1941) 657. —*Dracocephalum sibiricum* L. Syst. nat. ed. 10 (1759) 1104. —Ic.: Pall. Fl. Ross. tab. 113; Bot. Mag. tab. 2185.

Described from cultivated specimen of Siberian origin. Type in London (Linn.).

Coastal meadows, meadows on hill slopes, shaded talus and foot of rocks, floors of creek valleys, coastal gravel beds in the mountain- and forest-steppe belts.

IA. **Mongolia:** *Mong. Alt.* (mountain slope in Botkhon gorge, in ravine, July 18, 1898—Klem.; Khatu river, Aug. 8; on way to Kobdo town along Botogoin-Gol river, sandy-rocky soil, Sept. 1—1899, Lad.; Katu river, Aug. 7, 1909—Sap.; Khara-Dzarga mountain range, Shutyn-Gol river valley, deep rocky gorge, on pebble bed, Aug. 28; Khasagtu-Khairkhan mountains, emergence of Dundu-Seren-Gol river onto rocky trail, Sept. 15; same site, around Undur-Khairkhan mountain, along coastal pebble bed, Sept. 16—1930, Pob.: "entre Oulioun-Gour et Kobdo, Sept. 19, 1895, Chaff."—Danguy, l.c.) *East. Mong.* (Ourato, Près des ruisseaux, July 1866—David; Muni-Ula mountains, June 23, 1871—Przew.) *Bas. Lakes* (near Ubsa lake, Kharkhira river, in ditch with water, July 9; Telin-Gol river, south of Ulangom, on brim of ditch, Sept. 4—1879, Pot.; in Bogden-Gol valley, below Ulyasutai town, in a deciduous grove, July 7, 1894; sandy bank of Bogden-Gol river, between tall pea shrubs, July 14, 1896—Klem.; Borig-Del' sand, nor.-west. margin of Baga-Nur lake, meadow, July 25, 1945—Yun.). *Gobi-Alt.* (Dundu-Saikhan mountains, on sand bed, July 4, 1909—Czet.; Gurbun Saikhan, rocky places, 1925—Chaney; Khalga pass, near foot of rocky slope, Aug. 3; Bain-Tsagan range, Khukhu-Daban creek valley, Aug. 11; Dundu-Saikhan mountains, near foot of rocky slope, Aug. 16; Dzun-Saikhan mountains, Yalo creek valley, under overhanging rocks on Tsagan-Gol river, Aug. 25; Barun-Saikhan mountain on rock along Gegetu river, Sept. 20—1931, Ik.-Gal.; Dundu- and Dzun-Saikhan ranges, mountain slopes and gorges of trails up to upper belt, July-Aug., 1933—M. Simukova; south. slopes of Bain-Tsagan range, along rocky slope, July 31, 1938—Luk'yanov; Dzun-Saikhan range, west. extremity of gorge in juniper belt, June 18, 1945—Yun., rocky steppe on south. slope of Dundu-Saikhan, July 20, 1950—Kal.). *Alash. Gobi* (Alashan mountains, in wet valleys and meadows, June 21, 1873—Przew.; same site, Yamato gorge, near foot of rock in humus soil, May 5; Khote-Gol gorge, near foot of large rocks, on humns soil, June 9—1908, Czet.; "Ho Lan Shan, at edge of woods, No. 1128, Ching"—Walker, l.c.).

General distribution: Jung.-Tarb. (Saur and Tarbagatai), Nor. Tien Shan; West. Sib. (Altay), East. Sib. (Sayans), Nor. Mong. (Hang.).

**N. spathulifera** Benth. in DC. Prodr. 12 (1848) 380; Pojark. in Fl. SSSR, 20 (1954) 427; Ikonnikov, Opred. rast. Pamira (1963) 212. —*N. reniformis* Briq. in Bot. tidsskr. 28 (1908) 236. —*N. fallax* Briq. l.c. 237. —Ic.: Briq. l.c. fig. 3 (sub nom. *N. reniformis* Briq.).

Described from Fore Asia environs of Kabul). Type in London (K).
Rocky and rubble slopes in alpine and subalpine belts, 2500–4000 m.

IIIC. Pamir: occurrence possible since it is known from adjoining regions of East. Pam.).
General distribution: East. Pam.; Fore Asia, Himalayas (west.).

15. **N. transiliensis** Pojark. in Bot. mater. Gerb. Bot. inst. AN SSSR, 15
(1953) 286; Pojark. in Fl. SSSR, 20 (1954) 315; Fl. Kazakhst. 7 (1964) 333. —
Ic.: Fl. Kazakhst. 7, Plate 39, fig. 1.

Described from Nor. Tien Shan (Transili Alatau). Type in Leningrad.
Rocky and rubble slopes in alpine belt.

IIA. Junggar. *Tien Shan* (Narat crossing on Kunges river, shaded slopes, 2300 m, Aug. 7,
1958—A.R. Lee (1959)).
General distribution: Nor. Tien Shan.
Note. This species was formerly reported only within the USSR.

16. **N. ucrainica** L. Sp. pl. (1753) 570; Benth. Labiat. gen. et sp. (1834)
487; Krylov, Fl. Zap. Sib. 9 (1937) 2310; Pojark. in Fl. SSSR, 20 (1954) 415; Fl.
Kirgiz. 9 (1960) 54; Fl. Kazakhst. 7 (1964) 342. —*Teucrium sibiricum* L. Sp.
pl. (1753) 564; Pall. Reise, 2 (1773) 269. —Ic.: Reichb. Ic. fl. Germ. tab.
1243, fig. 2.

Described from East. Europe. Type in London (Linn.).
Exposed rubble and rocky slopes in steppe groups.

IIA. Junggar: *Tien Shan* (Almaty river gorge [nor. Kul'dzha], April 20; same site, May 1;
along Agnaz river, 1500–1800 m, June 25—1878, A. Reg.).
General distribution: Aralo-Casp. (nor.), Fore Balkh., Jung.- Tarb., Nor. and Cent. Tien
Shan; Europe, Balk.-Asia Minor, Mid. Asia, West. Sib.

17. **N. yanthina** Franch. in Bull. Mus. nat. hist. natur. 3 (1897) 324;
Pampanini, Fl. Carac. (1930) 183. —? *N. vakhanica* auct. non Pojark.; Murata
in Kitamura, Pl. West Pakistan and Afghanistan (1964) 129.

Described from Tibet. Type in Paris. Isotype in Leningrad.

IIIB. Tibet: *Chang Tang* (Col. etre Pangong et Lokong, July 25, 1892—Rhins, isotypus!).
General distribution: Himalayas (west.).
Note. *N. yanthina* is closely related to *N. floccosa* Benth. According to the author of the
species, it differs from the latter only in very small size, shape of leaves and violet colour of
inflorescence. Unfortunately, adequate material is not available to verify critically all of these
characteristics compared to *N. floccosa*.
On the basis of phytogeographic concepts and taking into consideration the morphology of these species, in our opinion *N. yanthina* is one of forms of *N. floccosa*. The latter is
polymorphous but the extent of variability of its characteristics is not so significant as to say
about completely categories of species level. *N. floccosa* has been described from the region
between "Nako" and "Chango" and is widely distributed in West. Himalayas, i.e., entirely
covers the range of *N. yanthina*, which is known only from a single region. The ecology of
these species is the same, i.e., they grow on rubble talus in high-mountain areas.
The plant cited by Kitamura (l.c.) under the name *N. vakhanica* Pojark. from Karakorum
range is evidently *N. yanthina* since *N. vakhanica* is known only from the southern part of

West. Pamir and may probably be found in Vakhana in Afghanistan; this species has never been reported east of the latter.

18. **N. wilsonii** Duthie in Gard. Chron. 40 (1906) 334; Kudo in Mém. Fac. Sci. and Agr. Taihoku Univ. 2 (1929) 232. —*N. veitchii* Duthie, l.c.; Kudo, l.c. 232. —*N. prattii* Lévl. in Feddes repert. 9 (1911) 245. —*N. przewalskii* Pojark. in Fl. SSSR, 20 (1954) 341, nomen. —*Dracocephalum wilsonii* (Duthie) Dunn in Notes Bot. Gard. Edinburgh, 8 (1913) 166, 6 (1915) 171; Hand.-Mazz. in Symb. Sin. 7 (1936) 918. —*D. veitchii* (Duthie) Dunn in Notes Bot. Gard. Edinburgh, 6 (1915) 171. —*D. prattii* (Lévl.) Hand.-Mazz. in Acta Horti Gotoburg. 9 (1934) 79; Hao in Engler's Bot. Jahrb. 68 (1938) 634. —*D. sibiricum* auct. non L.; Dunn in Notes Bot. Gard. Edinburgh, 6 (1915) 171; Walker in Contribs U.S. Nat. Herb. 28 (1941) 656. — Ic.: Duthie, l.c. fig. (sub nom. *N. veitchii*).

Described from specimens grown from seeds brought by Wilson from West China. Type in London (K). Plate II, fig. 4.

Scrub, forest fringes, often in houses, 2000–4000 m.

IIIA. Qinghai: *Nanshan* (Nor. Tetung mountain range, in valley, Aug. 9, 1872; along Yusun-khatyma river, at 2700–3000 m, in scrub, humus soil, July 23; Nor. Tetungk mountain range, Aug. 7—1880, Przew.; environs of Ganchan-gombo monastery and Chortenton, height 2100 m, in scrub, on rubbish heaps in a house, Sept. 1901—Lad.; "auf dem östlichen Nanschan, um 2900 m"—Ho, l.c.; "Upper Shui Mo Kou, near Lien Ch'eng, on a shrub-covered slope, No. 384, Ching"—Walker, l.c.).

IIIB. Tibet: *Weitzan* (Mekong river basin, along Chokchu river at height 3600 m, in forest, Aug. 31, 1900, Dzhagyn-Gol river, ?—Lad.).

General distribution: China (Nor.-West, South-West).

Note. *N. wilsonii* and *N. veitchii* Duthie were described from specimens grown in a garden from seeds brought by Wilson from West. China (Sunnan town). *N. prattii* Lévl. was described from Sichuan (Dadzyanlu pass). The study of a large number of herbarium specimens from Nor. Sichuan and Qinghai showed that only one species, which should be called *N. wilsonii*, grows here quite extensively. *N. wilsonii* is closely related to *N. sibirica* L., differing in sessile upper and middle caulous leaves and different type of pubescence of leaves, calyx and corolla.

There is a reference in Hemsley's 'The Flora of Tibet or High Asia' to the occurrence of *N. supina* Steven in our territory "Near Rakas Tal, 4500–5100 m, Strachey et Winterbottom". Although we could not locate these materials, we are confident that it could not be *N. supina* since it is known only from Caucasus.

## 12. Glechoma L.
Sp. pl. (1753) 578.

1. **G. hederacea** L. Sp. pl. (1753) 578; Krylov, Fl. Zap. Sib. Sib. 9 (1937) 2316; Kuprian. in Bot. zh. 33 (1948) 237; id. in Fl. SSSR, 20 (1954) 437; Fl. Kirgiz. 9 (1960) 56; Fl. Kazakhst. 7 (1964) 445. —*Nepeta glechoma* Benth. Labiat. gen. et sp. (1834) 485; Franch. Pl. David. 1 (1884) 238; Forbes and

Hemsley, Index Fl. Sin. 2 (1902) 290; Simpson in J. Linn. Soc. London (Bot.) 41 (1913) 436. —Ic.: Hegi, Ill. Fl. Mittel.-Europ. V, 4, 2373; Kuprian. l.c. (1948) fig. 1.

Described from Nor. Europe. Type in London (Linn.).

Forests, shaded banks of rivers, meadows, marshes, houses.

IIA. Junggar: *Tien Shan* (Kök-Terek—Tekkes, junct. Tekkes valley, No. 658, June 2, 1930—Ludlow [BM]).

General distribution: Aralo-Casp., Jung.-Tarb., Nor. Tien Shan, Cent. Tien Shan (introduced); Europe, Mediterranean, Caucasus, Mid. Asia, West. Sib., East. Sib., Far East, North America (introduced).

## 13. Dracocephalum L.

Sp. pl. (1753) 594; Hiltebr. Monogr. Dracoceph. (1805).

1. Anthers pubescent ............................................... 17. **D. ruyschiana** L.
+ Anthers glabrous ............................................................................. 2.
2. Annual ............................................................................................ 3.
+ Perennial ........................................................................................ 4.
3. Stem 5–15 cm tall; leaves with scattered sessile glands beneath; plant fetid ...................................................... 4. **D. foetidum** Bunge.
+ Stem 15–50 cm tall; leaves pitted-glandular beneath; plant pleasant-smelling ........................................ 10. **D. moldavica** L.
4. Stamens exserted from corolla by more than twice length .............
.................................................. 18. **D. stamineum** Kar. et Kir.
+ Stamens not exserted from corolla or barely exserted .................. 5.
5. Calyx sharply bilabiate, upper lip incised up to 1/4–1/3 its length into 3 subsimilar ovate teeth; lower lip incised up to 1/4 of its length or almost up to base ................................................................. 6.
+ Calyx not sharply bilabiate, all teeth nearly equally long but midtooth of upper lip often considerably broader than others. ....8.
6. Leaves deeply pinnatisect, sometimes almost up to midnrib sometimes bipinnate ........................................... 1. **D. bipinnatum** Rupr.
+ Leaves uniformly dentate or entire ..................................................... 7.
7. Stems spreading or central stem (if present) erect, 5–18 cm tall; leaves uniformly dentate, subcordate at base ..............................
................................................................ 7. **D. heterophyllum** Benth.
+ Stems erect, 20–70 cm tall; leaves unevenly dentate, teeth acuminatearistate or leaves entire but invariably cuneately narrowed toward base ................................... 14. **D. peregrinum** L.
8. Leaves lanceolate, entire or with few teeth ................................. 9.
+ Leaves arbicular-ovate, obtusely dentate or pinnatifid or pinnatisect, sometimes bipinnate ....................................................... 10.
9. Tip of leaves and their lateral teeth subulately acuminate.............
.............................................................. 5. **D. fruticulosum** Steph.

+ Leaves invariably entire, acuminate at tip but without subulate cusp
..................................................................... 9. **D. integrifolium** Bunge.
10. Leaves pinnatisect, sometimes bipinnate ..................................................
............................................................ 19. **D. tanguticum** Maxim.
+ Leaves with obtuse teeth, sometimes not pinnatisect .................. 11.
11. Stems trailing, with ascending branches, forming compact mat. ...
.............................................................................................................. 12.
+ Stems usually erect, single or more together, not forming compact
mat ........................................................................................................ 14.
12. Bracts shorter than calyx or as long; midtooth of upper lip of calyx
1.5–2 times broader than long and 4–5 times broader than lateral
teeth. ................................................................ 3. **D. discolor** Bunge.
+ Bracts longer than calyx; midtooth of upper lip of calyx longer than
broad and 2–3 times broader than lateral teeth. ........................... 13.
13. Corolla 12–15 mm long; bracts usually bluish; all teeth of calyx
acuminatearistate; midtooth only slightly broader than others ......
.................................................... 13. **D. origanoides** Steph. ex Willd.
+ Corolla 20–22 mm long; bracts reddish-violet; midtooth of calyx
considerably broader than rest, usually obtuse ...............................
.................................................... 2. **D. bungeanum** Schischk. et Serg.
14. Bracts entire ................................................................ 12. **D. nutans** L.
+ Bracts dentate in upper half ................................................................. 15.
15. Lower and middle caulous leaves with petiole several times shorter
than blade ..................................................... 11. **D. nodulosum** Rupr.
+ Lower and middle caulous leaves with petiole as long as blade or
longer ......................................................................................................... 16.
16. Radical leaves orbicular-cordate, cuneate or orbicular-reniform, not
longer than wide ..................................................................................... 17.
+ Radical leaves long-elliptical or oblong-ovate ............................... 18.
17. Leaves with scattered long flattened white hairs, whitish beneath,
with distinct sunken glands and white, rather flat hairs arranged
mainly along veins; calyx teeth stiff, acuminate-mistate, almost invariably reddish; corolla bluish-violet; stem and petioles patently
pilose ................................................................ 16. **D. rupestre** Hance.
+ Leaves glabrous or slightly pilose, greyish-green beneath, densely
pubescent with short appressed hairs; teeth of calyx soft, acuminate aristate, bluish or green; corolla dark blue; stem and petioles
pubescent with appressed short hairs ........... 8. **D. imberbe** Bunge.
18. Corolla more or less uniformly pubescent with short appressed hairs
on outer surface; upper lip of corolla with insignificant pubescence
inside ................................................................ **D. wallichii** Sealy.
+ Corolla unevenly pubescent on outer surface; upper part of upper
lip or both upper and lower lips pubescent with long white flattened

hairs; upper lip of corolla lanate inside ........................................... 19.

19. Corolla 35–45 mm long, bright blue, upper part of upper lip pubescent with white flattened hairs on outer surface; anthers 3 mm long, oblong ................................................................ 6. **D. grandiflorum** L.

+ Corolla small, 20–25 mm long, bluish-violet, pubescence of white flattened hairs outside denser on upper part of upper and lower lips; anthers 1.5 mm long, elliptical .. 15. **D. purdomii** Smith W.W.

1. **D. bipinnatum** Rupr. in Mém. Ac. Sci. St.-Pétersb. VII sér. 14 (1869) 66; Persson in Bot. notiser (1938) 300; Schischk. in Fl. SSSR, 20 (1954) 464; Fl. Kirgiz, 9 (1960) 68; Fl. Kazakhst. 7 (1964) 356. —*D. ruprechtii* Regel in Acta Horti Petrop. 6 (1880) 563; Hook. f. Fl. Brit. Ind. 4 (1885) 666. —**Ic.:** Gartenfl. 29, tab. 1018 (sub nom. *D. ruprechtianum* Regel); Fl. SSSR, 20, Plate 26, fig. 3.

Described from Tien Shan. Type in Leningrad.

Rocky and rubble slopes of steppes, banks of mountain rivers, 1300–2500 m.

IB. Kashgar: *Nor.* (Kapsalyon Tal und Nebental Kesyl-sai, auch Plateauhohe von Karadschon, Aug. 2–3, 1907—Merzb.; along road to Bai, on exposed slope, 2450 m, No. 8223, Sept. 7; nor.-west of Pocheenzy, Muzarta valley, in Langere, moraine formations, 1900–2200 m, No. 8338, Sept. 12; 8 km north of Aksu, in exposed valley sites, 2500 m, No. 8488, Sept. 24—1958, A.R. Lee (1959); valley of Yu. Muzart river, 4 km south of Yangimallya settlement, along road to exit from gorge, dry steppe belt, in chee grass thickets, rare, Sept. 11, 1958—Yun. et al.).

IIA. Junggar: *Jung. Alt.* (Urtak-Sary, July [18–20], 1878—Fet.; south. Alatau slopes, Aug. 4; Borotala, 2550 m, Aug.—1878, A. Reg.; Ven'tsyuan', in gorge, No. 3455, Aug. 14, 1957—Kuan; between Sairam lake and Borotola river valley, 5–6 km nor.-west of Shuvutin-Daba, along road to Urtak-Sary from Sairam, subalp. belt, south. rocky slopes, Aug. 18, 1957—Yun. et al.). *Tien Shan* (on Dzhin river, 1877—Larionov; along Tekes river, about 2500 m, July 7, 1877—Przew.; south. bank of Sairam lake, July; Talki gorge, 1800–2100 m, Aug. 15—1877, A. Reg.; Maralty, near Muzart inlet, Tekes river valley, 1800 m, Aug. 1, 1877; Sairam, July 28; Sharabaguchi, Aug. 1878, Fet.; Chapchal pass, June 28, 1878; Tsagan-Tunge, 1500 m, June 8; Kash between Ulutau and Nilki, 900–1200 m, June 30—1879, A. Reg.; in Tekes river valley near Agiaz [1886]—Krasnov; Kok-su Valley, Thian-Shan in about 83°E and 43°N, Aug. 15-Sept. 15, 1901— Littledale [K]; upper Agiaz, Aug. 11–12, 1907—Merzbacher; in Dzhagistai mine, on slope, No. 3139, Aug. 7; 1 km south of Dzhagistai mine on rubble slope, No. 686, Aug. 7; 3 km north of Chzhaosu on south-east. slope, No. 882, Aug. 13—1957, Kuan; Ketmen mountain range, above Sarbushin settlement, 3–4 km along road from Ili to Kyzyl-Kure, steppe belt, on south. talus slope, Aug. 23; right rocky bank of Ili river, on floor of lateral gorge, Aug. 29—1957, Yun. et al.).

General distribution: Jung.-Tarb., North and Cent. Tien Shan; Mid. Asia, Himalayas (Kashmir).

2. **D. bungeanum** Schischk. et Serg. in Krylov, Fl. Zap. Sib. 9 (1937) 2322; Schischk. in Fl. SSSR, 20 (1954) 446; Grubov, Konsp. fl. MNR (1955) 235. —*D. origanoides* auct. non. Steph.; Bunge in Mém. Ac. Sci. St.-Pétersb. Sav. Etrang. 2 (1835) 562; Sapozhn. Mong. Alt. (1911) 381, p.p.

Described from Altay (Chui valley). Type in Leningrad.

Rocks, rocky slopes, talus and alluvium in alpine belt.

IA. Mongolia: *Mong. Alt.* (pass from Malaya Kairta to Malyi Ku-Irtysh, talus, July 17, 1908—Sap.). *Gobi-Alt.* (Ikhe-Bogdo mountain range, Bityuten-Ama creek valley, alpine zone, Aug. 12, 1927—M. Simukova; same site, upper creek valley of Artsatuin-Ama, among rocky talus, in upper limit of vegetation, alt. 3950 m, Sept. 8, 1943; same site, Narin-Khurimt creek valley, nor. rocky slope; same site, plateau in upper creek valley of Ikhe-Khurimt, sheep's fescue-cobresia high-montane steppe, on debris—June 28, 1945, Yun.; Narin-Khurimt gorge, east. bank, on rocks, July 28, 1948—Grub.).

General distribution: West. Sib. (Altay), Nor. Mong. (Fore Hubs.).

Note. Species closely related to *D. origanoides* Steph., differing in very large corolla and reddish-violet bracts; its midtooth considerably broader than rest and usually obtuse.

More northwards, within the Soviet Union, in West. Siberia and Dauria, species *D. pinnatum* L. quite close to *D. origanoides* and *D. bungeanum* is found. *D. pinnatum* L. has also been reported from Mongolia [see Fl. SSSR, 20 (1954) 444] but we did not detect the typical form of *D. pinnatum* in Mongolian territory.

3. D. discolor Bunge in Mém. Ac. Sci. St. Pétersb. Sav. Etrang. 2 (1835) 560; Sapozhn. Mong. Alt. (1911) 381; Krylov, Fl. Zap. Sib. 9 (1937) 2320; Schischk. in Fl. SSSR, 20 (1954) 447; Grubov, Konsp. fl. MNR (1955) 236; Fl. Kazakhst. 7 (1964) 348. —*D. paulsenii* Briq. in Bot. tidsskr. 28 (1907) 238; Schischkin Fl. SSSR, 20 (1954) 448; Fl. Kirgiz. 9 (1960) 61; Fl. Uzbek. 5 (1961) 309; Ikonnikov, Opred. rast. Pamira (1963) 213; Fl. Kazakhst. 7 (1964) 348. —*Dracocephalum* sp.: Pampanini, Fl. Carac. (1930) 184. —Ic.: Briq. l.c. Fig. 4 (sub nom. *D. paulsenii*); Ledeb. Ic. pl. fl. ross. 2, tab, 128 (sub nom. *D. origanoides*).

Described from Altay (Charysh river valley). Type in Leningrad.

Rubble and rocky slopes, rocks in mountain-steppe belt.

IA. Mongolia: *Mong. Alt.* (on slope boundary of Dolan-Nor in Altay, July 8, 1877—Pot.; crossing from Narin river to Senkul' river, July 24; on bank of Khartsiktei river, on rocky talus, July 27—1898, Klem.; southern slope of Chinese Altay, Urmogaity pass, 2900 m, 1903—Gr.-Grzh.; Dain-Gol lake, steppe slopes, July 5, 1906—Sap.; southern slope of Tamchi-Daba pass through main mountain range, mountain steppe, July 16; same site, south-west of Tamchi-Nur lake, on midportion of slope, July 17—1947, Yun.; 8 km north of Yusup-Bulaka, along road from Tsaganolom, feather grass forb mountain steppe on ridge, Aug. 31, 1948—Grub.). *Bas. Lakes* (environs of Ak-Karasuk river, on south. slope of Tannu-Ol, southern steppe slopes, July 6, 1892—Kryl.; beyond Khoto-Khuduk collective, on southern slope of mountain, June 30; south-eastern slope of mountain between Khobur-Bulak and Khoto-Khuduk springs; steppe between Maikhanei-Shemaga-Bulak and Khobur-Bulak springs; on northern slope of mountains between Khoto-Khuduk wells and Shanda-Daban pass—June 30, 1894, Klem.; Khan-Khukhei mountain range, lower third of southern trail, wormwood-forb steppe, July 21, 1945—Yun.).

IIA. Junggar: *Tien Shan* (Talkibash, July 20; Sairam, July—1877, A. Reg.; M. Yuldus plateau, 2250-2700 m, June 10, 1877—Przew.; Muzart pass [1886]—Krasnov; M. Yuldus ulas, dry alluvium, 2500 m, Nos. 6345, 6369, Aug. 2, 1958—Lee and Chu (1959); intermontane basin of B. Yuldus site of Khaiduk river discharge from Yuldus, mountain steppe, Aug. 7, 1958—Fedorovich). *Jung. Gobi* (from Tsitai town to Beidashan pasture, on sunny slopes, No. 5230, Sept. 28, 1957—Kuan).

General distribution: Jung.-Tarb., Nor. and Cent. Tien Shan, East. Pamir; Mid. Asia, West. Sib. (Altay), East. Sib. (south-west.), Nor. Mong. (Hang.), Himalayas (Kashmir).

Note. *D. discolor* was described by Bunge from West. Altay (Charysh river valley) where this species grows on exposed rocky slopes of knolls. Within Mongolia, in Tien Shan, Pamiro-Alay and Karakorum mountain range, it is found in alpine and subalpine belts in steppe areas, rocks, rubble and rocky slopes of mountains. A careful scrutiny of material available from all over its distribution range showed that this species is more or less constant in characteristics over a comparatively large expanse. The degree of tapering in teeth and glandulosity of calyx and bracts varies somewhat but the variability is so insignificant that it hardly justifies separating the race with less glandular calyx and bracts and less acuminate teeth of calyx and bract lobes into a distinct species, *D. paulsenii* Briq., especially since specimens that vary in degree of tapering of teeth of the calyx and bracts can be found within Pamiro-Alay from where *D. paulsenii* was described. The size of flowers is more or less constant at 10 to 15 mm, with specimens of 10–12 mm long flowers encountered more often.

4. **D. foetidum** Bunge in Ledeb. Ic.pl. fl. ross, 2 (1830) 27 et tab. 193; id. in Ledeb. Fl. alt. 2 (1830) 386; Danguy in Bull. Mus. nat. hist. natur. 20 (1914) 84; Krylov, Fl. Zap. Sib. 9 (1937) 2329; Schischkin Fl. SSSR, 20 (1954) 462; Grubov, Konsp. fl. MNR (1955) 236. —*D. moldavica* L. β. *asiaticum* Hiltebr. Monogr. Dracoceph. (1805) 43; Sapozhn. Mong. Alt. (1911) 382. — Ic.: Ledeb. Ic. pl. fl. ross. 2, tab. 193.

Described from Altay (Chui river valley). Type in Leningrad.

River banks, pebble beds, valleys and flanks of gullies and creek valleys, foot of rocks and talus, sandy steppes, rubble and rocky slopes of steppes.

IA. Mongolia: *Khobdo, Mong. Alt., Cent. Khalkha, East. Mong., Bas. Lakes, Val. Lakes, Gobi-Alt., East. Gobi, Alash. Gobi, Ordos.,*

General distribution: West. Sib., Nor. Mong. (Hent., Hang., Mong.-Daur.).

Note. This species, closely related to *D. moldavica* L., possibly represents its wild ancestor species brought into cultivation long ago and later, according to some researchers, becoming wild once again. This aspect calls for special investigations however. In *D. foetidum* Bunge, the stem at 5–15 cm is not tall, branching right from the base, lateral branches ascending and as long as the main stem, widespreading bush is formed; (in *D. moldavica* the main stem is 15–50 cm tall, branching either from base or from middle but lateral branches obliquely erect, not gorwing into main stem, and thus the bush is more or less compact); lower surface of leaves with scattered sessile glands (in *D. moldavica*, lower surface of leaves punctate-glandular); filaments with distinct hairs (in *D. moldavica*, filaments perceptibly pilose). Apart from these characteristics, *D. foetidum* has a very unpleasant odour which readily distinguishes it from *D. moldavica* with an agreeable bouquet. Given these considerations, in our opinion these 2 species could possibly be considered independent, especially since a review of rather sizable material demonstrated the stability of characteristics of *D. foetidum* throughout its fairly extensive range.

5. **D. fruticulosum** Steph. in Willd. Sp. pl. 3 (1800) 152; Hiltebr. Monogr. Dracoceph. (1805) 29, 65; Danguy in Bull. Mus. nat. hist. natur. 20 (1914) 84; Schischk. in Fl. SSSR, 20 (1954) 456; Grubov, Konsp. fl. MNR (1955) 236; Henelt und Davažamc in Feddes repert. 70 (1965) 54. —*Dracocephalum* sp.: Sapozhn. Mong. Alt. (1911) 382. —Ic.: Hiltebr. l.c. tab. 3.

Described from Siberia. Type in Berlin.

Desert steppes and rocky steppe slopes, rocks, dry talus, flanks and floor of gullies, dry gravel beds on banks.

**IA. Mongolia:** *Khobd.* (Burgassutai river, in Uryuk-Nor lake basin, on dry terrace, rocky soil, June 21; steppe on left bank of Kharkhira river, dry pebble bed, July 10; mountains on right bank of Kharkhira, dry mountain slopes, July 11; valley of Kharkhira river, July 21; Tenemyk mountains, Sept. 4—1879, Pot.; Kharkhira mountain group, Burtu area, July 16, 1903—Gr.-Grzh.; right bank of Kharkhira, talus and deciduous forest, July 27, 1916—Neiburg; 3–4 km west of Ulan-Daba pass, along road from Ulangoma to Tsagan-Nur, dry hil montane steppe, July 29, 1945—Yun.). *Mong. Alt.* (slopes of cliffs surrounding Ulan-Sair creek valley, above estuary, July 23, 1894—Klem; Buyantu river, rubble steppe around Kobdo town, July 18, 1906—Sap.; Bodkhon-Gol area, 35 km south of Tugrik-Sume, feather grass and stipa-like desert steppes, July 11, 1945—Yun.; "Altai, alt. 1500–2100 m, region de Kobdo, Sept. 22, 1895, Chaff."—Danguy, l.c.). *Cent. Khalkha* (south-east. Hangay foothills, around Kholt area, June 20, 1926—Gus.; along road from Ekhini-Kuduk to Targat somon, feather grass-stipa-like rock steppe, Aug. 4, 1941—Yun.), *Bas. Lakes* (Ulyassutai river, July 9; Unyugyutei-Dzyukhe, dry gorge, discharging into Kholbo-Nor lake basin, on borders of dry river-bed, July 27—1879, Pot.). *Val. Lakes* (on ridge of Tuin-Gol river bank, below confluence of Sharagoldzhyut, on rocks, July 8; on hill, on right bank of Tuin-Gol, July 9; on hill, on left bank of Tatsi-Gol, below Tatsy urton, July 18—1893; on cliff on left bank of Uta river, June 17, 1894—Klem.; rocky precipice on Tuin-Gol river, Aug. 30, 1924—Kondr.; 40 km south-west of Khairkhan-Dulan somon, along road to Bain-Khongor, feather grass-stipa-like steppe on rubble chestnut soils, June 27; 30 km west of Tatsiin-Gol on south-Hangay road, wheat grass-feather grass dry steppe, June 28; 15–20 km south-south-west of Barun-Ul'dzeitu somon, right bank of Tuin-Gol, feather grass desertified steppe, June 29—1941, Tsatsenkin; 5 km east of Tatsiin-Gol, along road to Arbai-Khere in Bain-Khongor, gully and gully fringes, July 9; same site, Taridain-Gol area, steppe, July 9—1947, Yun.). *Gobi-Alt.* (Ikhe-Bogdo, Dundu, Saikhan, Daun-Saikhan, Bain-Tsagan, Tostu ranges). *East. Gobi* (Ondai Sair and Dating Gol, Outer Mongolia, wash at edge of fan and on dry ridges, 1500–1700 m, No. 189; Inner Mongolia, grasslands, No. 606—1925, Chaney). *Alash. Gobi* (in gravelly valleys on Alashan mountains, July 7; same site, in gorges, July 21—1873, Przew.; Inchuan, 45 km south-west of town, rocky low mountains of Alashan mountain range, hilly semi-desert, July 5, 1957; 50 km along road from Inchuan to Bayan-Khoto town, south. part of Alashan mountain range, rocky mountain slopes, June 10, 1958—Petr.).

General distribution: East. Sib. (south.), Nor. Mong. (Hang. south.).

6. **D. grandiflorum** L. Sp. pl. (1753) 595, ex parte; Hiltebr. Monogr. Dracoceph. (1805) 48, 74; Forbes and Hemsley, Index Fl. Sin. 2 (1902) 291, quoad sp. Hancockii; Simpson in J. Linn. Soc. London (Bot.) 41 (1913) 436; Schischk in Fl. SSSR, 20 (1954) 451; Keenan in Baileya, 5 (1957) 33; Wu in Acta Phytotax. Sin. 8, 1 (1959) 22; Fl. Kirgiz. 9 (1960) 61; Fl. Kazakhst. 7 (1964) 349; Hanelt. und Davažamc in Feddes repert. 70 (1965) 54. —*D. altaiense* Laxm. Nov. Comm. Ac. Petrop. 15 (1770) 556; Sapozhn. Mong. Alt. (1911) 382; Dunn in Notes Bot. Gard. Edinburgh, 6 (1915) 169, p.p.; Krylov, Fl. Zap. Sib. 9 (1937) 2323; Grubov, Konsp. fl. MNR (1955) 235. —**Ic.:** Laxm. l.c. tab. 29, fig. 3 (sub nom. *D. altaiense*); Keenan l.c. tab. 12; Curtis's Bot. Mag. 25, ab. 1009.

Described from Siberia. Type in London (Linn.).

Meadows, rocky alluvium, moraine, bald peaks, banks of brooks in alpine and subalpine belts; forest fringes and glades in upper forest boundary.

IA. **Mongolia:** *Khobd.* (Tszusylan, above forest line, in alpine meadow, July 13, 1879—Pot.; south. group of Kharkhira, Boro-Borgosun river, tributary of Saksai, Kobdo river system, July 2, 1903—Gr.-Grzw.). *Mong. Alt.* (slope of pass from Tamylta to Angyrta, July 26, 1898—Klem.; Bulugun river basin, Kharagaitu-Khutul' pass, alpine meadow and rubble alluvium with snow patches, July 24, 1947—Yun.). *Gobi-Alt.* (Ikhe-Bogdo mountain range, plateau-like crest of mountain range, in nor.-west. half, 3600 m, alpine carpet protected by high rocks, July 29, 1948—Grub.).

IIA. **Junggar:** *Cis-Alt.* (Oi-Chilik, alpine belt, 1876—Pot.; upper Ustyugan river, tributary of Kran river, alpine meadow, July 1, 1908—Sap.; Koktogoi region, 2500 m, No. 1978, July 17, 1956—Ching). *Tarb.* (north of Dachen, on slope, 2250 m, No. 1570, Aug. 13, 1957—Kuan; Saur mountain range, southern slope, valley of Karagaitu river, right bank of Bain-Tsagan creek valley, subalp. meadow belt, June 23, 1957—Yun. et al.). *Jung. Alt.* (Toli district, on slope, No. 1194, Aug. 6, 1957—Kuan). *Tien Shan* (Bogdo mountain, 3000 m; Sumbe pass, 2700–3000 m, June 22; on Kassan river, 2100 m, June 22; in Dzhagastai hills, chapchal gorge, June 28; Sairam—1878, A. Reg.; Burkhantau [Kul'dzha] [June 1878]—Fet.; on Borboro-gussun river, 2700 m, June 15; Karagol tributary, 2700 m, June 16; confluence of Karagol with Nilka, 2400–2700 m, June 16; Aryslyn estuary, 2400–2700 m, July 8; Aryslyn, 2700–3000 m, July 16—1879, A. Reg.; Tien Shan slopes between Barkul and Khama, June 2, 1879—Przew.; on Tekes river, 1886—Krasnov; 10 km nor. of Chzhaos on slope, No. 3312, Aug. 15; Nilka to Karasu, on sunny slope, No. 1674, Aug. 30—1957, Kuan). *Jung. Gobi* (Temirtam, No. 916, Aug. 15, 1957—Kuan).

**General distribution:** Jung.-Tarb., Nor. and Cent. Tien Shan; Mid. Asia, West. Sib. (Altay), East. Sib., Nor. Mong. (Fore Hubs., Hent., Hang.), China (North).

**Note.** Keenan (l.c.) points out that in the Linnean herbarium, *D. grandiflorum* is represented by 3 specimens of which 2 are entirely typical while 1 is not. Linneus in Sp. pl. states that *D. grandiflorum* has been described from Siberia after I. Gmelin who later cites for this species a more accurate geographic distribution in 'Flora sibirica' (III, 1768, 234): "Intra valles rupium Bargusinensium et earum quae existunt ad Maiam et Indomam b. Stellerus observavit". Evidently, the non-typical specimen in the Linnean herbarium came from Maya and Indoma. Plants widely distributed in this region were described in 1805 by Hiltebrandt in a monograph on genus *Dracocephalum* as *D. stellerianum* Hiltebr. while the other 2 wholly typical specimens were evidently collected from Barguzin and should be treated as type of *D. grandiflorum. L.*

7. **D. heterophyllum** Benth. Labiat. gen. et sp. (1835) 738; Franch. Pl. David. 1 (1884) 239; Hook. f. Fl. Brit. Ind. 4 (1885) 665; Hemsley in J. Linn. Soc. London (Bot.) 30 (1894) 118; Kanitz in Szechenyi, Wissensch. Ergebn. 2 (1898) 724; Hemsley and Pearson in Petermanns Mitt. 28 (1900) 374; Deasy, In Tibet and Chin. Turk. (1901) 398; Hemsley, Fl. Tibet (1902) 195; Diels in Futterer, Durch Asien, 3 (1903) 19; Keissler in Annal. naturhist. Hofmus. 22, 1 (1907) 30; Diels in Filchner, Wissensch. Ergebn. 10, 2 (1908) 263; Danguy in Bull. Mus. nat. hist. natur. 14 (1908) 132; ibid. 17 (1911) 345; Dunn in Notes Bot. Gard. Edinburgh, 6 (1915) 169; Paulsen in Hedin, S. Tibet, 6, 3 (1922) 45, incl. var. *rubicundum* Paulsen; Kudo in Mém. Fac. Sci. and Agr. Taihoku Univ. 2 (1929) 242; Hand.-Mazz. in Österr. bot. Z. 79 (1930) 38; Pampanini, Fl. Carac. (1930) 183; Kashyap in Proc. Indian Sci. Congr. 19

(1932) 49; Rehder and Kobuski in J. Arn. Arb. 14 (1933) 31; Hao in Engler's Bot. Jahrb. 68 (1938) 634; Persson in Bot. notiser (1938) 300; Walker in Contribs U.S. Nat. Herb. 28 (1941) 655; Schischk. in Fl. SSSR, 20 (1954) 465; Grubov, Konsp. fl. MNR (1955) 236; Murata in Acta Phytotax. et Geobot. 16, 1 (1955) 13; Fl. Kirgiz. 9 (1960) 68; Ikonnikov, Opred. rast. Pamira (1963) 212; Fl. Kazakhst. 7 (1964) 357. —*D. kaschgaricum* Rupr. in Mém. Ac. Sci. St.-Pétersb. VII ser. 14 (1869) 65. —*D. gobi* Krassan. in Zap. Russk. geogr. obshch-va, 19 (1888) 340. —*D. pamiricum* Briq. in Bot. tidsskr. 28 (1907) 239. —*D. fragile* auct. non Turcz.: Henderson and Hume, Lahore to Jarkend (1873) 331; Pampanini, Fl. Carac. (1930) 183. —? *D. nodulosum* auct. non Rupr.: Hance in J. Bot. (London) 20 (1882) 292. —**Ic.:** Briq. l.c. fig. 5 (sub nom. *D. pamiricum*).

Described from East. Himalayas. Type in Paris.

Rocky-rubble mountain slopes in steppe belt, rocks, meadows in alpine belt, coastal sand, 2000–5000 m.

**IA. Mongolia:** *East. Mong.* ("Ourato, près de Maomingan, dans les lieux secs, No. 2848, July 1866, David"—Franch, l.c.). *Alash. Gobi, Khesi.*

**IB. Kashgar:** *Nor.* (Uch-Turfan, Airi gorge, June 2, 1908—Divn.). *West., East.* (mountain road from Bartu to timber plant at Khomot, 2160 m, No. 6970, Aug. 3; Nyuitszygen in Khomot, meadow on south-west. slope, 3100 m, No. 7658, Aug. 10—1958, Lee and Chu).

**IIA. Junggar:** *Tien Shan.*
**IIIA. Qinghai:** *Nanshan, Amdo.*
**IIIB. Tibet:** *Chang Tang, Weitzan, South.*
**IIIC. Pamir.**

General distribution: Nor. and Cent. Tien Shan, East. Pam.; Nor. Mong. (Hang.), China (North and North-West), Himalayas.

8. **D. imberbe** Bunge in Mém. Ac. Sci. St.-Pétersb. 2 (1835) 560; Sapozhn. Mong. Alt. (1911) 381; Simpson in J. Linn. Soc. London (Bot.) 41 (1913) 436; Danguy in Bull. Mus. nat. hist. natur. 20 (1914) 84; Krylov, Fl. Zap. Sib. 9 (1937) 2322; Persson in Bot. notiser (1938) 300; Schischk in Fl. SSSR, 20 (1954) 453; Grubov, Konsp. fl. MNR (1955) 236; Keenan in Baileya, 5 (1957) 34; Fl. Kirgiz. 9 (1960) 62; Fl. Kazakhst. 7 (1964) 349. —*D. laniflorum* Rupr. in Mém. Ac. Sci. St.-Pétersb. VII Sér. 14 (1869) 65. —*D. albertii* Regel in Acta Horti Petrop. 6, 2 (1880) 362. —*D. altaiense* auct. non Laxm.: Hiltebr. Monogr. Dracoceph. (1805) 50, 75. —**Ic.:** Hiltebr. l.c. tab. 13 (sub nom. *D. altaiense*); Keenan, l.c. fig. 13; Fl. Kazakhst. 7, Plate 40, fig. 4.

Described from Altay. Type in Leningrad.

Rocks, cliffs, rocky slopes, coastal pebble beds, moraines in alpine belt.

**IA. Mongolia:** *Khobd.* (south. Kharkhira peak, on rocks, July 23; same site, on pebble bed near river and on mountain slopes, July 24—1879, Pot.). *Mong. Alt.* (slopes of Shadzagain-Suburga pass, July 22, 1898—Klem.).

**IB. Kashgar:** *Nor.* (Dzhanart pass, June 14–17, 1903—Merzbacher; Uch-Turfan, Karayutlik gorge, June 18, 1908—Divn.; Muzart river valley, lateral right bank valley 3–4 km before Sazlik area on road to Oi-Terek area, *Leucopoa*-covered mountain steppe along nor. slope,

2750 m, Sept. 9, 1958—Yun. et al.). *West.* (Sarykol'sk mountain range to west of Kashgar, Bostan-Terek locality, July 11, 1929—Pop.; "Bostan-Terek, 2400 m, Aug. 5, 1934, Persson"—Persson, l.c.). *East.* (on Bogdoshan in Turfan, on exposed slope, 2800 m, No. 5703, June 17; near Urumchi-Kucha highway, No. 6066, July 21; from Bort to east. Taskhan canal in Khomot, on valley slope 2900 m, No. 7087, Aug. 5; Ustun canal in Khomot, on slope, 3100 m, No. 7139, Sept. 7—1958, Lee and Chu).

IIA. Junggar: *Jung. Alt.* (near Toli town, on slope, 2450 m, Nos. 1183, 1215, 2643, Aug. 6; 20 km south of Ven'tsyuan', in forest, No. 1550, Aug. 14; same site, No. 101, Aug. 26—1957, Kuan). *Tien Shan* (nor. and east.).

IIIC Pamir (Ulugtuz gorge, in Charlysh river basin, along bank of brook near irrigation ditch, June 26, 1909—Divn.; along Mia river gorge, height 4000 m, July 21, 1941; Ulugtuz river gorge, 3500–4000 m, Aug.-Sept. 1942—Serp.).

General distribution: Jung.-Tarb., Nor. and Cent. Tien Shan; West. Sib. (Altay), East. Sib. (south-west.), Nor. Mong. (Fore Hubs.).

## 9. D. integrifolium Bunge in Ledeb. Ic. pl. fl. ross. 2 (1830) 10, tab. 120; id. in Ledeb. Fl. alt. 2 (1830) 387; Sapozhn. Mong. Alt. (1911) 382; Danguy in Bull. Mus. nat hist. natur. 20 (1914) 84; Krylov, Fl. Zap. Sib. 9 (1937) 2325; Persson in Bot. notiser (1938) 300; Schischk, in Fl. SSSR, 20 (1954) 457; Grubov, Konsp. fl. MNR (1955) 236; Grubov in Bot. mater. Gerb. Bot. inst. AN SSSR, 19 (1959) 549; Fl. Kirgiz. 9 (1960) 65; Fl. Kazakhst. 7 (1964) 350. — Ic.: Ledeb. Ic. pl. fl. ross, 2, tab. 120.

Described from Altay. Type in Leningrad.

Steppe and clayey meadow and rocky-rubble slopes, larch forest fringes, 900–3000 m.

IA. Mongolia: *Khobd.* (between border and Khobdo town, 1870— Kalning).

IB. Kashgar: *Nor.* (near Abad, May 30, 1903—Merzbacher). *West.* (King-Tau mountain range, 3 km south-east of Kosh-Kulak, steppe belt, 2900 m, June 10, 1959—Yun.; along road to Kosh-Kulak from Upal, 1800–3000 m, Nos. 199, 260, June 10, 1959—Lee et al.; "Bostan-terek, ca 2400 m, Aug. 1934, Persson"—Persson, l.c.), *East.* (4 km north of Sansankho-Turfan, on exposed slope, 2400 m, No. 5639, May 15, 1958—Lee and Chu).

IIA. Junggar: *Cis-Alt.* (20 km nor.-west of Shara-Sume on Kran river, steppe belt, scrub-meadow steppe on south. slope, July 7, 1959—Yun.). *Jung. Alt.* (Urtaksary, July 19, 1879—Fet.; Dzhair mountain range, Yamata river gorge, rocky slopes, July 2, 1947—Shumakov; nor. of Karamai settlement, 1800 m, No. 10766, July 21, 1959—Lee and Chu). *Tien Shan* (nor. and east.).

General distribution: Jung.-Tarb., Nor. and Cent. Tien Shan; Mid Asia, West. Sib. (Altay).

## 10. D. moldavica L. Sp. pl. (1753) 595; Franch, Pl. David. 1 (1884) 239; Forbes and Hemsley, Index Fl. Sin. 2 (1902) 292; Krylov, Fl. Zap. Sib. 9 (1937) 2328; Kitag. Lin. Fl. Mansh. (1939) 378; Schischk. in Fl. SSSR, 20 (1954) 463; Fl. Kirgiz. 9 (1960) 67; Fl. Kazakhst. 7 (1964) 355. —*D. moldavica* L. α *europaeum* Hiltebr. Monogr. Dracoceph. (1805) 43. —Ic.: Hiltebr. l.c. tab. 9.

Described from East. Europe. Type in London (Linn.).

Sandy river valleys, rocks, meadows, grown in kitchen gardens; also as ruderal.

IA. Mongolia: *East. Mong.* (Sartchy, 1866—David; Narum-Daban, between Kerulen and Dolon-Nor, 1870—Lom.; sand facing Khekou town, Aug. 8, 1884—Pot.). *East. Gobi* (Shara

Murun, red lake bottom, No. 569, 1925—Chaney). *Ordos* (in Huang He river valley, 1871—Przew.; Ushkyun-Tokhum area, Aug. 29; meadow around Ulan-Tologoi monastery, Sept. 6—1884, Pot.).

General distribution: Nor. Tien Shan; Europe, Mid. Asia, West. Sib. (south), East. Sib. (south), Far East (south), China (Dunbei, North, North-West, North-East).

Note. Long cultivated as a melliferous plant and a surrogate for tea. Cultivated at present as an aromatic plant. Emits very pleasant fragrance (see note under *D. foetidum* Bunge).

11. **D. nodulosum** Rupr. in Mém. Ac. Sci. St.-Pétersb. VII sér. 14 (1869) 65; Danguy in Bull. Mus. nat. hist. natur. 20 (1914) 84; Schischk. in Fl. SSSR, 20 (1954) 460; Fl. Kirgiz. 9 (1960) 66; Fl. Kazakhst. 7 (1964) 354. —**Ic.:** Fl. SSSR, 20, Plate 27, fig. 3.

Described from Tien Shan. Type in Leningrad.

Rocky and turfed mountain slopes, talus and rocks, up to 3000 m.

IB. Kashgar: *West.* (Sulu-Sakal valley, 25 km east of Irkeshtam, upper part of juniper zone, rocky valley, 2800–2900 m, July 26, 1935—Olsuf'ev).

IIA. Junggar: *Tarb.* (Saur mountain range, south. slope in Khobuksk depression, Karagait river valley near its emergence on trail, on rocky southern slopes, June 23, 1957—Yun. et al.).

General distribution: Jung.-Tarb., Nor. and Cent. Tien Shan; Mid. Asia (south.).

12. **D. nutans** L. Sp. pl. (1753) 596; Hiltebr. Monogr. Dracoceph. (1805) 54, 76; Hook. f. Fl. Brit. Ind. 4 (1885) 665; Sapozhn. Mong. Alt. (1911) 382; Simpson in J. Linn. Soc. London (Bot.) 41 (1913) 437; Danguy in Bull. Mus. nat hist. natur. 20 (1914) 84; Pampanini, Fl. Carac. (1930) 183; Krylov, Fl. Zap. Sib. 9 (1937) 2325; Kitag. Lin. Fl. Mansh. (1939) 378; Schischk in Fl. SSSR, 20 (1954) 468; Grubov, Konsp. fl. MNR (1955) 236; Fl. Kirgiz. 9 (1960) 65; Fl. Kazakhst. 7 (1964) 353; Hanelt und Davažamc in Feddes repert. 70 (1965) 54. —*D. nutans* L. var. *alpinum* Kar. et Kir. in Bull. Soc. natur. Moscou, 15 (1842) 424. —**Ic.:** Pall. Fl. Ross. tab. 115; Fl. Kazakhst. 7, Plate 40, fig. 9.

Described from Siberia. Type in London (Linn.).

Coastal, forest and meadow slopes, coastal pebble beds, turfed alluvium; from forest to alpine belt.

IIA. Junggar: *Cis-Alt.* (Qinhe, No. 980, Aug. 4, 1956—Ching; Koktogoi, No. 10398, June 6, 1959—A.R. Lee et al. (1959); nor.-west of Shara-Sume on Kran river, scrub-meadow steppe, July 7; north of Koktogoi, right bank of Kairta river, Kugtdyn valley forest belt, Aug. 15—1959, Yun.). *Tarb.* (Saur mountain range, south. slope of Karagaita river, Bain-Tsagan creek valley subalpine belt, meadow on south. rubble slope, June 23, 1957—Yun. et al.). *Jung. Alt.* (Toli, Albakain mountain, No. 2641, Aug. 6; nor. bank of Sairam-Nur, on sunny slope, No. 2128, Aug. 26—1957, Kuan). *Tien Shan* (nor.; in east up to Barkul' lake). *Dzhark.* (Suidun, June 16, 1877—A. Reg.; Kutukchi, west of Kul'dzha, June 6, 1878—Fet.).

General distribution: Fore Balkh., Jung.-Tarb., Nor. and Cent. Tien Shan; Europe, Fore Asia, Mid. Asia, West. and East. Sib., Far East, Nor. Mong., China (Dunbei) Himalayas (Kashmir).

13. **D. origanoides** Steph. ex Willd. Sp. pl. 3 (1800) 151; Sapozhn. Mong. Alt. (1911) 381; Krylov, Fl. Zap. Sib. 9 (1937) 2321; Schischk. in Fl. SSSR, 20 (1954) 445; Grubov, Konsp. fl. MNR (1955) 236; Fl. Kirgiz. 9 (1960) 58; Fl.

Kazakhst. 7 (1964) 347.   —*D. pinnatum* L. α *altaicum* Bunge in Mém. Ac.
Sci. St.-Pétersb. Sav. Etrang. 2 (1835) 562.   —*D. pinnatum* L. var. *minus*
Ledeb. Fl. Ross 3 (1849) 384; Sapozhn. Mong. Alt. (1911) 381.   —*D. pinnatum*
L. var. *pallidiflorum* Kar. et Kir. Bull. Soc. natur. Moscou, 15 (1842) 422.   —
**Ic.:** Ledeb. Ic. pl. fl. ross. 5, tab. 445 (sub nom. *D. pinnatum*); Fl. Kazakhst. 7,
Plate 40, fig. 1.

Described from Siberia. Type in Berlin. Isotype in Leningrad.

Rubble and rocky slopes, rocks, talus in mountain-steppe and alpine
belts.

**IA. Mongolia:** *Khobd.* (Kharkhira river valley, on rock beds, July 20, 1879—Pot.). *Mong.
Alt.* (near Kobdo town, June 16, 1870—Kalning; on foothill slopes of Tsastu-Bogdo, upper
course of Uzun-Dayur river, among pebble-clayey-shale formations, July 31, 1896—Klem;
upper course of Tsagan-Gol river, old moraines between Khor-sola and Prokhodna, June 30,
1905—Sap.; Urtu-Gol valley, dry rubble mountain slope, Aug. 19, 1930—Pob.; Tolbo-Kungei
mountain range, high-mountain belt, rocky and semi-fixed alluvium, Aug. 5, 1945; Adzhi-
Bogdo mountain range, midcourse of Burgastyin-Daba river, along road from Indertiin-Gol,
mountain steppe, Aug. 6, 1947—Yun.). *Cent. Khalkha* (Dzhargalante river basin, source of
Kharukhe river, around Ulan-Khada mountains, rubble slopes, Sept. 18, 1925—Krasch. and
Zam.; Ul'dzeitu somon, road from Arbai-Khere to Ulan-Bator, 15 km east of camp in somon,
on rocky crest of knoll, June 19, 1948—Grub.). *Gobi Alt.* (Baga-Bogdo, on upper meadows,
2700 m, 1925—Chaney).

**IIA. Junggar:** *Tien Shan* (in steppe valley between Tien Shan and Mechin-01, north of
mountain passage through Tien Shan, June 12, 1877—Pot.; Irenkhabirg, Upper Taldy, 3000 m,
May 21; Nilki, June 30—1879, A. Reg.).

**General distribution:** Jung.-Tarb., Nor. and Cent. Tien Shan; Mid. Asia, West. Sib. (Altay),
East. Sib. (south-west.), Nor. Mong. (Hang.).

**14. D. peregrinum** L. Cent. pl. 2 (1756) 20; Hiltebr. Monogr. Dracoceph.
(1805) 37, 69; Sapozhn. Mong. Alt. (1911) 382; Danguy in Bull. Mus. nat.
hist. natur. 20 (1914) 84; Krylov, Fl. Zap. Sib. 9 (1937) 2330; Schischk in Fl.
SSSR, 20 (1954) 464; Grubov, Konsp. fl. MNR (1955) 237; Fl. Kazakhst. 7
(1964) 356.   —**Ic.:** Hiltebr. l.c. tab. 8; Pall. Fl. Ross. tab. 117.

Described from Siberia. Type in London (Linn.).

Rocky and rubble slopes in desert-steppes, rocks and talus; ascends to
upper forest boundary.

**IA. Mongolia:** *Khobd.* (between boundary and Khobdo; between Khobdo and Khatu
rivers, July 2, 1870—Kalning; mountains between Khobdo and Ukha rivers, on road, July 4,
1898—Klem.; at crossing of Ulan-Daban and Khobdo town, Oigur river, July 30, 1899—Lad.;
east. steppe slope with exposed conglomerates at descent from Ulan-Dabar crossing into Ubsa-
Nor lake, Sept, 19, 1931—Bar.; midcourse of Tsagan-Nurin-Gol, south. slope in lower mountain belt, Aug. 2, 1945—Yun.). *Mong. Alt.* (Tsitsiriin-Gol, dry pebble bed, June 10, 1877—Pot.;
east. part of Mong. Alt., not far from Borogod-Daban, on rocky talus, Aug. 18; Khara-Dzarga
mountain range, Sakhir-Sala river valley, on east. rubble slope of Imertsik mountain, Aug.
22—1930, Pob.; upper course of Tsinkir river, near Tsagan-Burgas canal, Oct. 6, 1930—Bar.;
Bain-somon, Khalyun area, nor. slope, forest belt, steep rocky south. slope, Aug. 24, 1943;
along road to Yusun-Bulak from Bain-somon, 3 km west of passage through Khan-Taishir,
valley floor, Aug. 11, 1947—Yun.).

IIA. Junggar: *Jung Alt.* (toward Kuzyun pass, tip of ridge, pebble bed, July 2, 1908—B. Fedtschenko; Toli, on knoll, No. 1400, Aug. 8, 1957—Kuan).

General distribution: Jung. Tarb., Nor. Tien Shan; West. Sib. (south), East. Sib. (southwest).

15. **D. purdomii** W.W. Smith in Notes Bot. Gard. Edinburgh, 9 (1916) 105; Keenan in Baileya, 5 (1957) 36; Wu in Acta Phytotax. Sin. 8, 1 (1959) 24. —*D. grandiflorum* L. var. *purdomii* (W.W. Smith) Kudo in Mem. Fac. Sci. and Agr. Taihoku Univ. 2 (1929) 241. —*D. truncatum* Sun in Wu in Acta Phytotax. Sin. 8, 1 (1959) 25. —*D. imberbe* auct. non Bunge: Kanitz, A növénytani (1891) 47; Forbes and Hemsley, Index Fl. Sin. 2 (1902) 292; Rehder and Kobuski in J. Am. Arb. 14 (1933) 31; ? Walker in Contribs U.S. Nat. Herb. 28 (1941) 656. —*D. rupestre* auct. non Hance: Wu in Acta Phytotax. Sin. 8, 1 (1959) 23, p. min. p. —**Ic.:** Keenan l.c. fig. 15; Acta Phytotax. Sin. 8, 1, tab. 2 (sub nom. *D. truncatum* Sun).

Described from North-West China (Gansu province). Type in Edinburgh.

Rubble and rocky slopes in upper hill belt.

IIIA. Qinghai: *Nanshan* (Nor. Tetung mountain range, July 8, 1872; Rako-Gol river, July 21, 1880—Przew.; "Lang Tzu T'ang Ko, in dense tussocks, on an exposed, moist, gravelly foothill, common No. 590, Ching"—Walker, l.c.). *Amdo* (Alp. belt of Dzhakhar-Dzhargyn hill, 3150–3450 m, June 24; on Yussun-Khatyma river, June 24—1880, Przew.; "in jugi Kvetehiensis latere bor. 2700 m, Aug. 1879, Szecheny"—Kanitz, l.c.).

General distribution: China (North-West).

Note. *D.purdomii* is one of representatives of peculiar group of high-montane species. It differs very distinctly from *D. imberbe* Bunge in pubescence of the corolla and shape of radical leaves. It was described from the southern regions of Gansu province (Min-Shan mountains and has been collected many times by various collectors in Gansu, Qinghai and west. regions of Shenxi. The species is comparatively uniform within its distribution range. Only the shape of calyx teeth, shape of corolla and length of petioles of radical leaves vary insignificantly but this variation is within intraspecific limits and is clearly inadequate to merit a distinct species, *D. truncatum* Sun.

*D. purdomii* stands somewhat between *D. grandiflorum* and *D. imberbe*. Having studied these species, the conclusion is spontaneous that differentiation evidently proceeded on one side from type *D. purdomii* to *D. grandiflorum* and, on the other side, to *D. imberbe* and *D. wallichii—D. bullatum*. See also note under *D. rupestre*.

16. **D. rupestre** Hance in J. Bot. (London) 7 (1869) 166; Hand.-Mazz. in Acta Horti Gothoburg. 9 (1934) 78; Kitag. Lin. Fl. Mansh. (1939) 378; Keenan in Baileya, 5 (1957) 34; Wu in Acta Phytotax. Sin. 8, 1 (1959) 23, p.p. —*D. grandiflorum* auct. non L.: Franch. Pl. David. 1 (1884) 238; Forbes and Hemsley, Index Fl. Sin. 2 (1902) 291, p. max. p.; Dunn in Notes Bot. Gard. Edinburgh, 6 (1915) 169, p.p.; Kudo in Mém. Fac. Sci. and Agr. Taihoku Univ. 2 (1929) 240, p.p.—**Ic.:** Keenan l.c. fig. 14.

Described from North China (Peking (Beijing) environs). Type in Paris. Found at 1300–2000 m.

IA. **Mongolia:** *East. Mong.* (Ourato, lieux pierreux des montagnes, July, 1866—David; in Muni-Ula mountains, July 7, 1871—Przew.; Pei t'ai, Ordos, July 1, 1922—Licent [K]; "Tatsingschan, Hala Ch'ing Kow, 1300 m, No. 2772, July 22"—Wu, l.c.).

**General distribution:** China (Dunbei, Nor.).

**Note.** The species is closely related to *D. grandiflorum* L. but well differentiated in shape of radical leaves and absence of pubescence on inside of upper corolla lip. Several investigators have attempted to merge these species but in our opinion, as well as Keenan's (l.c.), there is no justification at all to merge these 2 distinct species.

*D. rupestre* is also closely related to *D. purdomii* Smith but differences between these species are nonetheless sharp (see key). In our view, even the distribution ranges of these species are entirely distinct. *D. rupestre* is distributed in Hebei, Zhekhe and Shanxi provinces while *D. purdomii* is found only in Qinghai, Gansu and western regions of Shenxi. From this viewpoint, the report of *D. rupestre* from Qinghai (Wu, l.c.) causes doubt, all the more since Walker (l.c.) identified this specimen as *D. imberbe* Bunge. In our opinion, this plant which has given rise to such contradictory judgements is evidently nothing else but *D. purdomii.*

17. **D. ruyschiana** L. Sp. pl. 2 (1753) 595; Hiltebr. Monogr. Dracoceph. (1805) 33, 66; Franch. Pl. David. 1 (1884) 240; Forbes and Hemsley, Index Fl. Sin. 2 (1902) 292; Simpson in J. Linn. Soc. London (Bot.) 41 (1913) 437; Danguy in Bull. Mus. nat. hist. natur. 20 (1914) 84; Krylov, Fl. Zap. Sib. 9 (1937) 2331; Kitag. Lin. Fl. Mansh. (1939) 378; Schischk. in Fl. SSSR, 20 (1954) 472; Grubov, Konsp. fl. MNR (1955) 237; Fl. Kirgiz. 9 (1960) 70; Fl. Kazakhst. 7 (1964) 360. —*D. argunense* auct. non Fisch. ex Link: Kudo in Mém. Fac. Sci. and Agr. Taihoku Univ. 2 (1929) 243, p.p. —**Ic.:** Hiltebr. l.c. tab. 6; Fl. Kazakhst. 7, Plate 41, fig. 4.

Described from Siberia and Sweden. Type in London (Linn.).

Sparse larch and mixed forests and their fringes, meadow slopes, floodplains of hill brooks, rocky slopes and rock crevices.

IA. **Mongolia:** *Khobd.* (between boundary and Khobdo, 1870—Kalning). *East. Mong.* (Irekte station, wet creek valley, June 23, 1924—Skvortzov).

IIA. **Junggar:** *Tien Shan* (Agyaz river, 1500–1800 m, June 25, 1878; Aryslyn, 2400–2700 m, July 11, 1879—A. Reg.; nor. steppe slope of Tekes, between Sumbe-Khonokhai and Sumbe, June 24–25, 1907—Merzbacher; east of Chzhaos [Kolman-Kure], 26 km from Tekes, along road, on nor. slope, Aug. 17, 1957—Kuan; "Sairam-Nor, montagnes, alt. 2000 m, No. 1255, July 21, 1895, Chaff".—Danguy, l.c.).

**General distribution:** Fore Balkh., Jung.-Tarb., Nor. and Cent. Tien Shan; Europe, Caucasus, Mid. Asia (hills), West. Sib., East. Sib., Nor. Mong. (Hent., Hang., Mong.-Daur.), China (Dunbei), Japan.

**Note.** This extensively distributed species depicts more or less constant characteristics throughout its range, *D. argunense* Fisch., closely related but well distinguished from *D. ruyschiana*, is found in the Far East, East. Siberia, Korean peninsula, China (Dunbei) and Japan. *D. argunense* Fisch. has nowhere penetrated our territory under study.

18. **D. stamineum** Kar. et Kir. in Bull. Soc. natur. Moscou, 15 (1842) 423; Henderson and Hume, Lahore to Jarkand (1873) 331; Hook. f. Fl. Brit. Ind. 4 (1885) 666; Paulsen in Hedin, S. Tibet, 6, 3 (1922) 46; Pampanini, Fl. Carac. (1930) 183; Persson in Bot. notiser (1938) 300; Schischk. in Fl. SSSR, 20 (1954) 473; Ikonnikov in Dokl. AN Tadzh. SSR, 20 (1957) 55; Fl. Kirgiz. 9 (1960) 70; Ikonnikov, Opred. rast. Pamira (1963) 212; Fl. Kazakhst. 7 (1964) 360.

*—D. pulchellum* Briq. in Bot. tidsskr. 28 (1907) 241.    *—Fedtschenkiella staminea*
S. Kudr. in Bot. mater. Gerb. Bot. inst. Uzb. filiala AN SSSR, 4 (1941) 6.    —
Ic.: Briq. l.c. fig. 6; S. Kudr. l.c. fig. 2.
Described from Jung. Alatau (Sarkhan). Type in Leningrad.
Usually found in high-mountain belts in alpine and subalpine mead-
ows, on sandy-rocky slopes, gravel beds of rivers, 1700–5600 m.

IB. **Kashgar:** *Nor.* (near Bedel' pass [Aug. 17, 1886]— Krasnov; around Kukurtuk pass,
June 26–July 2, 1903—Merzbacher; Uch-Turfan, Krargaily gorge, river bed, June 17, 1908—
Divn.). *West.* (Sulu-Sakal valley, 25 km east of Irkeshtam, July 26, 1935—Olsuf'ev; "Bostanterek,
ca. 2400 m, Aug. 5, 1934, Persson"—Persson, l.c.). *East.* (in Bogdoshan, Turfan, 2000 m, valley
terrace, No. 5741, June 18, 1958—Lee and Chu).

IIA. **Junggar:** *Jung. Alt.* (nor. bank of Sairam lake, 1920 m, July; Borotala, 2550 m, Aug.—
1878, A. Reg.; Borotala river basin, south. slope below Koketau pass, July 21, 1909—Lipsky).
*Tien Shan* (Kum-Daban, 2700 m, May 28; Kumbel', 2700–3000 m, May 31; Kash, June 9;
Monguto, 3000–3300 m, July 4; Yultu-Aryslyn, 2100–2400 m, July 7; Aryslyn, 2700 m, July 17;
Kunges, 2700 m, Aug. 22—1879, A. Reg.; Karagaitas pass, upper course of Khaidyk-Gol river,
July 24, 1893—Rob.; Irisu, on water, 1700 m, No. 1887, July 17; 28 km south-east of
Nyutsyuan'tsz, on bank of Ulausu river, in meadow, No. 184, July 18; Daban to Danu, on
slope, 2400 m, No. 2002, July 19; 30 km south of Nyutsyuan'tsz, on Ulausu bank, No. 322, July
19; 0.5 km south of Danu, on rocky slope, No. 360, July 21—1957, Kuan; Manas river basin,
Ulan-Usu river valley, 8–9 km beyond confluence of Koisu river with it, forest belt, on gravel
floor of small creek valley, July 18; same site, Danu Gol river valley, steep flank of glacial
trough, on rubble talus, July 21—1957, Yun. et al.).

IIIB. **Tibet:** *Chang Tang* (Chang-lá and Másinik Pass, 4575–4875 m, July 10–15, 1870—
Henderson: "Versante sett.: fra il Campo Remo nord, 5075 m ed in Campo Oltreremo, 5280 m,
Dainelli e Marinelli"—Pampanini, l.c.).

IIIC. **Pamir** (Tagdumbasch-Pamir, in angustiis Pistan jugi Sary-Kol, in detritu lapidoso,
3900–4200 m, July 15, 1901—Alexeenko; Kok-Muinak pass, on rocky-sandy slope, July 8, 1909—
Divn.; Pil'nen river, 3000–4000 m, June 30; Shorluk river gorge, 4000–5500 m, July 28; Kulan-
aryk district, between Zaz settlement and Tash-Ui river, 3500–3800 m, July 29—1942, Serp.;
"Mustag-ata, Jam-bulak-bashi, 4439 m, Aug. 14, 1894, Hedin"—Paulsen, l.c.; "Kungur mountain range, on slope, right Kok-sel "moraine glacier, upper course of Kenshirber-Su river,
4100 m, Aug. 16, 1956—Pen Shu-li, Skorobogatov; nor.-west. slope of Kungur mountain range,
near glacier on moraine, 4500 m, Aug. 13–17, 1956, Dmitriev et al."—Ikonnikov, l.c. 1957).

**General distribution:** Jung.-Tarb., Nor. and Cent. Tien Shan, East. Pam.; Mid. Asia (hills),
Himalayas (Kashmir).

19. **D. tanguticum** Maxim. in Bull. Ac. Sci. St.-Pétersb. 27 (1881) 530;
Kanitz, A növénytani (1891) 47; Forbes and Hemsley, Index Fl. Sin. 2 (1902)
293; Diels in Futterer, Durch Asien, 3 (1903) 19; Dunn in Notes Bot. Gard.
Edinburgh, 6 (1915) 168; Marquand in J. Linn. Soc. London (Bot.) 48 (1929)
218; Kudo in Mém. Fac. Sci. and Agr. Taihoku Univ. 2 (1929) 238; Rehder
and Kobuski in J. Arn. Arb. 14 (1933) 31; Hand.-Mazz. in Acta Horti
Gothoburg. 9 (1934) 78; ej. Symb. Syn. 7 (1936) 917; Hao in Engler's Bot.
Jahrb. 68 (1938) 634; Hand.-Mazz. in Acta Horti Gothoburg. 13 (1939) 343;
Walker in Contribs U.S. Nat. Herb. 28 (1941) 656; Murata in Acta Phytotax.
et Geobot. 16, 1 (1955) 13; Keenan in Baileya, 5 (1957) 26; Wu in Acta

Phytotax. Sin. 8, 1 (1959) 21.  —? *D. hookeri* Clarke in Hook. f. Fl. Brit. Ind. 4 (1885) 666; Hemsley, Fl. Tibet (1902) 195.  —Ic.: Keenan, l.c. fig. 9. Described from Qinghai. Type in Leningrad. Plate I, fig. 1.

Clayey and pebble slopes and rocks in grass-forb steppe, spruce forests, 2500–3600 m.

IA. **Mongolia:** *Khesi* ("T'ai Hua, on an exposed, dry roadside of hard clay, common, No. 535, 1923, Ching"—Walker, l.c.).

IIIA. **Qinghai:** *Nanshan* (exposed slopes of mountain range along Tetung river near Chertynton temple, July 8, 1872; high-mountains between Nanshan and Don-Kyr, along Raka-Gol river, 3000–3300 m, July 21, 1880; Nor. Tetung mountain range, Aug. 14, 1880—Przew.; Shibanguku area, south. slope, middle belt, Aug. 4, 1909—Czet.; on grassy mountains of Kumbum, south of Hsining, alt. 2700 m, No. 13249, Sept. 10; on loess banks between Hsining and Tankar, alt. 2700 m, No. 13260, Sept. 10—1925, Rock [K]; 66 km west of Xining, rocky slopes of knolls, grass-forb mountains steppe, 2800 m, Aug. 5, 1959—Petr.; "in jugi Kvetehiensis, latere merid. 2400 m supra Kázsán, init. Aug. 1879, Szecheny"—Kanitz, l.c.). *Amdo* (Mudzhik-kh 2700–2850 m, June 29, 1880—Przew., lectotypus!; Dabasun-Gobi, on clay, 3000 m, Aug. 13–15, 1901—Lad.; "in der Nähe des Klosters Ta-schiusze, 4000 m, Sept. 9, No. 1187; auf der Ebene Schalakutu, nicht selten, um 3400 m, No. 857, Aug. 18; auf dem östlichen Nan-Schan, um 2900-3200 m, Aug. 3, Nos. 791, 823; auf dem Plateau Dahoba, um 4000 m, Aug. 28, No. 1051; ebendort, im Tälchen eines Flusses, Sept. 7, No. 1178"—Hao, l.c.).

IIIB. **Tibet:** *Weitzan* (Yantszytszyan river basin, environs of Kabchzhi-Kamba village, 3630 m, July 20; Mekong river basin, along Chokchu river, in humus and spruce forest, 3600 m, Aug. 31; Dzhagyngol river—1900, Lad.) *South.* (Mt. Everest Exped., Shekar, alt. 4200 m, No. 348, July 9, 1922—Morton [K]; mountains above Lhasa, Aug. 1904; Karo La Pass, about 4800 m, 1904—Walton [K]; Everest Exped., 1921—Wollaston [K]; Atsa Tso, 4200–4500 m, Aug. 26, 1924—Ward [K]; Tzang Po, alt. 3600 m, No. 118A, Aug. 30, Shigatze, alt. 3800 m, Nos. 63, 82, Sept. 3—1935, Cutting and Vernay [K]; mountain behind Drepung N. West of Lhasa, alt. 5400 m, No. 36, Sept. 27, mountain behind Gyantze, alt. 4500 m, No. 857, Aug. 13; above Singma Khangchung, alt. 3400 m. No. 200 Aug. 21—1936, Chapman [K]: "Lhasa, Neesal temple, Sept. 24; Laten temple Oct. 14; circa Shigatse, Tsatan village, Aug. 1, 1914, Kawaguchi"—Murata, l.c.).

General distribution: China (North-West, South-West), ? Himalayas (Sikkim).

**Note.** In 1885, Clarke described *D. hookeri* Clarke from Sikkim. This species is closely related to *D. tanguticum* Maxim. and should possibly be regarded as a variety of *D. tanguticum*. However, a thorough study of adequate herbarium material from Sikkim and south. Tibet is necessary for a final solution to this problem. The few specimens from south. Tibet which studied help at present only to express doubts about the independent status of *D. hookeri*.

Within Qinghai and Tibet (Weitzan and south.), *D. tanguticum* is quite constant in its characteristics; comparatively small-flowered specimens, difficult to differentiate from *D. hookeri* were found only from Lhasa region but specimens with corolla typical of *D. tanguticum* are, however, found from the same region; the extent of pubescence of corolla varies somewhat over the entire distribution range of the species.

While describing *D. tanguticum*, there is only reference to the collection of N.M. Przewalsky (1872 and 1880) without citation of specimens. Only some have been preserved in the herbarium of the Komarov Botanical Institute of Academy of Sciences of the USSR (see geographic distribution). All of these specimens have been attested by K.I. Maximowicz. Since the type specimen had not been identified, I suggested specimen No. 409 as lectotype: "on Mudzhik-Khe river, 2700–2850 m, June 29, 1880, Przewalsky".

D. wallichii Sealy in Curtis's Bot. Mag. 164 (1944) sub tab. 9657, in adnot.; Keenan in Baileya, 5 (1957) 38.   —*D. calanthum* C.Y. Wu in Acta Phytotax. Sin. 8, 1 (1959) 24.   —*D. speciosum* auct. non Sweet: Benth in Wall. Cat. (1829) No. 2128 et in Wall. Pl. Asia Rar. 1 (1830) 65; Benth. Labiat. gen et sp. (1834) 494; Hook. f. Fl. Brit. Ind. 4 (1885) 665; Kudo in Mém. Fac. Sci. and Agr. Taihoku Univ. 2 (1929) 241.   —Ic.: Curtis's Bot. Mag. 33, tab. 6281 (sub nom. *D. speciosum*).

Described from East. Himalayas. Type in London (BM).

IIIB. Tibet: *South.* (occurrence possible since it is known from the adjoining Himalayan regions).

General distribution: Himalayas (west., east.).

Note. Within Yunnan province, *D. bullatum* Forest, closely related to *D. wallichii*, has been reported; the former differs from the latter in very low stem, strongly rugose leaves with deep purple petiole and nerves as well as shape of corolla. Unfortunately, we did not have good herbarium specimens from Yunnan province for critical evaluation of the aforesaid distinctive features. *D. wallichii* is distributed in the Himalayas from Simla to Sikkim and Bhutan while *D. bullatum* is found only in Yunnan. Considering all these aspects, all reports from Yunnan and Sikang cited for *D. calanthum* C.Y. Wu (= *D. wallichii* Sealy) need critical review.

## 14. Lallemantia Fisch. et Mey.
Ind. sem. hort. Petrop. 6 (1839) 52.

1. **L. royleana** (Benth.) Benth. in DC. Prodr. 12 (1849) 404; Hook. f. Fl. Brit. Ind. 4 (1885) 667; Gorschkova in Fl. SSSR, 20 (1954) 486; Fl. Kirgiz, 9 (1960) 71; Fl. Kazakhst. 7 (1964) 361.   —*Dracocephalum royleanum* Benth. in Wall. Pl. Asia Rar. 1 (1830) 65.   —*D. inderiense* Less. ex Kar. et Kir. in Bull. Soc. natur. Moscou, 15 (1842) 423.   —*Nepeta cordiifolia* Boiss. Diagn. ser. 1, 5 (1844) 24.   —Ic.: Fl. Kazakhst. 7, Plate 41, fig. 9.

Described from West. Himalayas. Type in London (K?).

Dry steppes, semi-deserts, foothill steppes, often as ruderal.

IIA. Junggar: *Tien Shan* (Borborogussun, 900–1200 m, April 27 [1879]—A. Reg.; Urumchi, Khunshan'river, on slope May 30; Manas, Khunshan'tszui, on slope, June 6; Savan district, Syaedi, June 20; Kuitun, near well, June 28—1957, Kuan). *Dzhark.* (Ili bank east of Kul'dzha, May 14; near Kul'dzha, May—1877, A. Reg.; same site, 1906—Muromsky).

General distribution: Aralo-Casp., Fore Balkh., Jung.-Tarb., Nor. and Cent. Tien Shan; Europe (southern Europ USSR), Fore Asia (Iran), Caucasus, Mid. Asia, West. Sib. (south-west.), Himalayas (west.).

## 15. Prunella L.
Sp. pl. (1753) 600.   —*Brunella* Moench, Meth. (1794) 414.

1. **P. vulgaris** L. Sp. pl. (1753) 600; Hance in J. Bot. (London) 20 (1882) 38; Franch. Pl. David. 1 (1884) 241; Hook. f. Fl. Brit. Ind. 4 (1885) 670; Forbes

and Hemsley, Index Fl. Sin. 2 (1902) 299; Dunn in Notes Bot. Gard. Edinburgh, 6 (1915) 177; Kudo in Mém. Fac. Sci. and Agr. Taihoku Univ. 2 (1929) 248; Rehder and Kobuski in J. Arn. Arb. 14 (1933) 300; Krylov, Fl. Zap. Sib. 9 (1937) 2333; Hao in Engler's Bot. Jahrb. 68 (1938) 634; Borissova in Fl. SSSR, 20 (1954) 495; Fl. Kirgiz. 9 (1960) 73; Fl. Kazakhst. 7 (1964) 364. —*P. parviflora* Gilib. Fl. Lithuan. 2 (1781) 88. —*P. officinalis* Güldenst. in Gmel. Reise, 1 (1774) 495. —*P. japonica* Makino in Tokyo Bot. Mag. 28 (1914) 158. —*P. asiatica* Nakai in Tokyo Bot. Mag. 49 (1930) 19, p.p. —*P. vulgaris* L. var. *japonica* (Makino) Kudo in J. Coll. Sc. Univ. Tokyo, 43, 8 (1921) 23. — *Brunella vulgaris* Moench, Meth. (1794) 414; Danguy in Bull. Mus. nat. hist. natur. 20 (1914) 85. —**Ic.:** Fl. Kazakhst. 7, Plate 42, fig. 1.

Described from Europe. Type in London (Linn.).

Meadows, forest fringes, river valleys, irrigation ditches, houses, gardens and kitchen gardens.

**IIA. Junggar:** *Tien Shan* (Talki pass, July 19, 1877—A. Reg.; same site, 1900 m, July 10, 1878—Fet.). *Dzhark.* (Kul'dzha, June 15, 1877—A. Reg.; Dzhagistai, sand, Aug. 6, 1957—Kuan).

**General distribution:** Aralo-Casp., Fore Balkh., Jung.-Tarb, Nor. and Cent. Tien Shan; Europe, Mediterranean, Balk.-Asia Minor, Fore Asia, Caucasus, Mid. Asia, West. Sib., East. Sib., Far East, China, Himalayas, Japan.

## 16. Eremostachys Bunge

in Ledeb. Fl. alt. 2 (1830) 414; Popov in Novye Mem. Mosk. obshch. ispyt. prir. 19 (1940) 48.

1. Calyx limb rotate, membranous, broad, reticulately nerved; root tuberous; all leaves undivided ........... 1. **E. moluccelloides** Bunge.
+ Calyx not enlarged upward, without limb; root fibrous, with tuberous thickening at tips of rootlets; radical leaves pinnatisect ...........
................................................................................ 2. **E. speciosa** Rupr.

1. **E. moluccelloides** Bunge in Ledeb. Fl. alt. 2 (1830) 415; Sapozhn. Mong. Alt. (1911) 382; Krylov, Fl. Zap. Sib. 9 (1937) 2337; Popov in Novye Mem. Mosk. obshch. ispyt. prir. 19 (1940) 134; Knorr. in Fl. SSSR, 21 (1954) 55, p.p.; Grubov, Konsp. fl. MNR (1955) 237; Fl. Kazakhst. 7 (1964) 386, p.p.; Pazij in Bot. mater. Gerb. Inst. bot. AN UzbSSR, 18 (1968) 16. —**Ic.:** Ledeb. Ic. pl. fl. ross, 5, tab. 437.

Described from Altay. Type in Leningrad.

Desert-steppe rocky slopes and rocks, fixed and coastal sand and sandy floors of ravines.

**IA. Mongolia:** *Khobd.* (between border and Khobdo, 1870—Kalning).

**IIA. Junggar:** *Cis-Alt.* (Ch. Irtysh valley, rocky slope between Shal'chigai and Ulasta river, June 3, 1903—Gr.-Grzh.; Kran river valley near Shara-Sume, June 28, 1908—Sap.; Quinhe region, 20 km south-east of Kalabuergun', on dune sand, No. 10331, June 1; Koktogoi, north-west of Ukhalashan' mine, No. 10433, June 10—1959, Lee et al.). *Zaisan* (50–55 km east of Zimunai settlement, along road along northern trail of Saur to Khobuk, desert, on floor of broad sand-covered ravine, July 10, 1959—Yun. et al.). *Dzhark.* (Aktyube near Kul'dzha, 900 m, May 13, 1877—A. Reg.). *Jung. Gobi.* (in steppe valley of Ch. Irtysh, Aug. [23–27] 1876—Pot.; along Urungu river, June 22, 1876—Pev.; on cliffs near Kyup spring, Aug. 8, 1898—Klem.; 30 km south of Sosotsyuan' source, Dinshan', sand, May 28, 1959—Lee et al.).

General distribution: Fore Balkh., Jung.-Tarb.

Note. *E. isochila* Pazij et Vved. described from west. Tien Shan is closely related to this species and is well differentiated from *E. moluccelloides* Bunge in equal corolla lips and profuse admixture of stellate hairs on under surface of radical and caulous leaves. V.K. Pazij and A.I. Vvedensky have suggested *E. isochila* somewhat doubtfully even for Kul'dzha region [see Bot. mater. Gerb. Inst. bot. AN UzbSSR, 18 (1968) 16]. Plants from this region available to us were defective, with poorly preserved leaves and corolla. Nevertheless, their study helped us to conclude that they cannot be placed among typical *E. isochila* Pazij et Vved. We treat these specimens as *E. moluccelloides* Bunge, all the more since we did not detect typical *E. isochila* in adjoining regions of Kazakhstan.

2. **E. speciosa** Rupr. in Mém. Ac. Sci. St.-Pétersb. VII sér. 14 (1869) 68; Popov in Novye Mem. Mosk. obshch. ispyt. prir. 19 (1940) 99; Knorr. in Fl. SSSR, 21 (1954) 27; Fl. Kirgiz. 9 (1960) 85; Fl. Kazakhst. 7 (1964) 376. —*E. transiliensis* Regel in Acta Horti Petrop. 9 (1886) 556. —*E. laciniata* auct. non Bunge: Regel in Acta Horti Petrop. 9 (1886) 552, p.p. —**Ic.:** Fl. Kirgiz. 9, Plate 7; Fl. Kazakhst. 7, Plate 53, fig. 4.

Described from Tien Shan (Shamsi pass). Type in Leningrad.

Rocky slopes in foothills and lower mountain belt.

IIA. **Junggar:** *Tien Shan* (at Sumbe pass, 1800–2100 m, July 29, 1877—Fet.; Chapchal pass, south of Kul'dzha, 1500–2100 m, June 28, 1878; Borborogusun, 900–1200 m, April, 27, 1879—A. Reg.; "Khanakhai, 1200–1500 m, June 27, 1878, A. Reg. [*typus E. transiliensis*]"—Regel, l.c.).

General distribution: Jung.-Tarb., Nor. and Cent. Tien Shan; Fore Asia, Mid. Asia.

Note. Highly polymorphic species varying in size and division of leaves, pubescence of leaves and calyx. We could not find herbarium specimens of *E. transiliensis* Regel as its type was evidently lost. Judging from the description, *E. transiliensis* Regel should be regarded as a race of *E. speciosa* Rupr. in which leaves are almost undivided, sometimes with sparsely lobed or lyrate leaves at base.

## 17. **Lamiophlomis** Kudo
In Mém. Fac. Sci. and Agr. Taihoku Univ. 2 (1920) 210.

1. **L. rotata** (Benth.) Kudo in Mém. Fac. Sci. and Agr. Taihoku Univ. 2 (1929) 211; Wu in Acta Phytotax, Sin. 8, 1 (1959) 33. —*Phlomis rotata* Benth. ex Hook. f. Fl. Brit. Ind. 4 (1885) 694; Hook. f. in Kew Bull. (1896) 214; Hemsley, Fl. Tibet (1902) 195; Dunn in Notes Bot. Gard. Edinburgh, 6 (1915) 185; Kashyap in Proc. Indian Sci. Congr. 19 (1932) 49; Rehder and Kobuski in J. Arn. Arb. 14 (1933) 31; Murata in Acta Phytotax et Geobot. 16, 1 (1955) 16.

Described from East. Himalayas (Sikkim). Type in London (K).

Grasslands on high-mountain areas, rocks and clayey-rocky slopes.

IIIA. Qinghai:*Amdo* (alpine meadows of mts south of Radja, southern slopes of Yangtse river, alt. 3600 m, No. 14172, June 1926—Rock; alpine region between Radja and Jupar range: Wajo valley south of Woti La, on grassy slopes, alt. 3900 m, No. 14406, July 1926—Rock [K]).

IIIB. Tibet: *Chang Tang* (Gooring Valley [30° 12 lat. 90° 25′ long.], about 5000 m, July–Aug. 1895—Litledale [K]), *Weitzan* (mountains along Talachu river, on humus-clayey slopes, 3900–4500 m, June 19; alpine region of mountains along Yangtse river, on bank, 3900–4500 m, June 20—1884, Przew.; on exposed clayey-rocky slopes of Purchekh-la pass, 3900 m, July 22; on humus in alpine meadows and above them on wet clay, Chamudug-la pass, 4700 m, July 26; in Ichu river valley and an around lake Rkhombo-mtso, in meadows and around lake, on humus, 3900 m, Aug. 1; in grasslands along valley and on mountain slopes, in humus, around Chzherku monastery, 3300 m, Aug. 8—1900, Lad.). *South.* (Dsaka Chu, 3900 m, No. 337, July 7, 1922—Morton [K]; Rongphar Valley, No. 65, July 2, 1924—Kingston [K]; Chongphu Chu, 4500 m, No. 170, July 5; Tinkye-la, 4800 m, No. 101, July 16—1933, Shebbeare [K]; Karo La, 4500–4800 m, No. 208, Aug. 18, 1936—Chapman [K]; mountains N. of Lhasa, 3750 m,, on peaty turf, No. 8679, June 10; mountains N. of Lhasa [Sarapu Valley], 4650 m, in damp ground surrounded by *Caragana* scrub , No. 8782, July 3—1942; Dechen Dzong, 3900 m, on open meadows, No. 9614, June 11, 1943—Ludlow and Sherriff [E]; "circa Shigatse, Linbun rivulet, No. SM 103268, July 29, 1914, Kawaguchi"—Murata, l.c.).

General distribution: China (South-West), Himalayas (east.).

## 18. Phlomis L.
Sp. pl. (1753) 584.

1. Stem glabrous or pubescent with simple hairs; bracts and calyx with only simple hairs ......................................................10. **Ph. tuberosa** L.
+ Stem, bracts and calyx pubescent with simple as well as stellate hairs ...........................................................................................................2.
2. Leafy bracts linear or linear-lanceolate, considerably longer than verticils ....................................................................................................3.
+ Leafy bracts broader, oblong or orbicular-ovate, shorter than or as long as verticils or even longer.............................................................4.
3. Stamens with filaments without appendages ......................................
............................................................................7. **Ph. oreophila** Kar. et Kir.
+ Stamens with filaments have well-developed appendages. ............
........................................................................... 3. **Ph. alpina** Pall.
4. Ovary at tip with dense brush of hairs.............................................5.
+ Ovary at tip glabrous or with scattered glands ..............................6.
5. Stem branched; small flat stellate and glandular hairs in pubescence of stem, calyx and bracts ................................... 2. **Ph. agraria** Bunge.
+ Stem simple, erect; stellate and bushy-stellate hairs with long central ray in pubescence of stem, calyx and bracts. Glandular hairs absent ............................................................. 6. **Ph. mongolica** Turcz.
6. Radical and lower caulous leaves suborbicular, not cordate at base
............................................................................ 11. **Ph. umbrosa** Turcz.
+ Radical and lower caulous leaves ovate, elliptical or oblong but never orbicular, truncate, cuneate or cordata at base ....................7.

7. Plant 40–60 cm tall, with large (12–17 cm long and 10–15 cm broad) radical and lower caulous cordate-ovate leaves; calyx with 10 sharply exserted coarse thickened nerves .................................................
.................................................. 8. **Ph. pratensis** Kar. et Kir.

+ Plant very low, 25–45 cm tall; radical and lower caulous leaves very small (3–13 cm long and 2–5 cm broad), oblong, elliptical or deltoid or ovate-cordate; calyx scarious with 5 nerves or with 10; in latter case 5 nerves distinct and 5 barely visible .................................... 8.

8. Radical and lower caulous leaves deltoid, cordate or ovate-cordate; calyx with 10 nerves..................................... 9. **Ph. similis** Tschern.

+ Radical and lower caulous leaves elliptical or oblong; calyx with 5 nerves ............................................................................................ 9.

9. Calyx, bracts and stem with densely appressed tomentose pubescence; pubescence consisting exclusively of stellate hairs with subidentical rays............................................. 4. **Ph. dentosus** Franch.

+ Calyx, bracts and stem with very loose pubescence comprising tiny stellate, simple, thickened multicellular, bushy hairs with long, erect central ray .......................................................................................... 10.

10. Leaves loosely pubescent on upper surface with simple hairs; calyx teeth with recurved, 4 mm long cusp. .........................................
..................................................... 1. **Ph. admirabilis** Tschern.

+ Leaves with stellate pubescence on both surfaces; calyx teeth with very short 2–3 mm long cusp.......................................................... 11.

11. Radical leaves oblong, with 7–13 cm long, 2.5–4 cm broad blade, densely pubescent with stellate hairs on upper and especially lower surface; corolla 16–17 mm long ............. 5. **Ph. kawaguchii** Murata.

+ Radical leaves elliptical or oblong, with 3–5 cm long, 2–3 mm broad blade and very loose stellate pubescence; corolla 15 mm long .......
..................................................... 12. **Ph. younghusbandii** Mukerjee.

1. **Ph. admirabilis** Tschern. sp. nova. —?*Ph. dentosa* Franch. var. *glabrescens* Danguy in Bull. Mus. nat. hist. natur. 17 (1911) 345.

Planta perennis. Caules 25–70 cm alt., tetragoni, vix ramosi, ramis lateralibus, caule plerumque brevioribus, pilis multicellularibus deorsum reflexis incrassatis laxis et pilis minute stellulatis solitariis obtecti, ad nodos densius pilosi; folia omnia supra viridia, pilis simplicibus laxe obsita, subtus cinerascentia, pilis tenuibus stellatis dense obtecta, radicalia et caulina inferiora elliptica, longe petiolata, petiolis 6–12 cm longis, lamina basi subtruncata vel vix cordata, 9–13 cm longa, 5–6 cm lata, margine crenata; caulina pauca, sensim deminuta, radicalibus conformia, sed brevius petiolata; floralia breviter petiolata, basi cuneiformiter attenuata, margine emarginato-dentata, oblonga 3–4 cm longa. Verticilli remoti 3–5, multiflori. Bracteae basi connatae, lineari-sub-ulatae, calyci aequilongae vel eo longiores, pilis multicellularibus incrassatis solitariis et minutis tenuibus stellatis dense obsitae. Calyx tubulato-campanulatus, 9–10 mm longus, quinquenervis, totus stellulato-pilosus, ad nervos pilis multicellularibus incrassatis longis obtectus, dentibus trilobatis, lobis lateralibus triangulariter acuminatis, brevibus, apice pilis albis fasciculatis ornatis, medio in acumen subulatum reflexum 4 mm longum pilis stellulatis et fruticulosis vestitum attenuato. Corolla rosea 16–17 mm longa, labio superiore galeiformi, extus dense piloso, intus margine dense barbato, labio inferiore superiori subaequilongo trilobato, lobo medio rotundato lateralibus majore, tubo extus glabro, intus annulo piloso ornato. Filamenta longe appendiculata; stigma inaequaliter lobatum. Ovarium glabrum.

Typus: China, Kansu, viciniae, opp. Kuan-gou-tschen, ad declivia montium et arva, in argillosis, rarius humosis 2300 m, No. 600, Sept. 17, 1901, Ladygin. In Herb. Inst. Bot. Acad. Sci. URSS (Leningrad) conservatur. Plate I, fig. 4.

Affinitas. Species nostra *Ph. younghusbandii* Mukerjee affinis est sed foliis majoribus , supra pilis simplicibus laxe obsitis, caule haud tomentoso, calycis dentibus longioribus bene differt.

On southern mountain slopes in forest belt, 2300–2700 m.

IA. Mongolia: *Khesi* (around Kuan'gouchen town, mountain slopes and ploughed fields, clay, more rarely humus, 2300 m, No. 600, Sept. 17, 1901—Lad., typus!).

IIIA. Qinghai: *Nanshan* (South Tetungsk. range, in lower forest belt, under rock, Aug. 6; North Tetungsk. range, in midbelt of forest, 2400–2700 m, Aug. 14—1880, Przew.; ? "Si-Ning-Fou, 2400 m, July 18, 1908, Vaillant"—Danguy, l.c.).

General distribution: endemic.

Note. Plant collected in 1908 from Xining town and named *Ph. dentosa* Franch var. *glabrescens* Danguy (Danguy, l.c.) possibly pertains to this species. We have questioned its inclusion, however, since no definite opinion can be expressed without a study of the plant.

2. **Ph. agraria** Bunge in Ledeb. Fl. alt. 2 (1830) 411; Regel in Acta Horti Petrop. 9 (1886) 589; Popov in Byull. Sredneaz. univ. 13 (1926) 137; Krylov, Fl. Zap. Sib. 9 (1937) 2342; Knorr. in Fl. SSSR, 21 (1954) 103; Grubov, Konsp. fl. MNR (1955) 237, p.p.; Fl. Kazakhst. 7 (1964) 397.    —Ic.: Ledeb. Ic. pl. fl. ross. 4, tab. 364; Regel. l.c. tab. 10.

Described from East. Kazakhstan. Type in Leningrad.

Steppe slopes, stagnant pools.

IA. Mongolia: *Khobdo* ("Ukhyn-Gol"—Grubov, l.c.).

IIA. Junggar: *Tien Shan* (Muzart gorge, 1600–2100 m, Aug. 15, 1877—A. Reg.; 20 km south of Nyutsyuanz, No. 240, July 18, 1957—Kuan).

General distribution: (?) Aralo-Casp., Fore Balkh., Jung.-Tarb., Nor. Tien Shan; West. Sib. (south-east.).

Note. This species was reported from Mongolia by V.I. Grubov (l.c.) from 2 sites. Plant from Kerulen-Gol turned out to be *Ph. tuberosa* L. and the other from Ukhyn-Gol (Mong. Altay) is evidently lost.

3. **Ph. alpina** Pall. in Acta Ac. Petrop. 2 (1779) 265; Regel in Acta Horti Petrop. 9 (1886) 583; Popov in Byull. Sredneaz. univ. 13 (1926) 138; Krylov,

Fl. Zap. Sib. 9 (1937) 2344; Knorr. in Fl. SSSR, 21 (1954) 92; Fl. Kazakhst. 7 (1964) 393. —Ic.: Pall. l.c. tab. 13; Regel, l.c. tab. 10.

Described from Altay. Type in Leningrad.

Meadows in forest and alpine belts of mountains.

IIA. Junggar: *Tien Shan* (Talki gorge, July 11; south. bank of Sairam lake, July 20—1877, A. Reg.; Kul'dzha, June 6; near Sairam lake, July 18—1878, Fet.).

General distribution: Jung.-Tarb.; West. Sib. (Altay).

Note. Closely affiliated to *Ph. oreophila* Kar. et Kir. but with a narrower distribution range, penetrating into the east. part of the range of *Ph. oreophila*.

4. **Ph. dentosus** Franch. in Nouv. Arch. Mus. hist. natur. 2 ser. 6 (1884) 243 [Pl. David. 1 (1884) 243]; Dunn. in Notes Bot. Gard. Edinburgh, 6 (1915) 187.

Described from East. Mongolia. Type in Paris. Isotype in Leningrad.

Dry mountain slopes.

IA. Mongolia: *East. Mong.* (Toumet, Sartchy, dans les montagnes et sur les coteaux secs, No. 2731, June 1866—David, typus!).

General distribution: endemic.

Note. Judging from the isotype (no other herbarium material was available), *Ph. dentosus* is very well distinguished from all other species of genera in section Phlomoides (Moench) Briq. found in our territory by its characteristic tomentose pubescence of calyx and stem comprising stellate hairs as well as scabrous upper surface of leaves due to dense bushy-stellate pubescence.

5. **Ph. kawaguchii** Murata in Acta Phytotax. et Geobot. 16, 1 (1955) 15. —Ic.: Murata, l.c. fig. 1.

Described from Tibet (environs of Lhasa). Type in Japan (KYO).

IIIB. Tibet: South. ("circa Lhasa, Pulunka temple, No. SM 103284, Aug. 20, 1914, Kawaguchi, typus"—Murata, l.c.).

General distribution: endemic.

Note. Judging from sketch and description, this plant is very similar to *Ph. younghusbandii* Mukerjee. Possibly, *Ph. kawaguchii* too should be treated as a synonym of the latter but this would call for a study of all specimens of this species. The species description mentions neither the presence of appendages of filaments nor pubescence of the ovary.

6. **Ph. mongolica** Turcz. in Bull. Soc. natur. Moscou, 24 (1851) 406; Franch. Pl. David. 1 (1884) 242; Forbes and Hemsley, Index Fl. Sin. 2 (1902) 306, p.p.; Kudo in Mém. Fac. Sci. and Agr. Taihoku Univ. 2 (1929) 216; Kitag. Lin. Fl. Mansh. (1939) 383; Walker in Contribs U.S. Nat. Herb. 28 (1941) 657. —*Ph. tuberosa* auct. non L.: Dunn in Notes Bot. Gard. Edinburgh, 6 (1915) 187, p.p.

Described from Nor. China. Type in Leningrad. Plate I, fig. 3.

Steppes, steppe slopes of mountains, floors of gorges, rocks.

IA. Mongolia: *East. Mong.* (Sartchy, Ourato, No. 2770, July 1866—David; Khukh-Khoto environs, Siustudzhao monastery, July 15, 1884— Pot.; between Ulan-Khada and Qindai, Aug. 10, 1898—Zab.; east. Mongolia, 1902—Campbell [BM]; east. Mongolia, Arban tolon ts'on, No.

7536, June 30, 1924—Licent [BM]; Shara Murun, No. 464; Chaptsur, No. 18—1925, Chaney; Shilin-Khoto town feather grass-wild rye steppe, 1960—Ivan.; between Sume-Khada and Muni-Ula, 1871—Przew.). *Alash. Gobi* ("Kuang Hsi Kou, at bottom of an exposed, moist, rocky gorge, common, No. 197; Chung Wei, on an exposed, bare, gravelly slope, No. 231—Ching"—Walker, l.c.).

General distribution: China (Dunbei, Nor., North-West).

7. **Ph. oreophila** Kar. et Kir. in Bull. Soc. natur. Moscou, 15 (1842) 426; Regel in Acta Horti Petrop. 9 (1886) 588; Sapozhn. Mong. Alt. (1911) 382; Popov in Byull. Sredneaz univ. 13 (1926) 138; Krylov, Fl. Zap. Sib. 9 (1937) 2343; Persson in Bot. notiser (1938) 300; Knorr. in Fl. SSSR, 21 (1954) 95; Grubov, Konsp. fl. MNR (1955) 237; Fl. Kirgiz. 9 (1960) 106 (excl. syn. *Ph. alpina*); Fl. Kazakhst. 7 (1964) 394. —*Ph. dszumrutensis* Afan. in Bot. mater. Gerb. Bot. inst. AN SSSR, 8 (1940) 110. —Ic.: Fl. SSSR, 21, Plate 6, fig. 1; Fl. Kirgiz. 9, Plate 12.

Described from Kazakhstan (Junggar Ala tau). Type in Leningrad.

Rocky steppes of slopes, rocks and moraines in mountain-steppe and forest-steppe belts; sometimes ascends up to alpine belt.

IA. Mongolia: *Mong. Alt.* (south. flank of Indertiin-Gol river valley, near summer camp in somon, mountain steppe, July 24; same site, subalp. steppe, July 25—1947, Yun.; "V. Kobdosskoe lake, Saksa-Gol, Shiverin-Gol"—Grubov, l.c.).

IB. Kashgar: *West.* ("Bostan-terek, No. 24a, July 20, 1921"—Persson, l.c.).

IIA. Junggar: *Cis-Alt.* (upper valley of Kran river, alp. meadow, July 11, 1908—Sap.; Burchum, No. 3200, July 16; Qinhe, No. 973, Aug. 4—1956, Ching). *Jung. Alt.* (Toli district, on east. slope of water divide, No. 1188, Aug. 6; nor. bank of Sairam lake, shaded slope, No. 2130, Aug. 8; Ven'tsyuan', on slope, No. 2067, Aug. 25—1957, Kuan). *Tien Shan* (nor. slope of Narat mountain range, Tsanma river valley, June 5–10, 1877—Przew.; Talki gorge, July 18; Sairam, July—1877; Dzhagastai range, 2400–3200 m, June 20, 1878—A. Reg.; 5–6 km south of Daban, No. 583, July 23; Chzhaosu district, Aksu region, Kantelek hills, on slope, No. 3529, Aug. 14—1957, Kuan; Narat mountains on Kunges, 2300 m, on shaded slope, Aug. 7, 1958—Lee and Chu; in Tsitai region, mountains, in undergrowth on sunny slope, 1800 m, No. 10731, July 21, 1959—Lee et al.; "Kirgis-at-davan, ca, 2600 m, No. 297, July 31, 1932"—Persson, l.c.).

General distribution: Fore Balkh., Jung.-Tarb., Nor. and Cent. Tien Shan; Mid. Asia (Pamiro-Alay), West. Sib. (Altay).

Note. Polymorphic species in which size of corolla and extent of pubescence of different plant parts are variable. Related to *Ph. alpina* Pall. from which it differs mainly in the absence of appendages of stamen filaments although vestiges of appendages could be detected at times.

Herbarium material from Mongolian Altay cited in V.I. Grubov's 'Konspekt fl. MNR' could not be located.

8. **Ph. pratensis** Kar. et Kir. in Bull. Soc. natur. Moscou, 15 (1842) 426; Regel in Acta Horti Petrop. 9 (1886) 590; Danguy in Bull. Mus. nat. hist. natur. 20 (1914) 86; Popov in Byull. Sredneaz. univ. 13 (1926) 137; Knorr. in Fl. SSSR, 21 (1954) 94; Grubov, Konsp. fl. MNR (1955) 237; Fl. Kirgiz. 9 (1960) 105; Fl. Kazakhst. 7 (1964) 394.

Described from Kazakhstan (Junggar Ala tau). Type in Leningrad.

Steppe slopes in thicket belt of mountains.

IA. Mongolia: *Mong. Alt.* (in upper Khatu-Gol near Kobdo town, June 20, 1870—Kalning).
IIA. Junggar: *Jung. Alt.* (ascent to Kuzyun' pass, descent to ravine, Aug. 2, (1908—B. Fedczenko). Tien Shan (fairly extensive).
General distribution: Fore-Balkh., Jung.-Tarb., Nor. and Cent. Tien Shan.

### 9. Ph. similis Tschem. sp. nova.

Planta perennis. Caules vix ascendentes, 40–45 cm alti; tetragoni patenter ramosissimi, pilis stellatis et fruticulosis radio centrali longiore ornatis obtecti. Folia supra viridia pilis fruticulosis suffultis dense vestita (caulina inferiora et radicalia supra laxe pilosa), omnia subtus ob pilos stellatos densos griseo-tomentosa; folia radicalia et caulina inferiora triangularia, cordata, 7–8 cm longa, 3–5 cm lata, petiolata, petiolis 7–9 cm longis, margine crenata; folia caulina pauca, diminuta, omnia petiolata, laminis oblongis, basi cuneatis, margine crenatis, folia floralia valde diminuta. Verticilli remoti, multiflori, bini-terni. Bracteae lineari-subulatae, calici aequilongae vel eo longiores, pilis minutissimis stellatis et multicellularibus longis incrassatis, rarius fruticulosis, radio centrali elongato ornatis obtectae. Calyx tubuloso-campanulatus, 8–9 mm longus, indistincte 10–nervis (nervis 5 bene distinctis manifeste elevatis), totus ob pilos minutos stellatos tenuiter tomentosus, ad nervos pilis multicellularibus longis incrassatis, rarius fruticulosis obtectus, dentibus trilobatis, lobis lateralibus triangularibus, abbreviatis, apice pilis longis albis fasciculatis ornatis, medio rigido, incrassato, in acumen reflexum subulatum 2–3 mm longum pilis minutis stellatis et fruticulosis tectum attenuato. Corolla pallide rosea 12–15 mm longa, labio superiore galeiformi, extus dense piloso, intus ad marginem dense barbato, inferiore superiore breviore, trilobato, lobo medio orbiculari, lateralibus majore, tubo extus glabro intus annulo pilorum ornato. Filamenta longe appendiculata, pilosa; ovarium glabrum, stigma inaequaliter lobatum.

Typus: China, Tsinghai, in fluxu superiore fl. Hoangho, oasis Guj-duj secus canales irrigatorios, 2100 m s.m., June 14, 1880, Przevalskij. In Herb. Inst. Bot. Acad. Sci. URSS (Leningrad) conservatur.

Affinitas. *Ph. younghusbandii* Mukerjee et *Ph. admirabili* Tschem. foliis radicalibus triangularibus basi cordatis et caulinis latis bene differt. A. *Ph. mongolica* Turcz. caulibus patenter ramosis vix ascendentibus et ovario glabro dignoscitur.

Steppe sections in lower belt of mountains, irrigation canals, more rarely in crops.

IA. Mongolia: *Khesi* (nor. foothills of Nanshan mountain range, near Dedzhin town, in crops, June 29; same site, in crops, among desert grasses, July 1—1872, Przew.; Godvoputan valley, in humus soil in lower hill belt, July 17; 1908—Czet.).
IIIA. Qinghai: *Amdo* (upper course of Huang He river, Guj-duj oasis, along irrigation canals, 2100 m, June 14, 1880—Przew., typus!).
General distribution: endemic.

10. **Ph. tuberosa** L. Sp. pl. (1753) 586; Sapozhn. Mong. Alt. (1911) 382; Danguy in Bull. Mus. nat. hist. natur. 20 (1914) 86; Dunn in Notes Bot. Gard. Edinburgh, 6 (1915) 187, p.p.; Krylov, Fl. Zap. Sib. 9 (1937) 2339; Kitag. Lin. Fl. Mansh. (1939) 384; Knorr. in Fl. SSSR, 21 (1954) 99; Grubov, Konsp. fl. MNR (1955) 237; Fl. Kazakhst. 7 (1964) 396; Hanelt und Davažamc in Feddes repert. 70 (1965) 55; Dashnyam in Bot. zh. 50 (1965) 1641.  —*Ph. agraria* auct. non Bunge: Grubov, Konsp. fl. MNR (1955) 237, p.p.  —*Ph. mongolica* auct. non Turcz.: Forbes and Hemsley, Index Fl. Sin. 2 (1902) 306, p.p.

Described from Siberia. Type in London (Linn.).

Steppe and meadow slopes of mountains, forest fringes, dry pebble beds of rivers and gully floors.

IA. **Mongolia:** *Cis-Hing., Cent. Khalkha, East. Mong.*

IIA. **Junggar:** *Cis-Alt., Jung. Alt., Tien Shan, Balkh.-Alak.*

**General distribution:** Fore Balkh., Jung.-Tarb., Nor. Tien Shan; Europe, Balk.-Asia Minor, Fore Asia, Caucasus, Mid. Asia, West. and East. Sib. (south.), Far East (south-east.), Nor. Mong., China (Dunbei).

11. **Ph. umbrosa** Turcz. in Bull. Soc. natur. Moscou, 13 (1840) 76; Franch. Pl. David. 1 (1884) 242; Forbes and Hemsley, Index Fl. Sin. 2 (1902) 306; Dunn in Notes Bot. Gard. Edinburgh, 6 (1915) 186; Kudo in Mém. Fac. Sci. and Agr. Taihoku Univ. 2 (1929) 211; Kitag. Lin. Fl. Mansh. (1939) 384.

Described from Nor. China. Type in Leningrad.

IA. **Mongolia:** *East. Mong.* (Muni-Ula mountains, July 13, 1871—Przew).

**General distribution:** China (Dunbei, Nor., Nor.-West., Cent., South-West.), Korean peninsula.

12. **Ph. younghusbandii** Mukerjee in Notes Bot. Gard. Edinburgh, 19, 95 (1938) 307.  —Ic.: Mukerjee, l.c.

Described from Tibet. Type in Calcutta. Isotype in London (K) and Edinburgh (E).

IIIB. **Tibet:** *South.* (Khambajong, No. 106, July 17, 1903—Younghusband, isotypus [K, E]; mountains above Lhasa, Aug.; Gyantse, Sept.—1904, Walton [K]; Mt. Everest Exped., N. Dsakachu, 3700 m, July 6, 1922—Morton [K]; Gyantse, Aug. 30, 1925—Ludlow [BM]; hills behind Gyantse, No. 863, Aug. 13, 1936—Chapman [K]; low ground across Kyl Chu, 2 miles south-east of Lhasa, 3600-3900 m, Sept. 1, 1936, Chapman [BM]; mountains north of Lhasa, 4000 m, sandy hillside, No. 8757, June 24; Reting, 60 miles N of Lhasa, 3900 m, among grass and rocks, No. 8896, July 27—1942, Ludlow and Sherriff [E, BM]; Reting, on dry open plain, No. 11035, July 20, 1944; Nyenchengtang-la, 4 days [sic] NW of Lhasa, 3900 m, on open meadows and slopes, No. 9646, June 21, 1943—Ludlow and Sherriff [E]; "Kambajong—Prain; Dochin Hills—Stewart"—Mukerjee, l.c.).

**General distribution:** Himalayas (east.).

**Note.** Plants with less pubescence of leaves and calyces and larger radical leaves are found in the environs of Lhasa. The presence of such races, which we treat as a manifestation of intraspecific variation, brings into question the independent status of *Ph. kawaguchii* Murata (see note under *Ph. kawaguchii*).

## 19. Eriophyton Benth.

Bot. Register, 15 (1829) in adnot ad tab. 1289 et in Wall. Pl. Asia Rar. 1 (1830) 63.

1. E. wallichii Benth. in Wall. Cat. (1829) No. 2070, nom. nud.; id. in Wall. Pl. Asia Rar. 1 (1830) 63; Dunn in Notes Bot. Gard. Edinburgh, 6 (1915) 188; Marquand in J. Linn. Soc. London (Bot.) 48 (1929) 219; Kudo in Mém. Fac. Sci. and Agr. Taihoku Univ. 2 (1929) 218; Wu in Acta Phytotax. Sin. 8, 1 (1959) 31.

Described from East. Himalayas. Type in London (K?). Plate II, fig. 2. High mountain areas.

IIIB. Tibet: *Weitzan* (in placers, above alp. plant boundary, on wet clay covered with rubble, Yantszytszyan river basin, along Khicha river, about 4500 m, July 12, 1900—Lad.). *South.* ("on slate screes at Atsa, 4800–5100 m, No. 6158, Aug. 1924, Ward"—Marquand, l.c.).
General distribution: China (South-West), Himalayas (east.).

## 20. Stachyopsis M. Pop. et Vved.

in Tr. Turk. nauch. obshch. 1 (1923) 120.

1. Lower and middle leaves on long petiole, cuneate at base, very insignificantly pubescent; calyx with short appressed hairs in upper half. ...................................3. S. oblongata (Schrenk) M. Pop. et Vved.

+ All leaves sessile or on very short petiole, insignificantly pubescent with invariably cordate base or at least with velvety pubescence on both surfaces and cuneate base; calyx lanate ...................................2.

2. Stem erect, unbranched, glabrous on margins, with short downward hairs only on ridges; all leaves sessile with cordate base, insignificantly pubescent ... 1. S. lamiiflora (Rupr.) M. Pop. et Vved.

+ Stem branched from base; leaves sessile or on short petiole, with cuneate base with velvety pubescence as in stem on both surfaces ...................................2. S. marrubioides (Regel) Ik.-Gal.

1. S. lamiiflora (Rupr.) M. Pop. et Vved in Tr. Turk. nauch. obshch. 1 (1923) 122; Kuprian. in Fl. SSSR, 21 (1954) 109; Fl. Kirgiz. 9 (1960) 110; Fl. Kazakhst. 7 (1964) 398. —*Stachys lamiiflora* Rupr. Sertum tianschan. (1869) 67. —*Phlomis lamiiflora* Regel in Acta Horti Petrop. 6 (1879) 374, 9 (1886) 593, p.p. —Ic.: Fl. Kazakhst. 7, Plate 46, fig. 1.

Described from Nor. Tien Shan (Kastek pass). Type in Leningrad. Meadows, from forest to subalpine belt.

IIA. Junggar: *Tien Shan* (Piluchi, April; Nilki, 2100 m, June 8; Aryslyn, 2400 m, July 8; same site, 2700 m, July 15—1879, A. Reg.; Ulastai-Yakou, on slope, Aug. 30; 1957—Kuan; Narat hills, on slope 2400 m, Aug. 7, 1958—Lee and Chu).
General distribution: Nor. and Cent. Tien Shan.

2. S. marrubioides (Regel) Ik.-Gal. in Izv. Glavn. bot. sada AN SSSR, 26 (1927) 72; Kuprian. in Fl. SSSR, 21 (1954) 110; Fl. Kazakhst. 7 (1964) 399. —*S. oblongata* var. *canescens* M. Pop. et Vved. in Tr. Turk. nauch.

obshch. 1 (1923) 122. —*Phlomis marrubioides* Regel in Acta Horti Petrop. 6 (1879) 375. —*Ph. oblongata* β. *canescens* Regel in Acta Horti Petrop. 9 (1886) 593.

Described from East. Kazakhstan (Junggar Ala tau). Type in Leningrad. Subalpine and alpine meadows.

IIA. Junggar: *Jung. Alt.* (Chubaty pass, Aug. 1878—A. Reg., typus! Ven'tsyuan', on slope, No. 3477, Aug. 14; same site, No. 4636, Aug. 25—1957, Kuan; near lake Sairam, July 18, 1878 Fet.).

General distribution: Jung.-Tarb. (Jung. Ala tau).

Note. Plants from Ven'tsyuan' differ somewhat from typical *S. marrubioides* in less-branched erect stems and laxer pubescence.

3. **S. oblongate** (Schrenk) M. Pop. et Vied. in Tr. Turk. nauch. obshch. 1 (1923) 121; Kuprian. in Fl. SSSR, 21 (1954) 110; Fl. Kirgiz. 9 (1960) 111; Fl. Kazakhst. 7 (1964) 399. —*Phlomis oblongata* Schrenk in Fish. et Mey. Enum. pl. nov. 1 (1841) 29; Regel in Acta Horti Petrop. 9 (1886) 591. —*Leonurus dschungaricus* Regel in Acta Horti Petrop. 6 (1876) 367. —Ic.: Regel, l.c. (1886) tab. 10, fig. 15, a, b; Fl. Kazakhst. 7, Plate 46, fig. 2.

Described from East. Kazakhstan (Dzhilkaragai). Type in Leningrad.

Meadows, among shrubs, on grassy slopes, mountains from middle to subalpine belt.

IIA. Junggar: *Tien Shan* (Muzart, 1650–2100 m, July 15; Talki gorge, July 18; Sairam, July; Dzhagastai, 1500–2100 m, Aug. 9—1877; Khorgos, 1500–1800 m, May; Khanakhai, 1500–2100 m, June 16; Agiaz river, 2100–2400 m, June 26—1878; Taldy, May 15; Bagaduslun-Bayanamyn, 1500–1800 m, June 4; Tsagan-Tunge, 1500–1800 m, June 8; Nilki environs, 2100 m, June 8; Aryslyn gorge, 2400–2700 m, July 10—1879, A. Reg.; Burkhantau, June 5; Kutukchi, June 6—1878, Fet.; Muzart and Tekes—1886, Krasnov; upper Tekes, Bayan-Gol river, among riverside bushes, June 24; Tekes river, among bushes, July 4—1893, Rob.; Kok-Tal [Temurlyk-tau], May 20–22; syrt (watershed upland) von Karabulak und Karadschon, July 30—1907, Merzb.; 28 km south-east of Nyutsyuan'tsz, on bank of Ulausu river, meadow, No. 191, July 18; Daban, on slope, 2300 m, No. 1959, July 18; Danu river, on slope, 2200 m, No. 2223, July 24; Dzhagistai, on slope, No. 3179, Aug. 8; Nilki-Tszinkho, on rocky slope, No. 4045, Sept. 1—1957, Kuan).

General distribution: Jung.-Tarb., Nor. and Cent. Tien Shan; Mid. Asia (Mountains).

## 21. Galeopsis L.
Sp. pl. (1753) 579.

1. **G. bifida** Boenn. Prodr. Fl. Monast. (1824) 178; Juzepczuk in Sorn. rast. SSSR, 4 (1935) 44; Krylov, Fl. Zap. Sib. 9 (1937) 2347; Hao in Engler's Bot. Jahrb. 68 (1938) 635; Juzepczuk in Fl. SSSR, 21 (1954) 119; Grubov, Konsp. fl. MNR (1955) 237; Fl. Kirgiz. 9 (1960) 112; Fl. Kazakhst. 7 (1964) 401. —*G. tetrahit* auct. non L.: Hook, f. Fl. Brit. Ind. 4 (1885) 677; Danguy in Bull. Mus. nat. hist. natur. 17 (1911) 345; ibid. 20 (1914) 85; Dunn in Notes Bot. Gard. Edinburgh, 6 (1915) 181; Kudo in Mém. Fac. Sci. and Agr. Taihoku

Univ. 2 (1929) 204; Rehder and Kobuski in J. Am. Arb. 14 (1933) 31; Kitag. Lin. Fl. Mansh. (1939) 379; Walker in Contribs U.S. Nat. Herb. 28 (1914) 656. —Ic.: Fl. SSSR, 21, Plate 7, fig. 2.

Described from Europe. Type evidently in Mümster.

River banks, forest fringes and meadow slopes of mountains as well as weed in crops, kitchen gardens, gardens, roadsides, fallow lands.

IIIA. Qinghai: *Nanshan* (South Tetung mountain range, July 25, 1872; high-mountain areas between Nanshan and Donkyr, up to Rako-Gol river, 3000–3300 m, July 22; up to Yusun-Khatym river, July 23—1880, Przew.; "Kokonor: auf dem östlichen Nan-Schan, 2900-3100 m, No. 751"—Hao, l.c.; "La Chiung Kou, forming dense, pure stands on exposed, grassy slopes, No. 629, Ching"—Walker, l.c.).

General distribution: Fore Balkh., Nor. Tien Shan; Europe, Caucasus, West. Sib., East. Sib., Far East, Nor. Mong. (Hent., Hang., Mong.-Daur.), China (Dunbei, Nor.-West., South-West.), Himalayas (east.), Korean peninsula, Japan, Nor. America (introduced?).

## 22. Lamium L.
Sp. pl. (1753) 579.

1. Upper leaves sessile, semi-amplexicaul, reniform, crenate; corolla tube usually thrice longer than calyx, glabrous inside; corolla purple or **pink** ........................................................ 2. **L. amplexicaule L.**

+ All leaves petiolate, ovate or cordate, sharply serrate along marginal; corolla tube nearly as long or slightly longer than calyx with oblique hairy ring inside; corolla yellowish-white or white or hoary pale-white ...................................................................... 1. **L. album L.**

1. **L. album L.** Sp. pl. (1753) 579; Franch. Pl. David. 1 (1884) 241; Forbes and Hemsley, Index Fl. Sin. 2 (1902) 302; Sapozhn. Mong. Alt. (1911) 382; Danguy in Bull. Mus. nat. hist. natur. 20 (1914) 85; Dunn in Notes Bot. Gard. Edinburgh, 6 (1915) 183; Krylov, Fl. Zap. Sib. 9 (1937) 2351; Persson in Bot. notiser (1938) 300; Kitag. Lin. Fl. Mansh. (1939) 380; Gorschkova in Fl. SSSR, 21 (1954) 134; Grubov, Konsp. fl. MNR (1955) 237; Fl. Kazakhst. 7 (1964) 403. —*L. turkestanicum* Kuprian. in Bot. mater. Gerb. Bot. inst. AN SSSR, 14 (1951) 346; Fl. Kirgiz. 9 (1960) 114. —Ic.: Sorn rast. SSSR, 4, fig. 393; Kuprian. l .c. fig. 1, 2 (sub nom. *L. turkestanicum*).

Described from Europe. Type in London (Linn.).

River banks, among coastal bushes, floors of shaded creek valleys, birch groves and as weed in gardens, roadsides and residences.

IA. Mongolia: *Cis-Hing.* (near Yaksha railway station, mountain slope, June 13, 1902—Litw.; environs of Manchuria railway station, 1915—E. Nechaeva; Khalkhin-Gol river, steppe, Sept. 4, 1928—Tug.). *East. Mong.* ("Vallé de Kailar, steppes, June 26, 1896, Chaff."—Danguy, l.c.).

IIA. Junggar: *Cis-Alt.* (in Koktogoi region, No. 2173, Aug. 18, 1956—Ching). *Tarb.* (Kos-Konus', 1550 m, No. 1555, Aug. 13; same site, No. 1643—1957, Kuan). *Jung. Alt.* (along road to Ven'tsyuan' from Syat, in forest along river bank, No. 1420, Aug. 13; Ven'tsyuan', Kurbanin mountains, near water, No. 4611, Aug. 25—1957, Kuan), *Tien Shan* (west.).

General distribution: Fore Balkh., Jung.-Tarb., Nor. and Cent. Tien Shan; Europe, Mediterranean, Balk.-Asia Minor, Fore Asia, Caucasus, Mid. Asia, West. Sib., East. Sib., Far East (south), Nor. Mong., China (Dunbei, Nor.-West., Cent., East.), Himalayas (west., Kashmir), Korean peninsula, Japan, Nor. America.

**2. L. amplexicaule** L. Sp. pl. (1753) 579; Forbes and Hemsley, Index Fl. Sin. 2 (1902) 303; Danguy in Bull. Mus. nat. hist. natur. 20 (1914) 85; Pampanini, Fl. Carac. (1930) 184; Krylov, Fl. Zap. Sib. 9 (1937) 2350; Walker in Contribs U.S. Nat. Herb. 28 (1941) 656; Gorschkova in Fl. SSSR, 21 (1954) 128; Fl. Kirgiz. 9 (1960) 113; Fl. Kazakhst. 7 (1964) 402. —**Ic.:** Sorn. rast. SSSR, 4, fig. 390; Fl. Kazakhst. 7, Plate 46, fig. 3.

Described from Europe. Type in London (Linn.).

Flooded meadows, forest fringes, rocky and rubble slopes of mountains and as weed in crops, gardens, kitchen gardens, roadsides and around residences.

**IIA. Junggar:** *Tien Shan* (environs of Urumchi, No. 577, June 2, 1957—Kuan).

**IIIA. Qinghai:** *Nan Shan* (up to Yusun-Khatyma river, 2700–3000 m, July 24, 1880 —Przew.; "La Chi Tzu Shan, in an exposed yard of rich, clay soil, No. 720, Ching"—Walker, l.c.).

**IIIB. Tibet:** *Weitzan* (Yantszytszyan river basin, Donra area of Khichu river, in old pasture corrals, 3900 m, July 16, 1900—Lad.).

**General distribution:** Fore Balkh., Jung.-Tarb., Nor. Tien Shan; Europe, Mediterranean, Fore Asia, Caucasus, Mid. Asia, West. Sib. (nor.-west.), East. Sib. (south.), Far East (introduced), China (Nor.-West., East., Cent., South-West., Taiwan), Himalayas (west., east.), Korean peninsula, Japan, Nor. Africa.

### 23. Chaiturus Willd.
Prodr. Fl. Berol. (1787) 200.

**1. Ch. marrubiastrum** (L.) Spenn in Nees, Gen. Fl. Germ. 2 (1843) tab. 31; Kuprian. in Fl. SSSR, 21 (1954) 144; Fl. Kazakhst. 7 (1964) 404. —*Leonurus marrubiastrum* L. Sp. pl;. (1753) 584; Benth. Labiat. gen. et sp. (1834) 520; Krylov, Fl. Zap, Sib. 9 (1937) 2359. —**Ic.:** Fedtsch. and Fler. Fl. Evrop. Ross. 3, fig. 373.

Described from West. Europe. Type in London (Linn).

Fallow land, river banks, roadsides.

**IIA. Junggar:** *Tien Shan* (Savan district, Datsyuan'gou, along ravine, July 4, 1957—Kuan). *Dzhark.* (environs of Kul'dzha [1876]—Golike)

**General distribution:** Aral-Casp., Fore Balkh. Jung.-Tarb., Nor. Tien Shan; Europe, Caucasus, Nor. America (introduced).

**Note.** Evidently introduced in our territory.

### 24. Leonurus L.
Sp. pl. (1753) 584; Gen. pl. (1754) 254.

1. Calyx indistinctly bilobiate, its 2 lower teeth not recurved but much longer than 3 upper; corolla 15–20 mm long, with straight tube not

enlarged near straight hairy ring, with galeate shaped upper and straight lower lip; latter 1/3 shorter than upper; leaves 3–lobed ... .................................................................................. 7. **L. sibiricus** L.

+ Calyx distinctly bilobiate, its 2 lower teeth declinate; corolla 8–12 mm long, with tube inflated above oblique hairy ring, with slightly concave upper and declinate lower lip; leaves 5-lobed .................. 2.

2. Whole plant glabrous. ............. 4. **L. mongolicus** Krecz. et Kuprian.

+ Plants pubescent ........................................................................... 3.

3. Plants with long patent hairs along edges and ribs of stems, petioles, blades of leaves and, on caly, or lanate on calyx and stem in inflorescence; lower part of stem and leaves with appressed pubescence ...................................................................................... 4.

+ Plants with short appressed hairs along ribs and edges of stems, petioles, blades of leaves and on caly, or latter and leaf blades glabrous ........................................................................................ 5.

4. All plants with long patent hairs; inflorescence long, with interrupted verticils .................................... 6. **L. quinquelobatus** Gilib.

+ Long patent pubescence only on calyces and stem in inflorescence, elsewhere plant pubescent with appressed short hairs; inflorescence short, with approximated verticils ........ 5. **L. panzerioides** M. Pop.

5. Lower caulous leaves divided up to centre into broad, oblong, large and unevenly dentate lobes; leaves, stem and calyx usually glabrous ........................................................ 1. **L. cardiaca** L.

+ Lower caulous leaves split for 2/3 or more into cuneate or narrowly cuneate lobes; leaves, stem and calyx usually pubescent or leaves and stem subglabrous but calyx invariably with appressed short hairs ................................................................................ 6.

6. Plant densely foliate; leaves split for 2/3 into broad cuneate lobes, in turn divided into broadly lanceolate lobes .................................... ................................. 8. **L. turkestanicus** Krecz. et Kuprian.

+ Plant not densely foliate; lower leaves shedding early so that lower part of stem leafless by anthesis; leaves split almost up to base into narrow cuneate lobes, in turn divided into lanceolate lobes. ........ 7.

7. Stem with appressed short pubescence throughout length, with tufts of long patent hairs only at nodes; upper lip of corolla densely hoary ................................................. 3. **L. glaucescens** Bunge.

+ Stem with appressed short pubescence throughout length and at nodes; upper lip of corolla weakly pubescent .................................... ................................................. 2. **L. deminutus** Krecz.

1. **L. cardiaca** L. Sp. pl. (1753) 584; Sapozhn. Mong. Alt. (1911) 382; Persson in Bot. notiser (1938) 300; Kuprian. in Fl. SSSR, 21 (1954) 148; Grubov, Konsp. fl. MNR (1955) 237. —Ic.: Reichb. Ic. fl. Germ. (1856) 1213; Hegi, Ill. Fl. 5, 4, fig. 2392, 3265.

Described from Europe. Type in London (Linn.).
Weed on roadsides and in residences.

IA. Mongolia: *Mong. Alt.* ("Tsagan-Gol, Dain-Gol"—Sapozhn. l.c.).
IIA. Junggar: *Tien Shan* ("Kunges, ca. 1130 m, No. 425, Aug. 23, 1932"—Persson, l.c.).
General distribution: Europe, Balk.-Asia Minor.
Note. This is a rare weed, evidently introduced in our territory. V.I. Grubov reports in "Konsp. fl. MNR" the occurrence of this species in Mong. Altay, evidently based on V.V. Sapozhnikov's list (l.c.) as these specimens are lacking in the Herbarium.

2. **L. deminutus** Krecz. in Bot. mater. Gerb. Bot. inst. AN SSSR, 11 (1949) 136; Kuprian. in Fl. SSSR, 21 (1954) 153; Grubov, Konsp. fl. MNR (1955) 237; Dashnyam in Bot. zh. 50 (1965) 1641.

Described from East. Sib. (environs of Verkhneudinsk). Type in Leningrad.

Dry steppe rocky slopes, talus, rubbish and pasture corrals.

IA. Mongolia: *Cen. Khalkha* (in Bombutei-Ama creek valley, on bank of spring, July 25, 1893—Klem.; Dzhargalante river basin, Uber-Dzhargalante river, between sources and Agit mountain, deris shoots in valley, Aug. 14, 1925—Krasch. and Zam.; environs of Ikhe-Tukhum-Nor lake, Khashatyn-Ama gorge; same site, Nomgani-Ama gorge—1926, Zam.). *East. Mong.* (Dashnyam, l.c.), *Gobi-Alt.* (Dzun-Saikhan mountain range, under overhanging rocks in Yalo creek valley, Aug. 21, 1931—Ik.-Gal.).
General distribution: East. Sib. Nor. Mong. (Hent., Hang., Mong.-Daur.).

3. **L. glaucescens** Bunge in Ledeb. Fl. alt. 2 (1830) 409; Krylov, Fl. Zap. Sib. 9 (1937) 2355; Kuprian. in Fl. SSSR, 21 (1954) 151; Grubov, Konsp. fl. MNR (1955) 238; Fl. Kazakhst. 7 (1964) 405.  —**Ic.:** Ledeb. Ic. pl. fl. ross. 2, tab. 179.

Described from Altay. Type in Leningrad.

Mountain slopes, rocks and talus, river banks, residences and roadsides.

IA. Mongolia: *Mong. Alt.* (Tsagan-Gol, upper camp near cold spring, rocks, July 3, 1905—Sap.). *East. Mong.* (Ar-Zhargalant somon, Tsog-Delger mountain, peak, north of somon, among rocks and stones, June 26, 1956—Dashnyam).
General distribution: Aral-Casp., Fore Balkh., Jung.-Tarb.; Europe (Europ. USSR), Caucasus, Mid. Asia, West. Sib.

4. **L. mongolicus** Krecz. et Kuprian. in Bot. mater. Gerb. Bot. inst. AN SSSR, 11 (1949) 137; Kuprian. in Fl. SSSR, 21 (1954) 154; Grubov, Konsp. fl. MNR (1955) 238; Dashnyam in Bot. zh. 50 (1965) 1641.

Described from Mongolia (environs of Ulan Bator, Songino area). Type in Leningrad.

River banks, among coastal bushes, floors of creek valleys and on mountain slopes, as well as weed.

IA. Mongolia: *Cent. Khalkha* (in scrub along slope of Tszakh hill, in Suchzh river valley, July 10, 1924—Pavl.; environs of Ikhe-Tukhum-Nor lake, Uberbumain-Ama trough, July 24, 1926—Zam.). *East. Mong.* (Dashnyam, l.c.).
General distribution: East. Sib. (south-east), Nor. Mong. (Mong.-Daur.).

5. **L. panzerioides** M. Pop. in Fl. SSSR, 21 (1954) 650, 154; Fl. Kirgiz. 9 (1960) 116; Fl. Kazakhst. 7 (1964) 406. —*L. oreades* Pavl. in Vestn. AN KazSSR, 8 (1954) 134.

Described from Tien Shan (Talassk Ala tau). Type in Tashkent (TAK).
Rocky-rubble slopes and meadows in alpine belt.

IA. **Mongolia:** *Mong. Alt.* (upper Bayan-Gol river along road to summer camp in somon, steppe slope, July 23, 1947—Yun.).

IIA. **Junggar:** *Cis-Alt.* (Quinhe, No. 1260, Aug. 3; south of Koktogoi, Aug. 11—1956, Ching; 15 km north-west of Shara-Sume genuine spirea scrub steppe, July 7; 20 km nor. of Koktogoi, right bank of Kairta river, Kuidyn river valley, forest belt, on south. steppe slope, July 15—1959, Yun. et al.).

General distribution: Mid. Asia (montane).
Note. Formerly regarded as endemic in USSR.

6. **L. quinquelobatus** Gilib. in Usteri, Delect. op. bot. 2 (1793) 321; Kuprian. in Fl. SSSR, 21 (1954) 148. —*L. cardiaca* auct. non L.: Krylov, Fl. Zap. Sib. 9 (1937) 2358. —*Cardiaca quinquelobata* Gilib. Fl. lithuan. (1781) 85. —Ic.: Fl. SSSR, 21, Plate 8, fig. 2.

Described from East. Europe. Type in Paris.
Rubbish heaps near residences. Forms scrub.

IIA. **Junggar:** *Tien Shan* (along Kunges river, June 16, 1877—Przew.).
General distribution; Europe, Caucasus, West. Sib. (all regions except Altay).
Note. Introduced plant in our territory.

7. **L. sibiricus** L. Sp. pl. (1753) 584; Franch. Pl. David. 1 (1884) 244; Forbes and Hemsley, Index Fl. Sin. 2 (1902) 302; Krylov, Fl. Zap. Sib. 9 (1937) 2356; Hao in Engler's Bot. Jahrb. 68 (1938) 635; Kitag. Lin. Fl, Mansh. (1939) 380; Walker in Contribs U.S. Nat. Herb. 28 (1941) 656; Kuprian. in Fl. SSSR, 21 (1954), 157; Grubov, Konsp. fl. MNR (1955) 238. —*L. sibiricus* L. f. *albiflorus* Wu et Li in Acta Phytotax. Sin. 10, 2 (1965) 163. —*L. manshuricus* Yabe *f. albiflorus* Nakai et Kitagawa in Rep. First Sci. Exp. Manch. sect. 4, 1 (1934) 47; Kitag. Lin. Fl. Mansh. (1939) 380. —Ic.: Amman, Stirp. rar. 48, tab. 8.

Described from Siberia. Type—figure cited: Amman, l.c. tab. 8.

Rocky and turfed steppe slopes of mountains and ravines, dry river-beds, among granite outliers, banks of springs, more rarely on semi-over-grown sand and as weed in crops.

IA. **Mongolia:** *Cis-Hing.* (Khalkhin-Gol river, Symbur area, Sept. 1, 1928—Tug.). *Cent. Khalkha, East. Mong., East. Gobi* (40 km nor. of Dzamyn-Ude, Motonge mountains, sandy valleys and rocky slopes, Aug. 30, 1931—Pob.). *Ordos* (5 km north-west of Ushin town, semi-overgrown sand, Aug. 3, 1957—Petr.). *Khesi* (between Lan'chzhou and Pinfansyan' towns, July 11—Pias.; 67 km north of Lan'chzhou town, in wheat field, in valley, June 29, 1957—Petr.).

IIIA. **Qinghai:** *Nanshan* (North Tetungsk mountain range, 2250 m, July 11, 1880—Przew.; Chortynton temple, in ploughed fields, on river banks and mountain slopes on clay and humus, 2100 m, Sept. 8, 1901—Lad.; Paba-tson valley, on Tetung river, on banks of irrigation canals in humus soil, July 24, 1908—Czet.). *Amdo* (resort near Guidui town, June 14, 1880—Przew.).

General distribution: West. Sib. (Altay), East. Sib. (south-east.), Nor. Mong. (Hent., cent. Hang., Mong.-Daur.), China (Dunbei, Nor.-West., South-West.).

Note. In 1957, Guan' Ke-syan' collected a plant on the nor. slope of East. Tien Shan, south of Sykeshu, which evidently turned out to be the same as *L. heterophyllus* Sweet. This report calls for confirmation, however, since only the upper part of the stem is preserved in the herbarium. *L. heterophyllus*—a species closely related to *L. sibiricus*—differs from it in appressed short pubescence of calyx, much smaller corolla and almost undivided leaves in inflorescence. *L. heterophyllus* is distributed in Far East China, Japan, Korean peninsula, America. No report from our territory.

8. **L. turkestanicus** Krecz. et Kuprian. in Bot. mater. Gerb. Bot. inst. AN SSSR, 11 (1949) 134; Kuprian. in Fl. SSSR, 21 (1954) 152; Fl. Kirgiz. 9 (1960) 116; Fl. Kazakhst. 7 (1964) 406. —**Ic.**: Fl. SSSR, 21, Plate 8, fig. 1.

Described from Mid. Asia (Fergana). Type in Leningrad.

Rocky and melkozem slopes, river banks, among shrubs.

IIA. Junggar: *Tarb.* (nor. of Chuguchak, Aug. 13, 1957—Kuan). *Jung. Alt.* (Dzhair mountain range, north of pass and 4–5 km south of Yamat picket, steppe belt, rocky flank of gorge, on short trail adjoining overhanging rock, together with *Urtica cannabina*, Aug. 4, 1957—Yun. et al.; Toli district, near water, Aug. 6; same site, on shaded slope, Aug. 12—1957, Kuan). *Tien Shan* (Talki gorge, July 16; same site, July 18—1877, A. Reg.; south-east of Urumchi, in swamp, July 14, 1956—Ching; Sin'yuan', Nanshan, on slope, Aug. 22; Fukan district, in Nyan'chi lake region, on slope, Sept, 19—1957, Kuan).

General distribution: Jung.-Tarb., Nor. and Cent. Tien Shan; Mid. Asia (montane).

## 25. Panzeria Moench.

Meth. pl. (1794) 402; Kuprian. in Bot. mater. Gerb. Bot. inst. AN SSSR, 15; (1953) 349.

1. Plant green, with sparse lax pubescence of short spaced hairs and distinctly visible glands on leaves and calyx teeth; verticils mostly interrupted ........................................................ 1. P. canescens Bunge.
+ Plant white-tomentose, only upper surface of leaf blade green, tomentose pubescence fairly dense, glands almost not perceptible; verticils mostly approximate ........................ 2. P. lanata (L.) Bunge.

1. **P. canescens** Bunge in Del. sem. hort. Dorpat. (1839) 15; Kuprian. in Bot. mater. Gerb. Bot. inst. AN SSSR, 15 (1953) 361; id. in Fl. SSSR, 21 (1954) 159; Grubov, Konsp. fl. MNR (1955) 238. — *Leonurus bungeanus* Schischk. in Krylov, Fl. Zap. Sib. 9 (1937) 2358.

Described from Altay (Katun' river). Type in Leningrad.

Rocky slopes, rocks, talus, steppe valleys of mountain rivers, coastal pebble beds in mountain-steppe and desert-steppe belts.

IA. Mongolia: *Khobdo* (beyond Kuku-Khoshun, on stones, July 30; Ukhi valley, Aug. 6—1899, Lad.). *Mong. Alt.* (mountains between Uruktu and Kobdo rivers, July 2, 1898—Klem.; Khara-Dzarga mountain range, Sakhir-Sala river valley, along coastal pebble bed, Aug. 21, 1930—Pob.; Khan-Taishiri mountain range, 15–18 km south-east of Yusun-Bulak, arid feather grass-wormwood-wheat grass steppe, July 14; 3–4 km south-west of Tamcha lake, knoll peak,

July 17; lower part of slope along road to Turgen'-Gol river from Munku-Sardyk hill [left bank tributary of Bulugun], desert steppe along gully, on pebble bed, July 26; 25–30 km south of Tamchi-Daba pass, midcourse of Bidzhi-Gol river, left bank slope of rocky valley, birch grove on slope, Aug. 10—1947, Yun.). *Bas Lakes* (Kharkhira river valley south of Ulangom, Aug. 6; left bank of Kharkhira, Sept. 4—1931, Bar.; granite knoll 3 km south of Ulangom, among rocks, July 28, 1945—Yun.). *Gobi-Alt.* (Ikhe-Bogdo mountain range, Bityuten-Ama creek valley, slopes, Aug. 12, 1927—M. Simukova).

IIA. Junggar: *Jung. Gobi* (105 km south of Ertai along road to Guchen from Altay, saxaul desert along hillocky area, along gullies, July 16, 1959—Yun. et al.).

General distribution: West. Sib. (Altay).

Note. Species closely related to *P. lanata* (L.) Bunge, differing from latter mainly in absence of tomentose pubescence on all parts (except lower part of stem where tomentose pubescence is more often seen) and presence of glands on calyx, under surface of leaf blade and stem. The distribution range of *P. canescens* somewhat wedges into the fairly broad range of *P. lanata* and the 2 species readily hybridise, resulting in a very large number of intergrades. These species sometimes grow together, as for example south of Ulangom.

2. P. lanata (L.) Bunge in Ledeb. Fl. alt. 2 (1830) 410; Sapozhn. Mong. Alt. (1911) 382; Kuprian. in Bot. mater. Gerb. Bot. inst. AN SSSR, 15 (1953) 357; id. in Fl. SSSR, 21 (1954) 159; Grubov, Konsp. fl. MNR (1955) 238; Hanelt und Dava ž amc in Feddes repert. 70 (1965) 55.—*P. albescens* Kuprian. in Bot. mater. Gerb. Bot. inst. AN SSSR, 15 (1953) 362.  —*P. alaschanica* Kuprian. ibid. 15 (1953) 363.  —*P. argyracea* Kuprian. ibid. 15 (1953) 364; Kuprian. in Fl. SSSR, 21 (1954) 158.  —*P. parviflora* Wu et Li in Acta Phytotax. Sin. 10, 2 (1965) 164.  —*P. kansuensis* Wu et Li, l.c. 165.  —*P. alaschanica* Kuprian. var. *minor* Wu et Li, l.c. 165.  —*Ballota lanata* L. Sp. pl. (1753) 582.  — *Leonurus lanatus* Pers. Syn. pl. 2 (1807) 126; Franch. Pl. David. 1 (1884) 244; Danguy in Bull. Mus. nat. hist. natur. 20 (1914) 85; Walker in Contribs U.S. Nat. Herb. 28 (1941) 656.  —Ic.: Gmel. Fl. Sib. 3, tab. 54; Kuprian., l.c. (1953) figs. 5 and 6 (sub nom. *P. lanata, P. alaschanica* et *P. argyracea*).

Described from Siberia. Type in London (Linn.).

Dry rubble and rocky slopes, foot of rocks, talus, flanks and floors of dry creek valleys and gullies, sandy steppes, small sand-covered ravines and sand-dunes.

IA. Mongolia: *Mong. Alt., Cent. Khalkha, East. Mong.* (near Bichikte, Aug. 6; along road to Yaman-Ikhe-Dulan-Khosun, Aug. 20—1928, L. Shastin). *Bas. Lakes, Val. Lakes, Gobi-Alt., East. Gobi, Alash. Gobi, Ordos, Khesi* ("Ku-lang, 1900 m, Oct. 19, 1958, Tsinghai—Kansu exped., 1958, typus *P. kansuensis*", Wu et Li, l.c.).

General distribution: West. Sib. (Altay, East. Sib. (south-west.), Nor. Mong. (Hent., Hang., Mong.-Daru.), China (Altay, Nor.).

Note. *P. lanata*, the most widely distributed species of the genus, is quite consistent in its main characteristics in our territory. Attempts to separate the more pubescent forms into independent species are, in our opinion, irrational since the characteristics used for separation are not localised even geographically. A study of the material from the entire distribution range of the species shows marked variation of the degree of pubescence of stem and under surface of leaf blade. Plants with lax tomentum as well as with very dense tomentose under surface of leaf grow in the same region. Corolla and calyx teeth show some variation in size.

However, within its entire range, regions in which one or the other form of the species is most frequently found can be distinguished. Thus, for example, plants with lax pubescence, greyish underside of leaf blade, long acuminate calyx teeth (*P. lanata* var. *typica*) are found mainly within the east. part of Angara-Sayan region and Dauria, Altay (more rarely) and Mongolia (quite extensively).

More pubescent plants with comparatively broader and short acuminate teeth of calyx are found in Altay, west. part of Angara-Sayan region and Ubsa Nur lake region. These plants were described as *P. argyracea* Kuprian. but, in our opinion, can only be treated as a variety of *P. lanata*, i.e., *P. lanata* (L.). Bunge var. *argyracea* (Kuprian.) Tschern. comb. nova [*P. argyracea* Kuprian. in Bot. mater. Gerb. Bot. inst. AN SSSR, 15 (1953) 364].

Plants very similar to *P. lanata* var. *argyracea* but with denser white tomentose pubescence on calyx, stem and underside of leaf blade grow within Ordos and Alashan. Leaves of these plants are divided into narrow lobes (3–4 mm boad at base) although in Alashan (and evidently in Kansu—*P. kansuensis* C.Y. Wu et H.W. Li, l.c.) plants are found with leaves in which lateral lobes are very broad (4–10 mm broad at base) and leaf divided only up to half length. These plants were described as *P. alaschanica* Kuprian. We, however, consider them *P. lanata* (L.) Bunge var. *alaschanica* (Kuprian.) Tschern. comb. nova [*P. alaschanica* Kuprian. in Bot. mater. Gerb. Bot. inst. AN SSSR, 15 (1953) 363].

*P. parviflora* Wu et Li, l.c., is indicated for Sinkiang (Altay). It has a fairly small corolla (2–2.2 cm long) but in all other respects does not differ from *P. lanata;* this is probably a small-flowered race of *P. lanata.*

## 26. Lagochilus Bunge[1]
in Benth. Labiat. gen. et sp. (1834) 640.

1. Bracts (spines) lacking in axils of vegetative shoots. ....................2.
+ Bracts (spines) present in axils of fertile and vegetative shoots ...4.
2. Calyx teeth nearly equal; bracts and calyx tube densely pilose due to articulate hairs ...................................... 4. **L. hirtus** Fisch. et Mey.
+ Calyx teeth unequal; bracts with sparse articulate hairs or glabrous; calyx tube invariably glabrous .........................................................3.
3. Leaves subsessile, stiff, coriaceous, cuneate-rhomboid in profile, 3–5-toothed, teeth spine-like acuminate; corolla pale yellow ..............
........................................................ 5. **L. ilicifolius** Bunge.
+ Leaves petiolate, slender, not coriaceous, broadly ovate or deltoid, pinnatifid into obtuse or short, acuminate lobes.; corolla pale pink
.......................................................... 1. **L. bungei** Benth.
4. Calyx teeth oblong, lanceolate, longer, as long or slightly shorter than tube ........................................................................................5.

[1]In our view, there is no justification to separate some species of section Inermes Fisch. et Mey. of this genus into genus *Lagochilopsis* Knorr. A study of the characteristics of this quite artificial section of the genus revealed groups some related groups within it but their significance can be evaluated only after thorough treatment of the genus as a whole in the form of a monograph. Genus *Lagochilopsis* Knorr., however, combines species of various related groups while species closely related to them even in morphological features, more so in origin, remain in genus *Lagochilus* Bunge. Taking this aspect into consideration, we have adopted genus *Lagochilus* Bunge in its original limits.

+ Calyx teeth broadly deltoid or broadly ovate, invariably shorter than calyx tube ................................................................................. 7.

5. Leaves coriaceous, cuneate-rhomboid, 3–5-lobed on top, lobes emarginate-dentate with spine-like acuminate teeth; calyx teeth smooth, glabrous along margin; verticils 2-flowered; sheath of lower leaves with tomentose pubescence ....................................................
..................................................... 7. **L. lanatonodus** Wu et Hsuan.

+ Leaves broadly ovate, cuneately narrowed toward base, 3-lobed, lobes divided into oblong, short acuminate lobes; calyx teeth scabrous along margin due to papilliform bristles; verticils 6-flowered; sheath of lower leaves without tomentose pubescence .................. 6.

6. Calyx tube densely covered with stiff bristles; bracts loosely covered with stiff bristles; long articulate hairs never seen in pubescence of calyx, bracts and stem ..... 3. **L. grandiflorus** Wu et Husan.

+ Calyx tube either glabrous or with only stray articulate hairs; bracts glabrous or loosely pubescent with long articulate hairs; short simple hairs sometimes found in pubescence along with articulate hairs.
.................................................... 2. **L. diacanthophyllus** (Pall.) Benth.

7. Calyx teeth 2–2.5 times shorter than tube; calyx tube with bristly pubescence ...................................................... 6. **L. kaschgaricus** Rupr.

+ Calyx teeth as long or slightly shorter than tube; calyx tube with tomentose pubescence ............................. 8. **L. platyacanthus** Rupr.

1. **L. bungei** Benth. Labiat. gen. et sp. (1834) 641; Krylov, Fl. Zap. Sib. 9 (1937) 2362; Knorr. in Fl. SSSR, 21 (1954) 168; Fl. Kazakhst. 7 (1964) 412. —
*L. bungei* β *minor* Fisch. et Mey. Enum. pl. nov. (1841) 32. —*L. altaicus* Wu et Hsuan in Acta Phytotax. Sin. 10, 3 (1965) 215. —*Lagochilopsis bungei* (Benth.) Knorr. in Novosti sist. vyssh. rast. (1966) 200. —*Molucella grandiflora* auct. non Steph.: Bunge in Ledeb. Fl. alt. 2 (1830) 418, excl. syn. Pall. —**Ic.**: Ledeb. Ic. pl. fl. ross. 5, tab. 436; Acta Phytotax. Sin. 10, 3, tab. 41, fig. 1–6 (sub nom. *L. altaicus*). .

Described from East. Kazakhstan (Arkaul mountains). Type in Leningrad. Map 2.

Submontane plains and foot of rocky slopes, pebble beds in river valleys.

IIA. Junggar: *Cis-Alt.* (3–4 km south of Shara-Sume, on road to Shipati, on rocky slope of knoll, July 7, 1959—Yun.; in Qinhe region, No. 10366, June 4; Koktogoi, on alluvial-proluvial trail, 950 m, No. 10427, June 9—1959, Lee and Chu ); "Altai, July 1955, No. 2263"—typus *L. altaicus* Wu et Hsuan). *Tarb.* (Shatszagai, Khobuk river delta, No. 10287, May 18, 1959—Lee and Chu). *Jung. Gobi.* (Urungu river valley, June 22, 1876—Pev.; in mountains between Bulugun river and Yamatei mountain range, Aug. 4, 1898—Klem.; Bulugun river valley, near winter camp in somon, grain sowings on ploughed terrace, in alluvial pebble beds, Aug. 1949—Yun.; "mons Altai, in itinere a Barbagai, ad Bultsing, 540 m s.m. Sept. 11, 1956, No. 2985, Ching"—Wu et al. l.c.).

General distribution: Fore Balkh., Nor. Tien Shan; West. Sib. (Altay, south-west.).

Note. We did not have authentic herbarium specimens of species *L. altaicus* Wu et Hsuan but drawings and a fairly detailed description of the species. Furthermore a study of our collection from the growing region of *L. altaicus* convinced us that it is no different from *L. bungei* Benth. Our findings do not confirm the evidently erroneous reference in "Fl. SSSR" (21: 168–169) to the distribution of *L. pungens* Schrenk in Sinkiang.

2. **L. diacanthophyllus** (Pall.) Benth. Labiat. gen. et sp. (1834) 641; Fisch. et Mey. Enum. pl. nov. (1841) 28; Krylov, Fl. Zap. Sib. 9 (1937) 2361; Knorr. in Fl. SSSR, 21 (1954) 170; Fl. Kazakhst. 7 (1964) 413. —*L. leiacanthus* Fisch. et Mey. in Bong. et Mey. Verzeichniss Saissang-Nor Pfl. (1841) [246]; eisd. Enum. pl. nov. (1841) 29; Sapozhn. Mong. Alt. (1911) 382; Krylov, Fl. Zap. Sib. 9 (1937) 2361; Knorr. in Fl. SSSR, 21 (1954) 171; Grubov, Konsp. fl. MNR (1955) 238; Grubov in Bot. mater. Gerb. Bot. inst. AN SSSR, 19 (1959) 549; Fl. Kazakhst. 7 (1964) 414. —*L. obliquus* Wu et Hsuan in Acta Phytotax. Sin. 10, 3 (1965) 218. —*L. chingii* Wu et Hsuan in Acta Phytotax. Sin. 10, 3 (1965) 219. —*Moluccella diacanthophylla* Pall. in Nova Acta Acad. Petrop. 10 (1797) 380. —Ic.: Pall. l.c. tab. 11; Fl. Kazakhst. 7, Plate 47, figs. 6, 8, 9; Wu et Hsuan, l.c. tab. 42, Figs. 8–11 (sub nom. *L. obliquus* et *L. chingii*).

Described from East. Kazakhstan (Tarbagatai foothills). Type in Leningrad. Map 1.

Rocky and stony slopes, wormwood–forb steppes, flanks of gullies; from middle to low mountain belts and foothills.

IA. **Mongolia:** *Mong. Alt.* (Kobdo, 1870—Kalning. Bulugun river bank, above Dzhirgalanta estuary, July 25, 1898—Klem.; Ubur-Maltikhin-Ama creek valley, left bank tributary of Bulugun river, rocky slopes of hills, July 28, 1947—Yun.).

IIA. **Junggar:***Cis-Alt.* ("Fuyun, juxta Wukagou, 1100 m s.m. Aug. 9, 1956, Ching, No. 1753, typus *L. chingii* Wu et Hsuan"—Wu et al. l.c.). *Tarb.* (nor. foothill of Saur, steppe along Tasta-Bulak river, June 19, 1908—Sap.; 10 km north of Kosh-Tologoi settlement [on Khobuk river] along road to Altay from Karamai, desert steppe on hummocky area, July 4, 1959—Yun. et al.). *Jung. Alt.* (on Borotala river, 2550 m, Aug. 1878—A. Reg.; along road to Kuzyun' pass, rocky site, Aug. 2, 1908—B. Fedtsch; Borotala river basin, south. slope of Jung. Alatau, below Koketau pass, near picket, July 21, 1909—Lipsky; Dzhair mountain range, Yamata river gorge, rocky slopes, July 2, 1947—Shum.; dry Tuz-Agny ravine [Karamai], unflooded terrace 10 km beyond oasis, April 14, 1954—Mois.; Toli district, Uty,, intermontane plain, Nos. 869, 891, Aug. 3; same site, on slope, No. 2485, Aug. 4—1957, Kuan; Dzhair mountain range, its south-east. extremity 20 km from Aktam settlement, on road to Chuguchak from Urumchi, nanophyton desert belt on flanks of gullies, Aug. 3; south-west. margin of Maili mountain range 20 km nor.-east of meteorological station in Junggar exit, towards pass in Kozhikhe valley, desert belt, on rocky slope, Aug. 14—1957, Yun. et al., environs of Dzinkho, 20 km nor. of Sairam-Nur lake, wormwood-forb hill steppe, 1500 m, intermontane plain, Aug. 31, 1959— Petr.). *Tien Shan* (Bayandai, near Kul'dzha, July 2 1877; Bainamun to Dzhin, 1500–1800 m, June 5, 1879—A. Reg.; 10 km south of Sairam-Nur lake, knolls with wormwood forb steppe, 1900 m, Aug. 31, 1959—Petr.; Daban pass, arid slopes, 2000 m, No. 1581, Aug. 8, 1956—Ching, isotype *L. obliquus* Wu et Hsuan). *Jung. Gobi* (from Tien Shan-Laoba to Myaoergou, on slope, No. 2367, Aug. 3, 1957—Kuan; "Chingho, inter Ertai et Chingho [Gikultai], No. 750, July 29, 1956, Ching"—Wu et Hsuan, l.c.).

General distribution: Nor. Tien Shan, Jung.-Tarb., Fore Balkh.; West. Sib. (Altay).

Note. All authors have distinguished *L. diacanthophyllus* and *L. leiacanthus* Fisch. et Mey. based on extremely insignificant characteristics. Calyx tube and bracts in *L. diacanthophyllus* are faintly pubescent with articulate hairs; calyx tube sometimes also with bristly pubescence. More often, however, articulate hairs on bracts and calyx tube are stray, calyx teeth elongated, as long or slightly longer than tube, lateral lobes of leaves elongate, acuminate. Bracts and calyx tube of *L. leiacanthus* are glabrous, calyx teeth as long or slightly shorter than calyx tube, lateral lobes of leaves obtuse but, as in *L. diacanthophyllus*, with cusp. In both species, throat of calyx slightly oblique; calyx somewhat indistinctly bilabiate. *L. diacanthophyllus* has been described from Tarbagatai foothills and *L. leiacanthus* from environs of Zaisan lake. A study of the fairly large collection of these species showed that their differentiation is very difficult. *L. leiacanthus* is only a form of *L. diacanthophyllus* with totally glabrous bracts and calyx tube. These plants are very often found in the same region and ecologically also identical. Variability of pubescence of bracts and calyx tube is very high; we found more densely pubescent bracts as well as subglabrous bracts with stray hairs in plants collected from the same site; the degree of cleavarge of leaf blades and pubescence as well as cleavage of upper lip of corolla also vary. In our opinion, therefore, all these insignificant differences should be treated as intraspecific variability of *L. diacanthophyllus*. A study of the range of this species confirmed our assumption.

Separation of *L. obliquus* Wu et Hsuan and *L. chingii* Wu et Hsuan is inappropriate. We could study firsthand only *L. obliquus* whose isotype is preserved in Leningrad but the drawings and fairly detailed descriptions available convince us that the characteristics whereby these species have been distinguished are characteristic of *L. dicanthophyllus* (Pall.) Benth.

3. **L. grandiflorus** Wu et Hsuan in Acta Phytotax. Sin. 10, 3 (1965) 217. —Ic.: Wu et Hsuan, l.c. tab. 42, figs. 6, 7.

Described from Tien Shan (Teksen). Type in Peking (Beijing) (HP). Map 1.

IIA. Junggar: *Tien Shan* (Khanakhai brook valley, June 15, 1878—A. Reg.;"Teksen, June 17, 1956, No. 26710, typus" Wu et Hsuan, l.c.).

General distribution: endemic.

Note. Thus species is closely related to *L. diacanthophyllus* (Pall.) Benth. but differs in characteristic pubescence of stem (stem with short stiff bristles), bracts (with lax, stiff bristles) and calyx (calyx tube densely setose, teeth setose along margin and very loosely setose outside but densely glandular) as well as bipinnatipartite leaves.

4. **L. hirtus** Fisch. et Mey. Enum. pl. nov. 1 (1841) 32; Sapozhn. Mong. Alt. (1911) 382; Krylov, Fl. Zap. Sib. 9 (1937) 2362; Knorr. in Fl. SSSR, 21 (1954) 165; Fl. Kazakhst. 7 (1964) 411. —*L. brachyacanthus* Wu et Hsuan in Acta Phytotax. Sin. 10, 3 (1965) 216. —*Lagochilopsis hirta* (Fisch. et Mey.) Knorr. in Novosti sist. vyssh. rast. (1966) 200. —Ic.: Fl. Kazakhst. 7, Plate 42, fig. 5; Knorr. l.c. (1966) figs. 1 and 3; Wu et Hsuan, l.c. tab. 41, figs. 7–13 (sub nom. *L. brachyacanthus*).

Described from East. Kazakhstan (Zaisan lake). Type in Leningrad. Map 2.

Desert steppes, rubble slopes, pebble beds and gravelly floors of gullies.

IIA. Junggar: *Tarb.* (Khobuk river valley, between Saur and Semistei mountain ranges, 10–12 km south of Khobuk settlement, along road to Kosh-Tologoi, desert along trails of low hummocky ridges, on gravelly floor of gully, June 24, 1957—Yun. et al.). *Jung. Gobi* (Kobu valley, south. bank of Ulyungur lake, on pebble bed, Aug. 15, 1876—Pot.). *Zaisan* (rubble steppe between Kara-Irtysh river and Saur mountain range, Aug. 21, 1906—Sap.).

General distribution: Fore Balkh.

Note. A critical study of material showed that this species is quite frequently reported only in Zaisan lake region within USSR territory, its occurrence being possible in Tarbagatai foothills. The geography of the species given in "Fl. Kazakhst." (7, 1964, 411) calls for thorough review.

We are not aware where *L. brachyacanthus* Wu et Hsuan was reported in Sinkiang territory nor from where this species was described since the first description mentions only "Sinkiang regin". However, the available drawing and description of the species suggest that this species was described for no purpose since the plant depicted in the drawing differs in no way from *L. hirtus* Fisch. et Mey.

5. **L. ilicifolius** Bunge in Benth. Labiat. gen. et sp. (1834) 641; Franch. Pl. David. 1 (1884) 244; Sapozhn. Mong. Alt. (1911) 382; Danguy in Bull. Mus. nat. hist. natur. 17 (1911) 555; ibid 20 (1914) 85; Knorr. in Fl. SSSR, 21 (1954) 169; Grubov, Konsp. fl. MNR (1955) 238; Hanelt et Davažamc in Feddes repert. 70 (1965) 55.

Described from Mongolia. Type in Leningrad. Plate III, fig. 2; map 2.

Sandy and rubble desert steppes, rocky and stony slopes, pebble bed terraces, mountain trails, flanks and pebble bed floors of gullies.

IA. Mongolia: *Mong. Alt., Cent. Khalkha* (south-west.), *Bas. Lakes, Val. Lakes, Gobi-Alt., East. Gobi, Alash. Gobi* (east.), *Ordos* (Khangian, south of town, on top of ridge, red sandstone outcrops, Aug. 5, 1957—Petr.). *Khesi* (south).

IIIA. Qinghai: *Nanshan* (eastward course beyond pass through North Tetungsk mountain range, Aug. 14–15, 1880—Przew.).

General distribution: West. Sib. (Ang.-Sayan., south. extremity of Tuva basin), Nor. Mong. (Hang., south. slope of Hangay mountain range).

6. **L. kaschgaricus** Rupr. in Mém. Ac. Sci. St.-Pétersb. VII sér. 14 (1869) 67; Knorr. in Fl. SSSR, 21 (1954) 181.   —*L. platyacanthus* auct. non Rupr.: Fl. Kirgiz. 9 (1960) 125, p. min. p.

Described from Kashgar. Type in Leningrad. Plate III, fig. 3; map 1.

Steep, washed or rubble slopes, steppised desert formations, 2000–3000 m.

IB. Kashgar: *West.* (Toyun river valley, July 30, 1867—Osten-Saken, typus!; Tashkurgansk road, Upal—Kosh-Kulak, No. 588, June 9, 1959—Lee et al.; King-Tau mountain range foothills, nor. slope, 5–8 km before reaching Kosh-Kulak settlement from Upal oasis, sympegma desert on old moraine, June 9; same site, 2–3 km south-east of Kosh-Kulak settlement, steppe belt, on steep washed slope, 2600 m, June 10; 67–68 km west of Kashgar on road to Irkeshtam, steppised desert, rubble slope of knoll, June 17, 1959—Yun. et al.; along Kashgar-Ulugchat highway, 71 km north-west of Kashgar, in river gorge, 2250 m, No. 380, June 17, 1959—Lee et al.; Baikurt settlement, 83 km north-west of Kashgar, on Kashgar-Turugart highway, steppised desert, 2300 m, June 19, 1959—Yun. et al; 25 km from Baikurt settlement, on slope, 3000 m, No. 9716, June 20, 1959—Lee et al.).

General distribution: Cent. Tien Shan.

Note. In 'Fl. SSSR' (21), this plant is shown as endemic to USSR. In fact, however, it was described from Chinese Kashgar, Toyun river valley and is distributed mainly outside the USSR. Within the USSR, we found only 1 occurrence of this species with a highly restricted distribution range (see Map 1). It is well distinguished from the closely related species *L. platyacanthus* Rupr. in shape of teeth and pubescence of calyx tube, pubescence of stem, slightly different shape of leaf and very short corolla.

7. **L. lanatonodus** Wu et Hsuan in Acta Phytotax. Sin. 10, 3 (1965) 216. —Ic.: Wu and Hsuan, l.c. tab. 42, figs. 1–5.

Described from East. Tien Shan. Type in Peking (Beijing) (HP). Isotype in Leningrad. Map 1.

Rocky and clayey slopes in steppe belt of mountains, 1100–2500 m.

IB. **Kashgar:** *Nor.* (east. tributary of Khaptsagal near Karashar, 2400 m, Aug. 30, 1879—A. Reg.; Khaidyk-Gol river, Sorsyn area, on clayey steppe, Aug. 12, 1893—Rob.; Bagrashkul' lake region, hill road from Bortu to timber plant, 1825 m, No. 6950, Aug. 3, 1958—Lee and Chu (A.R. Lee (1959), *East.* ("Liaudon-tou-si, regione montium, July 2, 1931, Liou; Turfan, in itinere a Sam-shan-shi ad Shi-yao-tse, 2400 m, No. 5662, Li and Chu)" Wu and Hsuan, l.c.).

IIA. **Junggar:** *Tien Shan* (between Khami and Nanshan-kou picket, Aug. 31, 1875—Pias.; Kul'dzha, June 21, 1877—A. Reg.; Nanshan-kou, on rocks, in crevices, June 7, 1877—Pot.; Tien Shan slopes, in gorges, 1700–2500 m, June 5, 1879—Przew.; Algoi, 1800–2400 m, Sept. 12, 1879—A. Reg.; Torkul'sk valley, Aug. 25, 1895—Rob.; around Urumchi town, Khunmiodza upland, rocky slopes on top of upland, Sept. 10, 1929—Pop.; nor. spur of Tien Shan, hilly slopes, 1160 m, No. 1791, July 23; 70 km east of Dashitou, desert, Oct. 2—1957, Kuan; south-west. foothills of Bogdo-Ul, 18–20 km south-east of Urumchi, along road to Turfan, desertified smoothed mountain ranges, nanophyte desert, June 2; 30 km on highway from Guchen to Khami, worm-wood desert on proluvial trails nor. of Tien Shan, Oct. 5—1957, Yun. et al.; Urumchi to Karashar [Yantszy], on slope, 2300 m, No. 6216, July 22, isotypus!; Bagrash lake region, along road to M. Yuldus, 2550 m, No. 6265, Aug. 1—1958, Lee and Chu); south. slope of Tien Shan 25 km be-yond Balinte settlement, Khanga valley, along road to Yuldus from Karashar, steppe belt, lower part of nor. rocky slope, Aug. 1, 1958—Yun. et al.; "Sha-wan, Kiang-guo, 1150 m, Oct. 2, 1956, No. 3710, Ching"—Wu and Hsuan, l.c.). *Jung. Gobi* (between Guchen town and Barkul' lake, Sept. 9, 1875—Pias.).

General distribution: endemic.

8. **L. platyacanthus** Rupr. in Mém. Ac. Sci. St.-Pétersb. VII sér. 14 (1869) 68; Persson in Bot. notiser (1938) 300; Knorr. in Fl. SSSR, 21 (1954) 180; Fl. Kirgiz. 9 (1960) 125, p. max. p.; Fl. Kazakhst. 7 (1964) 417. —*L. affinis* Rupr. in Mém. Ac. Sci. St.-Pétersb. VII sér. 14 (1896) 68. —*L. macrodontus* Knorr. in Bot. mater. Gerb. Bot. inst. AN SSSR, 13 (1950) 236; Knorr. in Fl. SSSR, 21 (1954) 179. —*L. keminsis* K. Isak. in Fl. Kirgiz. 9 (1960) 208, descr. ross. — *L. iliensis* Wu et Hsuan in Acta Phytotax. Sin. 10, 3 (1965) 219.—*L. kaschgaricus* auct. non Rupr.: Fl. Kazakhst. 7 (1964) 418. —Ic.: Fl. Kazakhst. 7, Plate 48, fig. 1.

Described from Cent. Tien Shan (Dzhamandaban mountains). Type in Leningrad. Map 1.

Rubble-rocky and clayey slopes, outcrops of variegated rocks, along river valleys in steppe and thicket belts.

IB. **Kashgar:** *Nor.* (Kapsalion river valley and Kyzysaya lateral valley, Karadzhon high plateau, Aug. 2–3, 1907—Merzbacher). *West.* (17 km east of Irkeshtam, on rocky-sandy soil, around mountain, Aug. 11, 1913—Knorring; Ulugchat, mountains, July 2; near Nagra-Goldy village, conglomerates on Kyzyl-Su, July 2; west of Kashgar, Bostan-Terek locality, July 10— 1929, Pop.; "Bostan-Terek, ca. 2400 m, July 27, 1921; Aug. 3, 1934"—Persson, l.c.).

IIA. **Junggar:** *Tien Shan* (Khanakhai river, 1200 m, June 27, 1878—A. Reg.; Tekes river, on clay under rock, July 1893—Rob.; "Tsaibutsar, 1900 m, No. 100, typus *L. iliensis*"—Wu and Hsuan, l.c.).

**General distribution:** Cent. and Nor. Tien Shan; Mid. Asia.

**Note.** Highly polymorphic species but insignificant variability of pubescence of calyx tube, size and shape of calyx teeth and size of lateral lobes of leaf blade should be regarded as variability of characteristics within the same species.

A study of herbarium material from all over the fairly extensive distribution range led to the conclusion that races with very broad calyx teeth (*L. macrodontus* Knorr.), very small and narrow calyx teeth (*L. keminsis* K. Isak.) and with rhomboid-deltoid very small leaves and hirtellous stems (*L. iliensis* Wu et Hsuan) do not merit separation into distinct species since characteristics differentiating these races from typical *L. platyacanthus* are not geographically localised and are often found in plants growing in the same locality.

However, plants described from Transili Alatau spurs (Turaigyr mountains) under the name *L. pulcher* Knorr. are quite well distinguished from *L. platyacanthus* in glabrous or subglabrous calyx tube and slightly different pubescence of stem. Such plants are found only in Transili Alatau.

## 27. **Metastachys** Knorr.
in Fl. SSSR, 21 (1954) 652, 192.

1. **M. sagittata** (Regel) Knorr. in Fl. SSSR, 21 (1954) 193; Fl. Kirgiz. 9 (1960) 128; Fl. Kazakhst. 7 (1964) 419. —*Phlomis sagittata* Regel in Acta Horti Petrop. 6 (1879) 373. —*Ballota sagittata* Regel in Acta Horti Petrop. 9 (1886) 607.

Described from East. Tien Shan. Lectotype in Leningrad. Plate III, fig. 1.

Meadow-steppe formations and among shrubs from foothills to forest belt of mountains.

IIA. **Junggar:** *Tien Shan* (Aktyube, near Kul'dzha, May 13, 1877 —A. Reg., lectotypus!; Piluchi, July 1877; Khangakhai, 1200 m, June 27; upper Khanakhai, in Akburtash, 2700 m, June 28—1879, A. Reg.).

**General distribution:** Cent. and Nor. Tien Shan.

**Note.** Species description does not indicate type; I suggest plant with label "Aktyube near Kul'dzha, May 13, 1877, A. Regel" as the lectotype.

## 28. Stachys L.
Sp. pl. (1753) 580.

1.  Leaves ovate, cordate or subcordate at base, suborbicular ..........2.
+   Leaves narrower, oblong-lanceolate or oblong, narrowing towards base, orbicular, slightly truncated or subcordate ............................4.
2.  Leaves softly hairy on both surfaces, greyish, with slender bristle at tip ......................................................................... 4. S. setifera Mey.
+   Leaves greenish on both surfaces, diffusely pubescent (pubescence denser on lower surface along veins), usually acuminate or subobtuse towards tip but invariably without bristles ..................3.
3.  Leaves large, lower and middle caulous leaves 10–12 cm long and 6.5–7.5 cm broad, on long (4–7 cm) petioles; corolla red; verticils 6–8-flowered, interrupted; inflorescence long ..........6. S. silvatica L.
+   Leaves very small, lower and middle cauline leaves 3.5 cm long and 1.2–1.5 cm broad (more rarely up to 7 cm long and 3 cm broad), on 1.5–3.5 cm long petioles, corolla pink, without or with dark violet spots (especially on lower lip of corolla); verticils 4–6-flowered, more or less approximate; inflorescence very short ........................
    ...............................................................................5. S. sieboldii Miq.
4.  Leaves oblong-lanceolate, subcordate at base, greyish-green on upper surface, greyish beneath due to dense velutinous pubescence; corolla pale pink, small, corolla tube not exserted from calyx ........
    ................................................................... 1. S. oblongifolia Benth.
+   Leaves linearlanceolate; as a rule narrowing or truncate at base, green on both surfaces, with pubescence only along veins on lower surface or even densely pubescent on both surfaces with patent hairs but not velutinous; corolla purple-lilac, purple or pink-lilac, corolla tube exserted from calyx. ..........................................5.
5.  Stem and leaves on both surfaces with dense setose pubescence ...
    ......................................................................................2. S. palustris L.
+   Stem and leaves loosely pubescent with flat articulate hairs ..........
    ............................................................. 3. S. riederi Chamisso ex Benth.

1. S. oblongifolia Benth. in Wall. Pl. Asia Rar. 1 (1830) 64; ej. Labiat. gen. et sp. (1834) 545; Hook. f. Fl. Brit. Ind. 4 (1885) 676; Dunn in Notes Bot. Gard. Edinburgh, 6 (1915) 181; Kudo in Mém. Fac. Sci. and Agr. Taihoku Univ. 2 (1929) 187.  —S. modica Hance in J. Bot. (London) 20 (1882) 292; Kanitz, A. növénytani (1891) 48; id. in Széchenyi, Wissensch. Ergebn. 2 (1898) 725.

Described from East. Himalayas. Type in London (K).

IIIA. Qinghai: *Nanshan* ("Szining-fu, ad fin. June 1879, Széchenyi"—Kanitz, l.c.).
General distribution: Himalayas (east.), Korean peninsula, China (South, Taiwan).

2. S. palustris L. Sp. pl. (1753) 580; Benth. Labiat. gen. et sp. (1834) 542; Sapozhn. Mong. Alt. (1911) 382; Danguy in Bull. Mus. nat. hist. natur. 20

(1914) 85; Dunn in Notes Bot. Gard. Edinburgh, 6 (1915) 181; Krylov, Fl. Zap. Sib. 9 (1937) 2364; Knorr. in Fl. SSSR, 21 (1954) 216; Fl. Kirgiz. 9 (1960) 130; Fl. Kazakhst. 7, (1964) 422. —Ic.: Fl. Kazakhst. 7, Plate 48, fig. 6.
Described from West. Europe. Type in London (Linn.).

Banks of rivers and lakes, wet swamp meadows, along irrigation canals.

IIA. Junggar: *Tien Shan* (on way to Nilka from 4th division state farm, No. 1206, Aug. 24; Sin'yuan'-Nilka, near water, No. 3775, Aug. 24—1957, Kuan). *Jung. Gobi* (Dyurbel'dzhin, meadow, Aug. 24; Ch. Irtysh valley, Aug. 26—1876, Pot.; Usu-San'tszyaochzhuan, along canal margin, No. 1042, June 25; Savan district, near water, No. 1567, June 25; Manas river, 2 km north of Kuitun, in pasture, humid site, No. 1122, June 29; same site, No. 1143, June 29; 3 km north of Kuitun, swamped meadow, No. 366, July 6—1957, Kuan; Kuitun station (on old Shikho-Manas road), sasa zone, forb-sedge marshy meadow, June 30; same site, July 6—1957, Yun. et al.; east of Barbagai, Kran river floodplain, No. 10597, July 8, 1959—Lee and Chu); "bords de l'Irtich, Aug. 29, 1895, Chaff."—Danguy, l.c.).

General distribution: Aralo-Casp., Fore Balkh., Jung.-Tarb., Cent. and Nor. Tien Shan; Arct. Europe, Mediterranean, Balk.-Asia Minor, Fore Asia, Caucasus, Mid. Asia, West. Sib., East. Sib. (west.), China (Altay), Himalayas, Japan.

Note. Highly polymorphic species. Several investigators have differentiated a large number of races. Material on this species in our territory is comparatively uniform.

3. S. riederi Chamisso ex Benth. in Linnaea, 6 (1831) 570; Benth. Labiat. gen. et sp. (1834) 540; Kitag. Lin. Fl. Mansh. (1939) 387. —*S. baicalensis* Fisch. in Benth. Labiat. gen. et sp. (1834) 543; Walker in Contribs U.S. Nat. Herb. 28 (1941) 657; Knorr. in Fl. SSSR, 21 (1954) 219; Grubov, Konsp. fl. MNR (1955) 238. —*S. chinensis* Bunge in Benth. Labiat. gen. et sp. (1834) 544; Knorr. in Fl. SSSR, 21 (1954) 219. —*S. japonica* Miq. in Ann. Mus. Bot. Lugd.-Batav. 2 (1865) 111; Knorr. in Fl. SSSR, 21 (1954) 218. —*S. aspera* auct. non Michx.: Franch. Pl. David 1 (1884) 241; Forbes and Hemsley, Index Fl. Sin. 2 (1902) 300. —Ic.: Fl. SSSR, 21, Plate 12, fig. 1 (sub nom. *S. baicalensis*).
Described from Kamchatka. Type in Leningrad.

Wet meadows, marshes, along wet banks of rivers and lakes, irrigation canals in forest belt.

IA. Mongolia: *East. Mong.* (environs of Kuku-Khoto, on Chaodzhyunfyn knoll, along irrigation canals, near water or on wet clayey soil, July 18, 1884—Pot.; Khakhir river, in wet meadows, July 14, 1899—Pot. and Sold.; Shilin-Khoto, true steppe, 1959—Ivan.; "Fossés à Sartchy, David"—Franchet, l.c.). *Ordos* (in Huang He river valley, in meadows, in high-water sections of rivers, in clayey or sandy soil, Aug. 9; same site, Aug. 18—1871, Przew.).

IIIA. Qinghai: *Nanshan* (Khagomi area in upper Huang He, 2100 m, near springs on rich soils, July 4; South Tetung mountain range, in farms, forms thickets, Aug. 8—1880, Przew.; "Yao Chieh, forming dense patches, along exposed, very moist roadsides of rich soil, common, No. 270, Ching"—Walker, l.c.).

General distribution: East. Sib., Far East, Nor. Mong. (Hent., Hang., Mong.-Daur.), China (Dunbei, North), Korean peninsula, Japan.

Note. Highly polymorphic species. In our view, however, there is no need to separate races with less pubescent or glabrous leaves as species. The degree of pubescence of leaves is highly variable in this species. Adequate material is not available to verify the variability of characteristics of *S. aspera* Michx; material on this species may turn out to be identical with our material, in which case our species will have to be named *S. aspera* Michx.

4. **S. setifera** Mey. Verzeichn. (1831) 94; Knorr. in Fl. SSSR, 21 (1954) 211; Fl. Kazakhst. 7 (1964) 421. —Ic.: Fl. SSSR, 21, Plate 11, fig. 1.

Described from Transcaucasus (Talysh). Type in Leningrad.

Wet sites in river valleys and along irrigations ditches in foothills and lower mountain belt.

IIA. Junggar: *Balkh.-Alak.* (Churchutsu river, Aug. 10, 1840—Schrenk).

General distribution: Fore Asia, Mid. Asia (West. Tien Shan, Pamiro-Alay), Caucasus.

Note. Semi-weedy species. Evidently introduced in our territory.

5. **S. sieboldii** Miq. in Ann. Mus. Bot. Lugd.-Batav. 2 (1865) 112; Forbes and Hemsley, Index Fl. Sin. 2 (1902) 301; Hao in Engler's Bot. Jahrb. 68 (1938) 635. —*S. affinis* Bunge in Mém. Ac. Sci. St.-Pétersb. Sav. Etrang. 2 (1832) 125. —*S. baicalensis* auct. non Fisch.: Rehder and Kobuski in J. Arn. Arb. 14 (1933) 31. —Ic.: Honda, Mizushima, Suzuki, Ill. Man. fl. Japan (1964) fig. 1059.

Described from cultivated specimens from Japan. Type in Leiden.

Shaded sites, grassy slopes in scrub belt of mountains, as weed in farms and houses.

IIIA. Qinghai: *Nanshan* (Chortynton temple, in humus soil in forests, 2100–2400m, Sept. 8, 1901—Lad.; 60 km south of Chzhana, Nanshan high foothills, 2200 m, July 12, 1958—Petr.). *Amdo* (Yellow river valley near Radzha, alt. 3300 m, on grassy slopes, June 1926—Rok.).

IIIB. Tibet: *Weitzan* (Yantszytszyan basin, Nruchu area, on right bank of river, in clayey-rocky soil of ploughed field, on rubbish heaps and on clay in juniper forest, July 25, 1900—Lad.).

General distribution: China (South-West), Japan.

6. **S. silvatica** L. Sp. pl. (1753) 580; Benth. Labiat. gen. et sp. (1834) 541; Krylov, Fl. Zap. Sib. 9 (1937) 2366; Knorr. in Fl. SSSR, 21 (1954) 215; Fl. Kirgiz. 9 (1960) 130; Fl. Kazakhst. 7 (1964) 421. —Ic.: Hegi, Ill. Fl. 5, 4, tab. 227.

Described from Europe. Type in London (Linn.).

Shade of trees, among shrubs, tall-grass meadows in scrub belt of mountains.

IIA. Junggar: *Tien Shan* (forest belt of Tien Shan in upper Kunges, 1200 m, in meadow, June 23, 1877—Przew.; Kul'dzha region, Urtas-Aksu, June 17, 1878—Fet.; Borgaty, 1800–2100 m, July 6, 1879—A. Reg.).

General distribution: Jung.Tarb., Cent. and Nor. Tien Shan; Europe, Balk.-Asia Minor, Caucasus, West. Sib.

## 29. Chamaesphacos Schrenk.
in Fisch. et Mey. Enum. pl. nov. 1 (1841) 27.

1. **Ch. ilicifolius** Schrenk in Fisch. et Mey. Enum. pl. nov. 1 (1841) 28; Kuprian. in Fl. SSSR, 21 (1954); 244; Fl. Kazakhst. 7 (1964) 424. —Ch.

*longiflorus* Bornm. et Sint. in Mitth. Thür. Bot. Ver. N.F. 18 (1903) 51. —Ic.:
Fl. Uzbek. 5, Plate 32, fig. 2; Fl. Kazakhst. 7, Plate 48, fig. 8.
Described from Balkhash lake. Type in Leningrad.
Dry river-beds, overgrown, sand-dunes and fixed sands ridges.

IIA. Junggar: *Jung. Gobi* (south. extremity of Dzosotyn sand, along right bank of Manas
river, 10–15 km north of 21st regiment state farm, sand ridge, acc. *Haloxylon-Artemisia santolina*,
June 11; same site, 20–24 km north of 30th regiment state farm, sand ridges, acc. *Haloxylon
persicum-Aritstida pennata*, July 9, 1957— Yun. et al.; same site, on top of sand-dunes, No. 487,
July 9, 1957—Kuan).

General distribution: Aralo-Casp., Fore Balkh., Jung.-Tarb., Nor. and Cent. Tien Shan;
Fore Asia, Mid. Asia.

## 30. Salvia L.
Sp. pl. (1753) 23.

1. Calyx 7 mm long, upper lip with 3 distinct, approximate acuminate teeth, calyx tube glabrous inside; corolla tube without hairy ring inside ............................................................ 1. **S. deserta** Schang.
+ Calyx 10–20 mm long, upper lip undivided, calyx tube inside as well as limb with short pubescence; corolla tube invariably with hairy ring or pubescent inside throughout length. ........................ 2.
2. Corolla yellow ................................................................................3.
+ Corolla blue, violet or purple .......................................................4.
3. Perennial; inflorescence simple or with 1–2 pairs of branches; corolla 30–40 mm long .......................... 2. **S. nubicula** Wall. ex Sweet.
+ Annual or biennial; inflorescence branched; corolla 15 mm long ...
.................................................................... 5. **S. roborowskii** Maxim.
4. Leaves loosely pubescent on upper surface, greenish, gryishtomentose beneath; corolla 25 mm long ............................................
.................................................................... 4. **S. przewalskii** Maxim.
+ Leaves green on both surfaces, loosely pubescent; corolla 35–40 mm long ............................................................................................5.
5. Radical leaves ovate or subhastate, deeply cordate at base, 7–16 cm long ................................................... 6. **S. Wardii** Peter-Stibal.
+ Radical leaves oblong or ovate, subcordata at base, 4–7cm long ....
.................................................................... 3. **S. prattii** Hemsl.

1. **S. deserta** Schang. in Ledeb. Ind. sem. hort. Dorp. Suppl. 2 (1824) 6;
Pobed. in Fl. SSSR, 21 (1954) 346; Grubov in Bot. mater. Gerb. Bot. inst. AN
SSSR, 19 (1959) 549; Fl. Kirgiz. 9 (1960) 138; Fl. Kazakhst. 7 (1964) 430. —
*S. silvestris* auct. non L.: Danguy in Bull. Mus. nat. hist. natur. 20 (1914) 83;
Krylov, Fl. Zap. Sib. 9 (1937) 2371; Persson in Bot. notiser (1938) 300. —*S.
nemorosa* auct. non L.: Grubov, Konsp. fl. MNR (1955) 238. —**Ic.:** Fl.
Kazakhst. 7, Plate 49, fig. 5.

Described from Altay. Type not known.

Steppe mountain slopes, banks of rivers and brooks, as weed on roadsides, around houses, in fields.

IA. Mongolia: *Khobdo* (between border and Khobdo town, 1870—Kalning).

IB. Kashgar: *Nor.* (Kapsalyon Tal und Nebental, Kisyl-sai, auch Plateauhöhe von Karadschon, Aug. 2–3, 1907—Merzb.).

IIA. Junggar: *Cis-Alt.* (Altay [Shara-Sume], No. 2371, Aug. 11; same site, 900 m, Aug. 11; environs of Koktogoi, No. 2106, Aug. 18; Altay, No. 2656, Sept. 3; south of Altay, No. 2769, Sept. 6—1956, Ching; 15 km north-west of Shara-Sume, steppe belt, scrub steppe, No. 1031, June 7, 1959—Yun. et al.). *Jung. Alt.* (north-west. extremity of Dzhair mountain range, 22 km nor.-nor-west of Toli settlement, along road to Temirtam, on flank of arid river-bed, in wormwood desert, Aug. 5, 1957—Yun. et al.; Toli-Emel', along dry ditch, No. 2775, Aug. 9, 1957—Kuan). *Tien Shan* (along Kunges river, 3500 m, June 27, 1871—Przew.; Matymbak spring, 1874-Larionov; Talki, north of Kul'dzha, July 16, 1877; near Nilka, 1500 m, June 1879—A. Reg.; upper Ili, in clayey soli, on edge of steppe, Aug. 29, 1877—Przew.; left bank valley of Kunges river, 3 km south of Besh-Tyube settlement, serus mountain steppe, Aug. 28, 1957—Yun. et al.; nor. slope of Merzbacher mountain range, Kok-Sai valley [right tributary of Dalunkho river], high-water erosion terrace, near irrigation ditch, 1100–1200 m, Sept. 9, 1952—Mois.). *Jung. Gobi* (near Guchen, 1876—Pev.; between Santakh and Dzhimuch, near wheat farm, Aug. 19, 1898—Klem.; Savan, No. 3718, July 4; Urumchi, No. 450, July 20—1956, Ching; in Shikhetsz region, near oat farm, in desert, No. 656, June 6; Tsynitsyuan', in wet site, No. 1225, July 7; from Shichan to Savan, No. 493, July 23; Savan district, 2–3 km south of Shichen, No. 847, July 27; 26 km east of Chzhaos, along road to Tekes, on roadside, No. 980, Aug. 17; near Fukan, in wet site, No. 4193, Sept. 17; in Gan'khetaz region, edge of farm, No. 5096, Sept. 23—1957, Kuan). *Dzhark.* (near Kul'dzha, May 15; same site, May 31; Chimpansi, near Kul'dzha, June Pilyuchi, June 17—1877, A. Reg.; outside Kul'dzha town, June 1878—Larionov, Ili-Yamatu, near water, No. 3028, Aug. 4, 1957—Kuan).

General distribution: Aralo-Casp., Fore Balkh., Jung.-Tarb., Nor. and Cent. Tien Shan; West. Sib. (Altay), Caucasus, Mid. Asia.

2. **S. nubicola** Wall. ex Sweet, Brit. Fl. Garden, 2 (1825–1827) 140; Hedge in Notes Bot. Gard. Edinburgh, 23, 3 (1961) 206. —*S. glutionosa* auct. non L.: Strachey, Catal. (1906) 142; Peter-Stibal in Feddes repert. 39 (1936) 176. —Ic.: Sweet, l.c.

Described from Himalayas. Type—drawing.

Wet, humus-rich soil.

IIIB. Tibet: *South.* (Niti, 3450 m, No. 3, Aug.—Strachey and Winterbottom).

General distribution: Fore Asia (Afghanistan), Himalayas.

3. **S. prattii** Hemsl. in J. Linn. Soc. London (Bot.) 29 (1893) 316; Kudo in Mém. Fac. Sci. and Agr. Taihoku Univ. 2 (1929) 163; Rehder and Kobuski in J. Arn. Arb. 14 (1933) 31.

Described from South-West. China (Sichuan). Type in London (K).

Rocks and rocky slopes of mountains, along river banks, 3900–4000 m.

IIIB. Tibet: *Weitzan* (Yantszytszyan basin, upper course of Ichu river, on rocky banks of river, gorge slopes, rocks, 3900 m, July 29, 1900—Lad.; "Amnyi Machen range, No. 14426, Rock"—Rehder and Kobuski, l.c.).

General distribution: China (South-West, Sichuan).

4. **S. przewalskii** Maxim. in Bull. Ac. Sci. St.-Pétersb. 27 (1881) 526; Forbes and Hemsley, Index Fl. Sin. 2 (1902) 287; Kudo in Mém. Fac. Sci. and Agr. Taihoku Univ. 2 (1929) 163; Rehder and Kobuski in J. Arn. Arb. 14 (1933) 31; Hao in Engler's Bot. Jahrb. 68 (1958) 635; Walker in Contribs U.S. Nat. Herb. 28 (1941) 657.

Described from Qinghai. Type in Leningrad.

Meadows in forest belt in high mountain areas.

IIIA. Qinghai: *Nanshan* (in alpine belt of North Tetungsk mountain range, in forest meadows, rare, Aug. 13, 1872—Przew., typus!); "Shui Mo Kou, near Lien Cheng, on an exposed, wet stream bank, flowers bluish-purple, No. 355, Ching"—Walker, l.c.).

General distribution: China (North-West, South-West).

5. **S. roborowskii** Maxim. in Bull. Ac. Sci. St.-Pétersb. 27 (1881) 527; Forbes and Hemsley, Index Fl. Sin. 2 (1902) 287; Diels in Futterer, Durch Asien, 3 (1903) 19; Kudo in Mém. Fac. Sci. and Agr. Taihoku Univ. 2 (1929) 166; Rehder and Kobuski in J. Arn. Arb. 14 (1933) 31; Hao in Engler's Bot. Jahrb. 68 (1938) 635; Walker in Contribs U.S. Nat. Herb. 28 (1941) 657; Murata in Acta Phytotax. Geobot. 16, 1 (1955) 16.

Described from Qinghai. Type in Leningrad.

In rich meadow soils on mountain slopes, on and under rocks, often in old pasture corrals, coastal, pebble beds, 3000–4000 m alt.

IIIA. Qinghai: *Nanshan* (South-Tetungsk mountain range, July 24, 1872, typus!; Rako-Gol river, in humus of old pasture corrals, July 22; on high mountain areas of South Tetung mountain range, 3000–3600 m, July 31—1880, Przew.; Kuku-Nor lake, 3300 m, in old pasture corrals, Aug. 18, 1901—Lad.; Mon'yuan', Ganshig river valley, left tributary of Peishikhe river [joins Tetung river], 3350–3720 m, 1958—Petr.; "Kokonor: unweit der Stadt Tenkar, um 3000 m, No. 1300, Sept. 14, Hao"—Hao, l.c.; "Am Kükenor, Futterer"—Diels, l.c.).

IIIB. Tibet: *Weitzan* (on and under rocks, in humus and clay, sometimes on pebble bed, Yantszytszyan basin, Donra area, on Khichu river, 3900 m, July 16, 1900—Lad.). *South* ("circa Lhasa, Elba temple, Aug. 23, 1914, Kawaguchi"—Murata, l.c.).

General distribution: China (North-West, South-West), Himalayas.

6. **S. wardii** Peter-Stibal in Feddes repert. 39 (1936) 176. —*S. hians* auct. non Royle: Marq. et Shaw in J. Linn. Soc. London (Bot.) 48 (1929) 217.

Described from Tibet. Type in London (K).

Alpine rocky meadows, among scrub.

IIIB. Tibet: *South* ("Near Gyamda, 3600 m, No. 6172, Sept. 1, 1924, Ward, typus; Ata Kang La, 4200–4500 m, No. 10591, Ward"—Peter-Stibal, l.c.).

General distribution: endemic.

## 31. Perovskia Kar.

in Bull. Soc. natur. Moscou, 14 (1841) 15; Kudryashev, Rod [Genus] *Perovskia* Kar. (1936) 6.

1. Leaves undivided, irregularly serrate .......... **P. atriplicifolia** Benth.

+ Leaves bipinnatisect ............................................ 1. P. abrotanoides Kar.

1. P. abrotanoides Kar. in Bull. Soc. natur. Moscou, 14 (1841) 15; Henderson and Hume, Lahore to Jarkend (1873) 331; Pampanini, Fl. Carac. (1930) 184; Kashyap in Proc. Indian Sci. Congr. 19 (1932) 48; Kudryashev, Rod [Genus] Perovskia Kar. (1936) 10; Gorschkova in Fl. SSSR, 21 (1954) 375; Fl. Kirgiz. 9 (1960) 144; Fl. Kazakhst. 7 (1964) 433. —Ic.: Kar. l.c. tab. 1; Kudryashev, l.c. fig. 3.

Described from Mid. Asia (Balkhany mountains). Type in Leningrad.

Dry river-beds and banks of mountain rivers and brooks, talus and rocky-rubble slopes of mountains, up, to 2000 m alt.

IA. Mongolia: *Khobdo* (between boundary and Khobdo, 1870—Kalning).
IB. Kashgar: *West.* (Bakh village on Charysh river, on clayey slope, Aug. 6, 1909—Divn.).
General distribution: Nor. and Cent. Tien Shan; Fore Asia, Himalayas (west., Kashmir).

P. atriplicifolia Benth. in DC. Prodr. 12 (1848) 261; Hook. f. Fl. Brit. Ind. 4 (1885) 652; Pampanini, Fl. Carac. (1930) 185; Kudryashev, Rod [Genus] *Perovskia* Kar. (1936) 33. —Ic.: Bot. Mag. 8 (1912) tab. 8441; Kudryashev, l.c., fig. 13.

Described from Fore Asia (Afghanistan). Type in London (K).

At 2100–3600 m alt.

IIIB. Tibet:*Chang Tang* (occurrence possible since known from Karakorum pass— Pampanini, l.c.).
General distribution: Fore Asia, Himalayas (west., Kashmir).

## 32. Ziziphora L.
Sp. pl. (1753) 21.

1. Annual, with more or less elongated spicate inflorescence. ............ ................................................................................ 4. Z. tenuior L.
+ Perennial; inflorescence capitate ........................................................ 2.
2. Calyx whitish due to dense and continuous pubescence of long, patent hairs ........................................................ 3. Z. pamiroalaica Juz.
+ Calyx greenish or greyish, with more or less uniformly dense, short pubescence or long hairs seen in pubescence but not fully covering calyx. ................................................................................................ 3.
3. Calyx with pubescence of dense uniform short stiff hairs; leafy bracts horizontally declinate or appressed to calyx without cilia along margin; caulous leaves lanceolate, more than twice longer than wide ................................................................................ 1. Z. bungeana Juz.
+ Calyx pubescence consists of long hairs scattered all over surface; leafy bracts deflexed, with cilia along margin; caulous leaves ovate or suborbicular, their length not more than twice width. ................ ................................................................ 2. Z. clinopodioides Lam.

1. **Z. bungeana** Juz. in Fl. SSSR, 21 (1954) 664, 386; Fl. Kirgiz. 9 (1960) 147; Fl. Kazakhst. 7 (1964) 434. —*Z. clinopodioides* auct. non Lam.: Grubov, Konsp. fl. MNR (1955) 239, p.p.; Grubov in Bot. mater. Gerb. Bot. inst. AN SSSR, 19 (1959) 549. —**Ic.:** Fl. Kazakhst. 7, Plate 50, fig. 1.

Described from East. Kazakhstan (Zaisan lake). Type in Leningrad.

Rocky and rubble slopes in desert-steppe and mountain-steppe belts, intermontane basins, along low hills, occasionally in meadows in subalpine belt of mountains.

IA. **Mongolia:** *Khobd.* (between boundary and Khobdo, 1870—Kalning).

IB. **Kashgar:** *Nor.* (Kapsalyon Tal et Nebental, Kisyl-sai, auch Plateauhöhe von Karadschon, Aug. 2–3, 1907—Merzb.).

IIA. **Junggar:** *Cis-Alt.* (20 km south of Shara-Sume, 7000 m. No. 2704, Sept. 6, 1956—Ching). *Tarb.* (between Chuguchak and Churchutsu [Aug.] 1840—Schrenk; Kotbukha, rocks, Aug. 10, 1876—Pot.). *Jung. Alt.* (south-east. extremity of Dzhair mountain range, 8 km southwest of Boguta brook along road to Otu from Aktama and Chuguchak, desert-steppe belt, gully, Aug. 3, 1957—Yun. et al.; Toli district, Uty, low hill, No. 931, Aug. 4; along road to Chzhaos, on road side, No. 1585, Aug. 5; 1 km east of Chzhaosu, in intermontane basin, No. 798, Aug. 11; Chzhaosu district, Ven'tsyuan' [Arasan], No. 3451, Aug. 14—1957, Kuan). *Tien Shan* (Issykul'-Muzart, July; on Tekes, Aug. 13; same site, Aug. 25—1877; Kash, Dzhirumto, 1350 m, May 2; Tsagan-Tunge, June 8—1879, A. Reg.; Yuldus, Sept. 1878—Fet.; south of Dzhinkho, Bogda-Ola; nor. slope of Tien Shan mountain range, on Pichan meridian, mountains east of Kicha-Ulan-Su [Sept. 27, 1889]—Gr.-Grzh.; Urumchi, Nanshan, No. 546, July 21, 1956—Ching; Kul'dzha-Chapchal, mines on Dzhagastai, ravine in desert, No. 3118, Aug. 7, 1957—Kuan; Ketmen' mountain range, nor. slope, Sarbushin river valley, 1 km below Sarbushin settlement, along road to Ili from Kzyl-Kure, steppe belt on south. Rocky slope, Aug. 21, 1957; Tsanma valley, left bank tributary of Kunges, Dagdyn-Gol area, subalp. belt, burnet meadow, Aug. 8, 1958—Yun. et al.; 50 km nor. of Tsitszyaotszintsza mountain valley near pass, 1500 m, Oct. 3, 1958—Petr.). *Jung. Gobi* (west. extremity of Junggar desert, on Chipeitszy-Chuguchak road, 600 m, Aug. 5, 1951—Mois.; Savan district, east of Datszyamyao, 2000 m, Oct. 1, 1956—Ching; near Kuitun, No. 1158, June 29; Savan district, San'daokhetsza, on river bank, No. 349, July 4; Savan district, Datszymyao, on slope, No. 1270, July 8; from Tsitai to Meiyao, in Gobi, No. 5147, July 25; Savan district, 6 km south-west of Shichan, No. 2292, July 26; Fukan district, near Tyan'chi lake, near water, No. 4263, Sept. 20; 60 km north-east of Tsitai, No. 2313, Sept. 25; from Tsitai through pasture up to Beidashan', on shaded slope, No. 5233, Sept. 28—1957, Kuan). *Dzhark.* (Piluchi, near Kul'dzha, June 20; Kul'dzha, June—1877, A. Reg.).

General distribution: Aralo-Casp., Fore Balkh., Jung.-Tarb., Nor. Tien Shan; Mid. Asia, West. Sib. (south.).

Note *Z. clinopodioides* and *Z. bungeana*, very closely related species, are perhaps better treated as races of the same species. Their differences are very insignificant and both are found in the same regions of Cent. Asia. Some relationship of these races of different attitudes can be perceived, however. As a rule, *Z. clinopodioides* is found in the subalpine belt at 1200–3500 m, descending along pebble beds into desert steppe zone. *Z. bungeana* generally grows at a lower level in desert-steppe zone, in low hills, intermontane basins, at 600–1200 (2000) m, but occasionally ascends even up to subalpine meadows. Only a diligent field study of all races of *Z. clinopodioides* s. l. throughout its distribution range would reveal the variability of characteristics of this compelx and widely distributed group of plants.

2. **Z. clinopodioides** Lam. Illustr. I (1791) 63; Krylov, Fl. Zap. Sib. 9 (1937) 2374; Persson in Bot. notiser (1938) 300; Juzepczuk in Fl. SSSR, 21

(1954) 398; Fl. Kirgiz. 9 (1960) 149; Fl. Kazaskhst. 7 (1964) 435. —Ic.: Fl. SSSR, 21, Plate 20, fig. 4; Fl. Kazakhst. 7, Plate 50, fig. 2.

Described from Siberia. Type in Paris.

Gravelly banks of rivers and floors of gullies, on rocks, rocky slopes and talus in mountain-steppe and subalpine belts.

IIA. Junggar: *Cis-Alt.* (between Qinhe and Tsagankhe, No. 1513, Aug. 8; Shara-Sume, No. 2424, Aug. 26—1956, Ching; 80 km east-nor.-east of Burchum, along road to Shara-Sume, steppe belt, on granite eluvium, July 5; 15–20 km north-west of Shara-Sume, on Kran river, scrub meadow steppe on granite eluvium, July 7—1959, Yun. et al.; in Altay mountains, on exposed slope, No. 10660, July 16, 1959—Lee et al.) *Tarb.* (Saur mountain range, south. slope of Karagaitu river valley, Bain-Tsagan creek valley, subalp. belt, along slopes, June 23, 1957; nor. trail of Saur mountain range, 30 km west-nor.-west of Kheisangou settlement, along road to Karamai from Burchum, desert steppe, on pebble beds along gully, July 19, 1959—Yun. et al.). *Jung. Alt.* (Urtaksary, July 20, 1878—Fet.; ascent to Kuzyun' pass, rocky site, Aug. 2, 1908— B. Fedtschenko; Toli district, Myaoergou, No. 2420, Aug. 4; Barlyk mountain range, on slope, No. 986, Aug. 6—1957, Kuan). *Tien Shan* (in upper course of Kunges river, 3500 m, June 18, 1877—Przew.; Chapchal pass, 1500–1800 m, June 20, 1878—A. Reg.; Talki, 1950 m, July 10, 1878—Fet.; Mongoto, 3000–3300 m, July 4; Araslyn, 2400 m, July 8—1879, A. Reg.; Urumchi river, upper Tasenku river, Biangou locality, Sept. 25, 1929—Pop.; nor. spur of Tien Shan, Danu river, on plain, No. 1470, July 17; south of Nyutsyuan'tsze, on alp. meadows, No. 636, July 18; Ulastai, on slope. No. 3811, Aug. 26—1957, Kuan; "Thian-Shan, Kunges, ca. 1130 m, Aug. 24, 1932"—Persson, l.c.). *Jung. Gobi* (Baitak-Bogdo-Nuru mountain range, Takhiltu-Ula, Ulyastu-Gola gorge, 5 km beyond estuary, on main bank, Sept. 17, 1948—Grub.).

General distribution: Jung.-Tarb., Cent. Tien Shan; Mid. Asia, West. Sib. (Altay), East. Sib. (south-east.).

Note. See note under Z. *bungeana* Juz.

3. **Z. pamiroalaica** Juz. in Tr. Bot. inst. AN SSSR, ser. 1, 4 (1937) 328, nomen; Juzepczuk in Fl. SSSR, 21 (1954) 668, 399, descr.; Fl. Kirgiz. 9 (1960) 150; Fl. Kazakhst. 7 (1964) 436. —Z. *tomentosa* Juz. in Fl. SSSR, 21 (1954) 667, 399; Fl. Kirgiz. 9 (1960) 150. —Z. *pulchella* Pavl. in Vestn. AN KazSSR, 8 (1954) 133. —Z. *clinopodioides* Lam. var. *media* Benth. in Labiat. gen. et sp. (1833) 321; Sapozhn. Mong. Alt. (1911) 381. —Z. *clinopodioides* auct. non L.: Grubov, Konsp. fl. MNR (1955) 239, p.p.

Described from Mid. Asia (Alay valley). Type in Leningrad.

Rocky and stony slopes of knolls and mountains, floors of gullies.

IA. Mongolia: *Mong. Alt.* (Tsitsirin-Ama, on bank of hill brook, July 9; Tsitsirin-Gol, between stones on valley floor and on slopes, July 10; Taishiri-Ula, at foot of mountains, July 15—1877, Pot.; before reaching Dzasaktu-khana camp, along margin of dry river-bed, July 14; in steppe between Olonnor and Taishiri-Ola mountains in small ravine, July 17—1894; in mountains between Naryn and Shadzagai rivers, July 20; Tashilta river bank, righ tributary of Bulugun, July 25; hill slope near Dzurkhe river, one peak of Tsagan-Gol, slope covered with sparse larch forest, July 30—1898, Klem.; Khara-Dzarga mountain range, Sakhir-Sala river valley, on coastal pebble bed, Aug. 22, 1930—Pob.; exposed solonetzes in Bodonchi river valley, Sept. 25, 1930—Bar.; Khalyun area, nor. slope in forest belt, steep rocky slope, Aug. 24, 1943; nor. trail of Taishiri-Ul, 7–8 km east-south-east of ajmak centre [administrative territorial unit in Mongolia], upper part of trail, along sandy floor of gully, July 11, 1945; nor. trail of Bus-Khairkhan mountain range, small ravine on upper third of trail, July 17; south. flank of Indertin-Gol river valley, in summer camp in somon, hill steppe, July 24—1947, Yun.).

IIA. Junggar: *Tien Shan* (in Kokkamyr mountains, 2100–2400 m, July 31, 1878; Kara-Gol, in pass near Nilka, 3000 m, June 17, 1879 —A. Reg.). *Jung. Gobi* (Tukhumyin-Khundei valley, on knolls and small ravines, Aug. 9, 1947—Yun.; Baitak-Bogdo-Nuru mountain range, nor. slope, upper Ulistu-Gola gorge, 7 km from estuary, large creek valley on left, rocky and stony north-exposed slopes, Sept. 18, 1948—Grub.).

General distribution: Mid. Asia (West. Tien Shan, Pamiro-Alay).

4. **Z. tenuior** L. Sp. pl. (1753) 21; Krylov, Fl. Zap. Sib. 9 (1937) 2375; Juzepczuk in Fl. SSSR, 21 (1954) 408; Fl. Kirgiz. 9 (1960) 151; Fl. Kazakhst. 7 (1964) 437. —Ic.: Fl. SSSR, 21, Plate 20, fig. 2.

Described from Siberia. Type in London (Linn.).

Rocky and rubble slopes of mountains, steppefied-low-mountains areas, along river banks.

IIA. Junggar: *Tarb.* (nor. of Dachen [Chuguchak], south. spur of Tarbagatai mountain range, No. 1714, Aug. 14, 1957—Kuan). *Tien Shan* (Khanakhai 900–1200 m, June 15, 1879—A. Reg.).

General distribution: Aralo-Casp., Balkh. region, Jung.-Tarb., Nor. and Cent. Tien Shan; Europe, Balk.-Asia Minor, Fore Asia, Caucasus, Mid. Asia, West. Sib. (south.).

### 33. Antonina Vved.
in Bot. mater. Gerb. Inst. bot. AN UzbSSR, 16 (1961) 16.

1. **A. debilis** (Bunge) Vved. in Bot. mater. Gerb. Inst. bot. AN UzbSSR, 16 (1961) 16; Fl. Kazakhst. 7 (1964) 438. —*Thymus debilis* Bunge in Ledeb. Fl. alt. 2 (1830) 391. —*Calamintha annua* Schrenk in Bull. Ac. Sci. St.-Pétersb. 10 (1842) 353. —*C. debilis* (Bunge) Benth. in DC. Prodr. 12 (1848) 232; Borissova in Fl. SSSR, 20 (1954) 430. —*Satureja debilis* Briq. in Engler-Prantl, Naturl. Pflanzenfam. IV, 3a (1897) 302; Krylov, Fl. Zap. Sib. 9 (1937) 2378. —*S. annua* Briq. l.c. —Ic.: Ledeb. Ic. pl. fl. ross. 5, tab. 438 (sub nom. *Thymus debilis*); Fl. Uzbek. 5, Plate 34, fig. 2; Fl. Kazakhst. 7, Plate 50, fig. 4.

Described from Altay. Type in Leningrad.

Rocky slopes of mountains, rocks and talus, shade of rocks and trees in midbelt.

IIA. Junggar: *Tien Shan* (Issykul'-Muzart, 2100–2400 m, Aug. 1877; Khorgos, 1200–1500 m, May 15, 1878; Mongoto, 2700 m, July 4, 1879—A. Reg.). *Jung. Gobi* (Sulyugou area, 75 km north-east of Shatszge state farm [Khobuka lowland], along road to Din'syan, *Anabasis salsa-nanophyte* desert, July 12, 1959—Yun. et al.).

General distribution: Fore Balkh., Jung.-Tarb., Nor. Tien Shan; Caucasus (all regions excluding West. Transcaucasus), Mid. Asia (west. Tien Shan, Pamiro-Alay), West. Sib. (south.).

### 34. Hyssopus L.
Sp. pl. (1753) 569; Borissova in Bot. mater. Gerb. Bot. inst. AN SSSR, 12 (1950) 251.

1. Leaves, bracts and calyx teeth with long subulate cusp; corolla about 11 mm long ...................................................... 1. **H. cuspidatus** Boriss.

+ Leaves, bracts and calyx teeth without subulate cusp; corolla 13–15 mm long ...................................................... 2. **H. macranthus** Boriss.

1. **H. cuspidatus** Boriss. in Bot. mater. Gerb. Bot. inst. AN SSSR, 12 (1950) 256; Borissova in Fl. SSSR, 21 (1954) 455; Grubov, Konsp. fl. MNR (1955) 239; Fl. Kazakhst. 7 (1964) 441. —*H. cuspidatus* Boriss. var. *albiflorus* Wu et Li in Acta Phytotax. Sin. 10, 3 (1965) 229. —Ic.: Borissova, l.c. (1950) figs. 1 and 5; Fl. SSSR, 21, Plate 24, fig. 1; Fl. Kazakhst. 7, Plate 50, fig. 7.

Described from East. Kazakhstan (Kurchum river). Type in Leningrad.

Rocky and stony slopes in steppe belt, rocks, ravines and dry riverbeds, up to 2000 m.

IA. **Mongolia**: *Mong. Alt.* (in mountains between Moga and Bidzhiin rivers, Aug. 6, 1896; mountain slope near Dzurkhe river, one of Tsagan-Gol peaks, slope covered with sparse larch forest, July 30; on knoll, near Bulugun river bank, Aug. 4—1898, Klem.; demolished mountain between Adzhi-Bogdo and Altay, in ravine, Aug. 10, 1947—Yun.; Bulugun river valley above somon, complex meadows, July-Aug. 1947—Tarasov).

IIA. **Junggar**: *Cis-Alt.* (Qinhe, on exposed slope, No. 924, Aug. 1; 15 km from Qinhe, on dry rocky slopes, 1400 m, No. 1362, Aug. 1; same site, 1700 m, No. 865, Aug. 2; same site, No. 1257, Aug. 3—1956, Ching; 20 km north-west of Shara-Sume, scrub-meadow steppe on granite eluvium, July 7, 1959—Yun. et al.; in Altay mountain range, on exposed slope, 1200 m, No. 10652, July 16, 1959—Lee et al.). *Tarb.* (Dachen, Tarbagatai, on shaded slope, No. 2899, Aug. 12; same site, along irrigation canal, No. 1693, Aug. 14, same site, on shaded slope, No. 1700, Aug. 14—1957, Kuan). *Jung. Alt.* (ascent to Kuzyun' pass, rocky site, Aug. 2, 1908—B. Fedtschenko). *Jung. Gobi* (Oshigiin-Usu area, on smoothened granite hillocky area, July 30, 1947—Yun.; 17 km west-nor.-west of Ubchugiin-Gol source, on rocks in low isolated hills, Sept. 9; Barlagiin-Gol dry river-bed, near Bodonchiin-Baishing road, on peble-rocky floor, Sept. 9; Baitak-Bogdo-Nuru mountain range, Takhiltu-Ula, left creek valley of Ulyastu-Gol, 7 km from mouth of gorge, on nor. slope, 2000 m, Sept. 17; same site, nor. slope of Ulyastu-Gol gorge, 3–4 km from estuary, along flanks and terrace above floodplain, Sept. 18—1948, Grub.; on Shara-Sume—Karama road, in ravine, No. 10805, July 30, 1959—Lee et al.).

General distribution: Jung.-Tarb.; West. Sib. (Altay-Narym mountain range).

2. **H. macranthus** Boriss. in Bot. mater. Gerb. Bot. inst. AN SSSR, 12 (1950) 260; Borissova in Fl. SSSR, 21 (1954) 457; Fl. Kazakhst. 7 (1964) 442. —*H. ambiguus* (Trautv.) Iljin in Prokhorov and Lebedev, Dushistye rasteniya Altaya [Aromatic Plants of Altay] (1932) 35, p.p.; Krylov, Fl. Zap. Sib. 9 (1937) 2380, p.p. —*H. latilabiatus* Wu et Li in Acta Phytotax. Sin. 10, 3 (1965) 229. —*H. officinalis* auct. non L.: Danguy in Bull. Mus. nat. hist. natur. 20 (1914) 83. —*H. officinalis* var. *ambigua* auct. non Trautv.: Sapozhn. Mong. Alt. (1911) 381. —*H. cuspidatus* auct. non Boriss.: Grubov in Bot. mater. Gerb. Bot. inst. AN SSSR, 19 (1959) 549. —Ic.: Borissova, l.c. (1950) illustr. 3, fig. 1, a–e; Fl. SSSR, 21, Plate 25, fig. 2.

Described from Kazakhstan (Akmolinsk). Type in Leningrad.

Rocky and rubble slopes and steppe formations, rocks, coastal pebble beds and along dry river-beds.

IIA. **Junggar**: *Tarb.* (Urta-Ulasty river, nor. foothill of Saur, steppe, June 18, 1908—Sap.). *Jung. Alt.* (Taidzhal hills [Maili range], near brow of dry ravine, around 1200 m, July 19, 1953

—Mois.; Toli-Uty, intermontane plane, No. 894, Aug. 3, 1957—Kuan; south-east. margin of Dzhair mountain range, 8 km south-west of Boguta brook, along Otu and Chuguchak road to Aktama, desert-steppe belt, along gully, Aug. 3; south-west. Maili mountain range, 40–42 km from meteorologial station at Junggar exist to Karaganda-Daban pass, mountain-steppe belt, on rubble south. slope, Aug. 14—1957, Yun. et al.). *Tien Shan* (in Dzhergalan tributaries, 1800 m, 1874—Larionov). *Jung. Gobi* (west. extremity of Junggar desert, along Chineitszy-Chuguchak road, about 600 m, Aug. 5, 1951—Mois.; Tien Shan Laoba-Myaoergou, steppe, No. 2375, Aug. 3, 1957—Kuan [isotype *H. latilabiatus* Wu et Li]; "montagnes pres de l'Ebi-Nor, No. 1263, July 30, 1895, Chaff."—Danguy, l.c.).

General distribution: Aralo-Casp., Fore Balkh., Jung.-Tarb.; West. Sib. (south.).

## 35. Origanum L.
Sp. pl. (1753) 588.

1. **O. vulgare** L. Sp. pl. (1753) 590; Franch. Pl. David. 1 (1884) 235; Forbes and Hemsley, Index Fl. Sin. 2 (1902) 282; Danguy in Bull. Mus. nat. hist. natur. 20 (1914) 83; Kudo in Mém. Fac. Sci. and Agr. Taihoku Univ. 2 (1929) 90; Krylov, Fl. Zap. Sib. 9 (1937) 2381; Persson in Bot. notiser (1938) 300; Hao in Engler's Bot. Jahrb. 68 (1938) 635; Borissova in Fl. SSSR, 21 (1954) 464; Svenson in Brittonia, 8, 1 (1954) 58; Grubov, Konsp. fl. MNR (1955) 239; Fl. Kirgiz. 9 (1960) 159; Fl. Kazakhst. 7 (1964) 444. —Ic.: Syreishch. Fl. Mosk. gub. 3, 77; Hegi, Ill. Fl. 5, 4 tab. 229, 4, fig. 3219.

Described from Europe. Type in London (Linn.).

Rocky steppe slopes, forest fringes in upper mountain belt.

IA. Mongolia: *Khobd.* (between border and Khobdo, 1870—Kalning). *East. Mong.* ("Vallée du Kéroulen, June 1896, Chaff." Danguy, l.c.).

IB. Kashgar: *East.* (Turfan, Sept. 29, 1879—A. Reg.).

IIA. Junggar: *Cis-Alt.* (in Altay mountains, on exposed mountain slope, 1700 m, No. 10726, July 21, 1959—Lee et al., *Tarb.* (Kzyl-Kungei, Nos. 1524, 1542, Aug. 12; same site, No. 1698, Aug. 14—1957, Kuan). *Jung. Alt.* (Toli district, Albakzin mountains, on slope, No. 2501, Aug. 5; same site, Barktok—Arba-Kezen', Nos. 1176, 1334, Aug. 7—1957, Kuan). *Tien Shan* (forest zone in upper Kunges, 3500 m, Sept. 6, 1876; same site, June 29, 1877—Przew.; Talki gorge, July 19; along Muzart, 1650–2100 m, Aug. 15—1877; in Kokkamyr hills, 1800–2100 m, July 27 , 1878; near Nilka on Kash river, 2100 m, June 8; between Dzhirgalan and Aryslyn rivers, 2400 m, July 8—1879, A. Reg.; Urtas-Aksu, June 17; Talki, 1950 m, July 10; Sharaboguchi, Aug.—1878, Fet.; 1 km south of mines in Dzhagastai, No. 671, Aug. 7; Iliisk Dzhagastai, on slope, No. 3144, Aug. 7; same site, 1940 m, No. 733, Aug, 8; 3 km north of Chzhaosu, No. 874, Aug. 13; Chzhaosu-Tekes, on slope, No. 3596, Aug. 16; 20 km north of Ulastai, on slope, No. 3860, Aug. 28; Tsitai district, steppe, No. 4429, Sept. 22—1957, Kuan; south. slope of Boro-Khoro mountain range, near N. Ortai settlement on Ili—Sairam-Nur road, forest belt, grassy spruce thicket, Aug. 19; Ketmen' mountain range, Sarbushin pass on Ili—Kzyl-Kure road, flat trough of pass, steppe meadow, Aug. 23—1957; on Narat mountain along Kungesu, in shade on slope, 2400 m, Aug. 7; left bank of Kunges, Tsanma river valley, its tributary Dagdyn-Gol, subalp. belt, burnet meadow with *Salix* groves, Aug. 8—1958, Yun. et al.; "Kunges, Jailo, ca. 2200 m, Aug. 14, 1932"—Persson, l.c.).

General distribution: Aralo-Casp., Fore Balkh., Jung.-Tarb., Nor. and Cent. Tien Shan; Europe, Mediterranean, Fore Asia, Caucasus, Mid. Asia, West. Sib., Far East, China (Cent., East., South-West.). Himalayas (west.), Japan.

Note. Svenson (l.c.) refers to this species growing somewhere between Lhasa and Darjeeling without citing precise location. This plant was most likely collected on the south. slope of the Himalayas.

## 36. Thymus L.[3]
Sp. pl. (1753) 590.

1. Inflorescence elongated with distinctly separated verticils; leaves subsessile ............................................ 4. **Th. marschallianus** Willd.
+ Inflorescence capitate, leaves petiolate ............................................ 2.
2. Stem distinctly tetrahedral, pubescent only on 2 opposite sides alternately from one to another internode ............................................ ............................................................. 3. **Th. kitagawianus** Tschern.
+ Stem indistinctly tetrahedral, uniformly pubescent on all sides. . 3.
3. Leaves with long scattered articulate hairs on upper surface, glabrous beneath ........................................ 2. **Th. gobicus** Tschern.
+ Leaves glabrous on both surfaces, only bracts sometimes with short pubescence at base on upper surface ........................................ 4.
4. Plant bushy, totally devoid of procumbent vegetative shoots, with ascending, highly branched stems; stems with short pubescence throughout length; teeth on upper calyx lip with or without well-developed cilia along margin. ........................................ 5.
+ Plant not bushy, with decumbent stems and developed procumbent vegetative shoots; stems with more or less long pubescence under inflorescence; teeth of upper calyx lip invariably with short setae along margin ........................................ 7.
5. Teeth of upper calyx lip with well-developed cilia; petiole of caulous leaves invariably with cilia reaching base of blade ........................................ ............................................................. 6. **Th. petraeus** Serg.
+ Teeth of upper calyx lip without cilia; petiole of caulous leaves either totally devoid of cilia or with 1–2 pairs of cilia ........................ 6.
6. Caulous leaves obovate or spatulate, 4–9 mm long, 2–3.5 mm broad, broadest most often above middle, with few cilia along petiole; leafy bracts with cilia along margin reaching midleaf ........................................ ............................................................. 8. **Th. roseus** Schipcz.
+ Caulous leaves oblong or oblong-elliptical, 4–10 mm long, 1–2 mm broad, with/without pair of cilia along petiole; leafy bracts with/without 1–2 pairs of cilia. ........................ 7. **Th. rasitatus** Klok.

---

[3]Thymes of affinity *Th. serpyllum* s.l. are highly polymorphic and form a large number of closely related races in Mid. Asia as well as West. Sib. In our treatment, we have attempted to cover all diverse races reported in our territory. It was not possible for us, however, nor was it our objective at present, to compare and evaluate the entire range of variation of pubescence of stems, leaves and calyx teeth throughout the distribution range of the genus in order to critically relate them to known species.

7. All leaves ovate, with numerous cilia on petiole; leafy bracts with cilia reaching midblade ................... 1. **Th. altaicus** Klok. et Schost.

+ Middle caulous leaves oblong-lanceolate, lower smaller than middle leaves and with broader blade (heterophylly), petiole with 1–3 pairs of cilia; leafy bracts with cilia only at base ............................ .................................................5. **Th. mongolicus** (Ronnig.) Ronnig.

1. **Th. altaicus** Klok. et Schost. in Zhurn. Inst. bot. AN UkSSR, 10/18 (1936) 159; Klokov in Fl. SSSR, 21 (1954) 540; Fl. Kazakhst. 7 (1964) 451. — *Th. altaicus* Serg. in Sist. zam. Gerb. Tomsk. univ. 6–7 (1937) 12; Krylov, Fl. Zap. Sib. 9 (1937) 2387. —*Th. serpyllum* auct. non L.: Grubov, Konsp. fl. MNR (1955) 239, p.p.

Described from Altay. Type in Khar'kov. Isotype in Leningrad. Rubble and rocky slopes in steppes, coastal pebble beds.

IA. **Mongolia:** *Khobd.* (Burgassutai river, in Uryuk-Nor lake basin, on pebble bed near brook, June 21; same site, Kharkhira river valley, July 10—1879, Pot.; steppe valley on left bank of Khingel'tsik river near confluence of Turun river, Aug. 29, 1895—Klem.; Kharkhira mountain group, Kendulyun river, July 27, 1903—Gr.-Grzh.; descent into Ubsa-Nor lake basin from Ulan-Daba pass, conglomerate exposures, Sept. 19, 1931—Bar.).

IIA. **Junggar;** *Tarb.* (Khohuk river valley, desert-rubble slopes, July 20, 1914 —Sap.).

**General distribution:** West. Sib. (Altay).

Note. Characterised by densely foliated vegetative shoots; patently pilose stems under inflorescence; broadly ovate leafy bracts; cilia on both sides up to middle; patently pilose calyx; hirtellous teeth of upper calyx lip; leaves caulous ovate; petioles with long cilia on both sides reaching leaf blade, sometimes even on it.

2. **Th. gobicus** Tschern. sp. nova. —*Th. serpyllum* auct. non. L.: Grubov, Konsp. fl. MNR (1955) 239, p.p. —*Th. nerczensis* auct. non Klok.: Hanelt und Davažamc in Feddes repert. 70 (1965) 55.

Suffruticulus, trunculis procumbentibus, non radicantibus; rami floriferi toti regulariter patenter pilosi, sub inflorescentia tantum densius pilosi; folia petiolata, caulina et floralia oblongo-lanceolate, 10–12 mm lg., 2–3 mm lt., caulina inferiora minora, elliptica, omnia utrinque atro-glandulosa, subtus glabra, supra pilis multicellularibus longis solitariis per totam paginam dispersis vel in parte inferiore congregatis (in foliis surculorum sterilium praecipue) praedita, petiolis margine dense ciliatis. Inflorescentia capitata, pedicellis 1–2 mm lg., florum inferiorum ad 3 mm lg., regulariter breviter pilosis; calyx sub anthesi 4.4–5 mm lg., dense patenter pilosus, labio inferiore densius piloso, dentibus labii superioris triangularibus, margine dense ciliolatis; corollae 7 mm lg. roseae.

Typus: Jugum Gurban-Sajchan, zona montana inferior ad declivitatam lapidosam, in schistosis unacum *Artemisia procera*, Aug. 22, 1943, Junatov; in Herb. Inst. Bot. Acad. Sci. URSS (Leningrad) conservatur.

Affinitas. Ab omnibus speciebus *Thymi* in regionibus nostris obvenientibus foliis pilis longis multicellularibus supra per totam paginam dispersis, solitariis vel in parte inferiore congregatis differt.

Rocks, rocky and rubble slopes and talus from lower to upper mountain belt, desert steppes, limestone outcrops, coastal pebble beds, semifixed and fixed sand.

IA. Mongolia: *Mong. Alt.* (Khan-Taishiri mountain range, Khabchigiin-Daba pass, 2275 m, on Yusun-Bulak—Tonkhil'—Daba road, sheep's fescue-forb mountain steppe, Sept. 3, 1948—Grub.). *Cent. Khalkha* (in valley on right bank of Ongiin river, among rocks, July 25; same site, on fine rubble, July 27—1893; sand knolls in Baranchin area, June 23, 1895—Klem.; between Urga and Kherulen, hill near Dzhergalantu pass; same site, near Berlik mountain July 1899—Pal.; along Mandalyn-Gobi—Khara-Tologoi road, June 8, 1909—Czet.; water divide between Ara-Dzhirgalante and Ubur-Dzhirgalante rivers, rubble sections, Aug. 10; steppe zone on left bank of Ubur-Dzhirgalante river, rubble slopes of Agit mountain, Aug. 22; same site, in upper courses, sand of second terrace, Sept. 13—1925, Krasch. and Zam.; rocks and stony steppe near Choiren, July 1, 1926—Kondr.; Khukhu-khoshun, rocky slope, July 24, 1926—Lis.; south-east. extremity of Khangai mountain range, Kholt area and ridges, Aug. 16, 1926—Glag.; environs of Ikhe-Tukhum-Nor lake, Mishik-Gun, Ulain-Khoshu hill, 1926; from Choiren to Naran area, on Sair-Usinsk road, Sept. 7, 1927—Zam.; Choiren, July 22, 1928—Tug.; 5 km east of Choiren-Ul, rubble feather-grass steppe, Aug. 22, 1940—Yun.; Ubur-Dzhirgalante area, rocky slope exposed westward, July 15, 1925—Gus.; Arbai-Khere—Ulan-Bator road, 15 km east of Ul'dzeit somon, on rocky crest of knoll, June 19; Ongiin-Gol river floodplain near crossing on Arbai-Khere—Khuchzhirte road, on pebble bed, June 20; Ulan-Bator—Tsetserlik road, sand on south. extremity of Tsagan-Nor lake lowland, June 25—1948, Grub.; midcourse of Ara-Dzhirgalante-Gol, hilly sheep's fescue-forb steppe, July 2; water divide of Ubur and Ara-Dzhir-galante rivers, hummocky sand overgrown with willow grove, extending onto swampy meadow, July 2—1949, Yun.; "Offene Schotterfelder innerhalb der Anwiesen am Ongin-gol nördl. Arbaicher"—Hanelt und Davažamc, l.c.). *Bas. Lakes* (Tutu river, on limestone exposures, July 18; Khudzhirte, on granite rocks, July 19; Shuryk river, on rocks and in dry depressions, July 23—1877, Pot.; on Tuz-tag hill, desert rocky slopes, July 2, 1892—Kryl.). *Val. Lakes* (Tui river, near Sharagol'dzhyut post office, Sept, 9, 1886—Pot.; on small hillock on right bank of Sharagol'dzhyut, among rocks, July 5, 1893; on hills between Ut and Baidarik rivers, June 15, 1894—Klem.; 30 km west of Tatsain-Gol on south. Hangay road, wheat grass-chee grass dry steppe, June 28; 5–7 km south of Targatu somon, along road to Guchin-Usu somon, chee grass desert steppe, July 4—1941, Tsatsenkin). *Gobi-Alt.* (south. slope of Ubten-Daban pass, Aug. 30, 1886—Pot.; Dundu-Saikhan hills, on rock ridges, July 7, 1909—Czet.; Baga-Bogdo, upper terrace and rocky canyon borders at 1800 m, 1925—Chaney; rocks of Ikhe-Bogdo mountain range spurs, June 28, 1926—Kozlova; Ikhe-Bogdo, Bityuten-Ama, mountain slopes, Aug. 12, 1927—M. Simukova; Bain-Tsagan mountains, Subulyur creek valley, Aug. 5; near rocks on Dundu-Saikhan mountain slope, Aug. 16—1931, Ik.-Gal.; Dundu-and Dzun-Saikhan mountain ranges, slopes of hills and gorges from trail to upper mountain belt, July–Aug. 1933—M. Simukova; pass between Dundu-and Dzun-Saikhan, steep rocky slope to gully, 1 km north of pass, July 22; Gurban-Saikhan mountain range, lower hill belt, rocky slope, talus with *Artemisia procera* shrubs, Aug. 22, 1943—Yun., typus!; Ikhe-Bogdo mountain range, south. slope, lower part of Narin-Khurimt creek valley, rubble talus slope, Sept. 5, 1943; Dzun-Saikhan mountain range, middle and lower mountain belts, June 19; Ikhe-Bogdo mountain range, upper part of Bityuten-Ama creek valley, steppe belt, juniper thickets on south. rocky slope, July 1—1945, Yun.; Ikhe-Bogdo mountain range, south-east. slope of Narin-Khurimt gorge, east. flank, on rocks about 2900 m, July 28, 1948—Grub.; rock slide on south. slope of Dundu-Saikhan mountain range, July 20, 1950—Kal.).

General distribution: Nor. Mong. (Hang., Mong.-Daur.).

Note. Characteristic Mongolian species distributed quite extensively. Varies in leaf size as well as degree of pubescence of calyx but variability of these characteristics is not significant.

In Gobi Altay, *Th. gobicus* is found high in mountains on rocky talus and rocks; more northward, within Cent. Khalkha, it grows on fixed and semi-fixed sand, pebble beds along rivers, in sections of rubble steppe. Might even be found on limestone exposures. All of this points to fairly broad ecological adaptability of the species. Of all thymes found in our territory, this species is well distinguished in leaves pubescent on upper surface with lax multicellular long hairs, sometimes even randomly scattered all over leaf blade surface.

In Cent. Khalkha, eastern part of the distribution range, plants falling betwen *Th. gobicus* and *Th. kitagawianus* are found; the leaves in these plants are narrower than in *Th. gobicus;* stem pubescence is very similar to that of *Th. kitagawianus* but without distinct succession of pubescent by non-pubescent faces from one to another internode.

**3. Th. kitagawianus** Tschern. nom. nov. —*Th. asiaticus* Kitag. Lin. Fl. Mansh. (1939) 388, non Serg. (1937). —*Th. serpyllum* L. var. *asiaticus* Kitag. in Rep. First Sci. Exped. Manchoukhudo, 4, 4 (1936) 92. —*Th. serpyllum* var. *angustifolius* auct. non Ledeb.: Danguy in Bull. Mus. nat. hist. natur. 20 (1914) 83.

Described from China (Dunbei). Type in Tokyo (TI).

Thin and fixed sand, sand steppes, sand-covered slopes of knolls in steppes.

IA. **Mongolia:** *East. Mong.* (between Kulusutaevsk settlement and Dolon-Nor, 1870—Lom.; Kulun-Buir-nor plain, Elisyn-Khuduk collective, dry sandy soil, June 5; Khaligakha area, dry sandy soil, June 23; syrts on Abdar river, June 25; syrt on Bilyutai hill, July 4—1899, Pot. and Sold.; Khailar, July 6, 1901—Lipsky; near Kharkhonte railway station, sand, June 7, 1902—Litw.; Manchuria station, steppe, Aug. 26; Khailar station, pine grove on sand-dunes, Aug. 28—1925, Gordeev; Khailar town, Sishan' hills, on sand, June 7; Manchuria station, Beishan' hill, on rubble, June 23—Li S.H. et al. (1951); Khailar town, sand knoll, 1954—Wang; Khailar town, true steppe, 1959—Ivan.; "Kailar, steppe sablonneuse, 700 m, June 20, 1896, Chaff."—Danguy, l.c.).

General distribution: Nor. Mong. (Hent., Mong.-Daur.—very rare), China (Dunbei).

Note. Our plant is evidently the same as *Th. asiaticus* Kitag. described from border regions of Dunbei. However, since the epithet *asiaticus* has already been used in this genus, we have renamed this species. It is closely related to *Th. disjunctus* Klok. but the leaves in the latter are dentate. The ecology of these species also differs. *Th. disjunctus* is found on rocks, more rarely on dunes, while our species is mainly confined to sand or sandy steppes.

**4. Th. marschallianus** Willd. Sp. pl. 3 (1800) 1141; Krylov, Fl. Zap. Sib. 9 (1937) 2391; Klokov in Fl. SSSR, 21 (1954) 511; Fl. Kirgiz. 9 (1960) 161; Fl. Kazakhst. 7 (1964) 447. —**Ic.:** Fl. Kazakhst. 7, Plate 51, fig. 1.

Described from East. Europe (south. Ukraine). Type in Berlin.

Steppe and meadow slopes, coastal meadows and pebble beds, tugais in foothills and midbelt of mountains.

IIA. **Junggar:** *Cis-Alt.* (Kaba river, near Kaba village, tugai, June 16, 1914—Schischk; Altay [Shara-Sume], No. 2428, Aug. 26; same site, in wet lowland, No. 2687, Sept. 3; 120 km nor. of Burchum, No. 3098 , Sept. 14—1956, Ching). *Tarb.,* (nor. of Dachen, on slope, No. 2903, Aug. 12, 1957—Kuan). *Tien Shan* (Arshan river, June [1875]—Larionov; Tekes river bank, 1350–1500 m, Aug. 13, 1877; nor. of Chapchal pass, 1500 m, June 23; Chapchal pass, 1500–1800 m, June 28; Talki—1878, A. Reg.; Urtas-Aksu, June 27, 1878—Fet.; 1 km south of Dzhagastai, 1790 m, No. 669, Aug. 7; Chzhaosu, environs of horse-breeding centre, intermontane basin, No.

3242, Aug. 11; Syate in Chzhaosu district, No. 1327, Aug. 11; same site, Nos. 776, 846, 867—Aug. 11, 1957, Kuan). *Balkhash-Alak.* (Chuguchak, No. 1521, Aug. 12, 1957, Kuan).

General distribution: Aralo-Casp., Fore Balkh., Jung.-Tarb., Nor. and Cent. Tien Shan; Europe, Balkans (?), Caucasus, Mid. Asia, West. Sib.

Note. Hybrids are found between *Th. marschallianus* Willd. and *Th. mongolicus* (Ronnig.) Ronnig. These plants have somewhat interrupted inflorescence, fairly large petiolate leaves broader than in *Th. marschallianus*, and ascending or suberect vegetative shoots. Such plants are found in upper Kunges and along Tsanma river (1877, Przew.), Kurdai pass (1907, Merzbacher) and in Urumchi in the upper reaches of the Tasengou river (1929, Popov) in Tien Shan.

5. **Th. mongolicus** (Ronnig.) Ronning. in Acta Horti Gotoburg. 9 (1934) 99. —*Th. serpyllum* L. var. *mongolicus* Ronning. in Notizbl. Bot. Gart. und Mus. Berlin-Dahlem, 10 (1930) 890; Rehder and Kobuski in J. Arn. Arb. 14 (1933) 31; Walker in Contribs U.S. Nat. Herb. 28 (1941) 657. —*Th. asiaticus* Serg. in Sist. zam. Gerb. Tomsk. univ. 6–7 (1937) 1; Krylov, Fl. Zap. Sib. 9 (1937) 2286; Klokov in Fl. SSSR, 21 (1954) 535; Fl. Kazakhst. 7 (1964) 450. —*Th. mongolicus* Klok. in Bot. mater. Gerb. Bot. inst. AN SSSR, 16 (1954) 311. —*Th. serpyllum* var. *angustifolia* auct. non Ledeb.: Franch. Pl. David. 1 (1884) 235. —*Th. serpyllum* auct. non L.: Forbes and Hemsley, Index Fl. Sin. 2 (1902) 282, p.p.; Hao in Engler's Bot. Jahrb. 68 (1938) 635. —Ic.: Fl. Kazakhst. 7, Plate 51, fig. 4 (sub nom. *Th. asiaticus* Serg.).

Described from East. Kazakhstan (Balkhash region). Lectotype in Vienna. Isolectotype in Leningrad.

Sandy steppes, thin sand, sandy-pebble river banks and gully floors, rocky-rubble slopes of knolls in steppes, low hills and midbelt of mountains.

IA. Mongolia: East. Mong. (in locis subarenosis Mongoliae Chinensis, 1831 (? Kuznetzov], from Turchaninov herbarium; Sume-Khada mountains June 7; Muni-Ula mountains, June 30—1871; same site, July 6, 1873—Przew.; Dzhungor royal camp, Aug. 15, 1884—Pot.; 10 km nor. of Khukh-Khoto town, nor. foothill of Datsin'shan', dry chee grass-lyme grass forb steppe, 1300 m, June 4, 1958—Petr.; Ourato, June 1866—David; "Oulachan, No. 2654, June, David"—Franch, l.c.). *Alash. Gobi* (Alashan mountain range, Tszosto gorge, on dry river-bed, pebble bed, May 11; same site, Khote-Gol gorge, in sandy soil of small ravines, May 11; same site, river-bed, among thickets, June 18—1908, Czet.; ? "Ho Lan Shan, in dense *Picea* forests, Ching"—Walker, l.c.). *Ordos* (70 km south of Khanchin town, on ridge peak, red sandstone exposures, Aug. 5; 35 km south-east of Khanchin town, sandy-pebble upland plain, Aug. 7—1957, Petr.).

IIA. Junggar: *Cis-Alt.* (Altay [Shara-Sume], 1100 m, No. 2433, Aug. 26, 1956—Ching; 20 km north-west of Shara-Sume, scrub meadow steppe on granite eluvium, July 7, 1959—Yun. et al.; in Altay hills, on exposed slope, 1200–1500 m, No. 10653, July 16, 1959—Lee et al.). *Tarb.* (Saur mountain range, south. slope, Karagaitu river valley 3 km beyond its discharge on the trail, poplar valley forest on pebble beds, June 23, 1957—Yun. et al.; Chuguchak region, 2250 m, No. 1589, Aug. 13, 1957—Kuan). *Jung. Alt.* (nor. bank of Sairam lake, exposed slope, 2800 m, No. 2129, Aug. 28, 1957—Kuan; ?oberstes Dunde-Kelde Tal und oberstes Chustai Tal, July 5–6, 1908—Merzb.). *Tien Shan* (Sairam lake, July 1877—A. Reg.; same site, July 1878—Fet.; Kash river, near Nilka, 2100 m, June 8; Kara-Gol river, pass near Nilka, 3000 m, June 17, on Aryslyn river, 2400–2700 m, July 8; same site, July 10; same site, July 15—1879, A. Reg.; Urumchi, Nanshan, No. 539, July 21, 1956—Ching; in M. Yuldus ulas, pebble bed floodplain, 2500 m,

No. 6327, Aug. 2, 1958—Lee and Chu; Myaoergou district, on slope, No. 2435, Aug. 4; same site, on peak of hill, No. 1086, Aug. 6; same site, on slope, 2400 m, No. 1219, Aug. 6; same site, No. 2645, Aug. 6—1957, Kuan).

General distribution: Fore Balkh., West. Sib. (south-east.), China (North-West).

Note. *Th. mongolicus* was described from several specimens without type identification. As lectotype of this species, we selected the plant labelled: "In subalpinis Alatau ad fl. Lepsa, Sarchan et Aksu, No. 1814, 1841, leg. Karelin et Kirilov".

Under *Th. mongolicus*, Ronniger has combined plants differing in the following characteristics; stem with uniform short pubescence of deflexed hairs; leaves of flowering branches petiolate, lower ones oblong-spatulate, with 8–9 mm long, 3 mm broad petiole, upper ones very large, 11–12 mm long, 3-3.5 mm broad, oblong; leaves of vegetative branches oblong, up to 11 mm long, 4 mm broad; all leaves glabrous and except for leafy bracts, slightly pubescent on upper surface, with non-profecting veins (venation camptodrome, with 3–4 pairs of lateral veins), with few cilia at base; calyx 4 mm long, glabrous on back, but pubescent in ventral half, teeth of upper lip shortly deltoid, with fine hairs along margin. Very similar plants were later described by L.P. Sergievskaja under *Th. asiaticus* Serg.

In East. Mongolia, Alash. Gobi and Ordos, plants with somewhat narrower leaves are found; such were probably the plants that M.V. Klokov had in view when he described his *Th. mongolicus* Klok.

6. **Th. petraeus** Serg. in Sist. zam. Gerb. Tomsk. univ. 2 (1937) 5; Krylov, Fl. Zap. Sib. 9 (1937) 2390; Klokov in Fl. SSSR, 21 (1954) 590; Fl. Kazakhst. 7 (1964) 461. —**Ic.:** Fl. Kazakhst. 7, Plate 52, fig. 2.

Described from Altay. Type in Tomsk.

Rocks, granite outcrops, rocky and rubble slopes in steppe belt of mountains.

IIA. Junggar: *Jung. Alt.* (Dzhair mountain range, between Otu and Sardzham, 1500–2000 m, July 2, 1947—Shum.; same site, 1–2 km north of Otu, on road to Dzhair pass and later into Chuguchak, steppe belt, on smoothened granite knolls, Aug. 4; same site, pass on Chuguchak-Aktam road, mountain steppe belt, pillow granites, Aug. 9—1957, Yun. et al.).

General distribution: Fore Balkh., Jung.-Tarb.; West. Sib. (Altay).

Note. Among closely related species, distinguished by comparatively small, dull green leaves and presence of cilia on teeth of upper lip of calyx.

7. **Th. rasitatus** Klok. in Bot. mater. Gerb. Bot. inst. AN SSSR, 16 (1954) 315; Klokov in Fl. SSSR, 21 (1954) 562; Fl. Kazakhst. 7 (1964) 456.

Described from East. Kazakhstan (Bektau-Ata mountains). Type in Leningrad.

Rocky steppe slopes, rock crevices, granite exposures.

IIA. Junggar: *Jung. Alt.* (Ven'tsyuan', 2400 m, No. 1445, Aug. 14, 1957—Kuan). *Tien Shan* (Chzhaosu district, Aksu region, on slope, No. 3488, Aug. 14, 1957—Kuan).

General distribution: Fore Balkh., Jung.-Tarb. (Junggar Alatau); West. Sib. (south-east.).

Note. Species closely related to *Th. roseus* Schipcz., differing primarily in very narrow oblong-lanceolate leaves.

8. **Th. roseus** Schipcz. in Bot. mater. Gerb. Gl. bot. sada, 2, 24–25 (1921) 95; Krylov, Fl. Zap Sib. 9 (1937) 2389; Klokov in Fl. SSSR, 21 (1954) 564; Fl. Kazakhst. 7 (1964) 456. —*Th. serpyllum* auct. non L.: Grubov, Konsp. fl. MNR (1955) 239, p.p. —*Th. serpyllum* var. *vulgaris* auct. non Ledeb.: Sapozhn. Mong. Alt. (1911) 381. —Ic.: Fl. Kazakhst. 7, Plate 52, fig. 3.

Described from East. Kazakhstan (Kandygatai mountains). Type in Leningrad.

Granite rocks, rocky slopes and talus.

IA. **Mongolia:** *Mong. Alt.* (Dain-Gol lake, south-west. bank, July 29, 1908—Sap.).
IIA. **Junggar:** *Cis-Alt.* (Korunduk hill, vicinity of Kurtu river, June 16, 1903—Gr.-Grzh.), *Jung. Gobi* (Qinhe district, Chzhunkhaitsza, 2550 m, Nos. 1133, 1438, Aug. 5–6, 1956—Ching).

General distribution: Fore Balkh., Jung.-Tarb. (Junggar Alatau); West. Sib. (south).

Note. Small subshrub with intensely branched, procumbent, strongly woody stems, short flowering branches, oblong-ovate or spatulate leaves with petioles nearly devoid of cilia or rarely with 1 pair of cilia and glabrous teeth on upper lip of calyx.

## 37. Lycopus L.

Sp. pl. (1753) 21; Henderson in Amer. Midl. Naturalist, 68, 1 (1962) 95.

1. Leaves lanceolate–elliptical or lanceolate; only lower ones pinnatifid at base ..................................................................... 1. **L. europaeus** L.
+ Leaves ovate or elliptical; all leaves deeply pinnatifid. ...................
.............................................................................. 2. **L. exaltatus** L. f.

1. **L. europaeus** L. Sp. pl. (1753) 21; Forbes and Hemsley, Index Fl. Sin. 2 (1902) 282; Krylov, Fl. Zap. Sib. 9 (1937) 2393; Volkova in Fl. SSSR, 21 (1954) 595; Fl. Kirgiz. 9 (1960) 164; Henderson in Amer. Midl. Naturalist, 68, 1 (1962) 133; Fl. Kazakhst. 7 (1964) 462. —Ic.: Henderson, l.c. Fig. 22; Fl. Kazakhst. 7, Plate 53, fig. 2.

Described from Europe. Type in London (Linn.).

Banks of rivers and lakes, near springs, in floodplain meadows, among shrubs, along fringes of swamped forests and in clearances.

IB. **Kashgar:** East. (nor. extremity of Khami desert, Bugas area, near river, Aug. 20; same site, Aug. 23—1895, Rob.).
IIA. **Junggar:** *Jung. Alt.* (Uchtyube, 1200 m, Aug. 1878—A. Reg.). *Jung. Gobi* (Dachuan area, swamp, Aug. 20, 1898—Klem.). *Zaisan* (Belezeka river lower course, tugai, June 18, 1914—Schischk). *Dzhark.* (right bank of Ili river west of Kul'dzha, May 26; Kul'dzha, July 5—1877, A. Reg.).

General distribution: Aralo-Casp., Balkh. region, Jung.-Tarb., Nor. and Cent. Tien Shan; Europe, Balk.-Asia Minor, Caucasus, Mid. Asia, West. Sib., East. Sib., China (North), Himalayas (Kashmir), Japan, Nor. America (introduced).

2. **L. exaltatus** L. f. Suppl. (1781) 87; Krylov, Fl. Zap. Sib. 9 (1937) 2394; Volkova in Fl. SSSR, 21 (1954) 593; Henderson in Amer. Midl. Naturalist, 68, 1 (1962) 130; Fl. Kazakhst. 7 (1964) 461. —Ic.: Henderson, l.c. fig. 20; Fl. Kazakhst. 7, Plate 53, fig. 1.

Described from Italy. Type in London (Linn.).

Along banks of rivers and lakes, in coastal thickets, tugais, more rarely forest fringes.

IIA. **Junggar:** *Cis-Alt.* (Kran river, Aug. 29, 1876—Pot.; Altay [Shara-Sume], 700 m, No. 770, Sept. 6, 1956—Ching). *Tien Shan* (near Kash river, 600–1200 m, Sept. 15, 1878—A. Reg.). *Jung. Gobi* (Savan district, Datsyuan'gou, in ravine, No. 1195, July 4, 1957—Kuan; 3–4 km north of Kuitun settlement along old Shikho--Manas road, sasa zone, forb-sedge marshy meadow, July 6, 1957—Yun. et al.). *Dzhark.* (Kul'dzha, June 1877—A. Reg.). *Balkh.-Alak.* (in Chuguchak region on marsh, No. 2812, Aug. 10; 4 km east of Chuguchak, No. 1482, Aug. 10—1957, Kuan).

**General distribution:** Aralo-Casp., Fore Balkh., Jung.-Tarb.; Europe, Fore Asia, Caucasus, Mid. Asia, West. Sib., East. Sib. (south-west.).

## 38. Mentha L.
Sp. pl. (1753) 576.

1.  Verticils in axils of nearly all caulous leaves or those in upper half of stem; leaves longer than verticils; upper leaf bracts similar to caulous leaves .......................................................... 1. **M. arvensis L.**
+   Verticils aggregated in terminal spicate inflorescences; upper leaf bracts strongly reduced. ................................... 2. **M. asiatica** Boriss.

1. **M. arvensis L.** Sp. pl. (1753) 577; Henderson and Hume, Lahore to Yarkend (1873) 331; Franch. Pl. David. 1 (1884) 235; Forbes and Hemsley, Index Fl. Sin. 2 (1902) 281; Danguy in Bull. Mus. nat. hist. natur. 17 (1911) 345, 20 (1914) 83; Pampanini, Fl. Carac. (1930) 186; Krylov, Fl. Zap. Sib. 9 (1937) 2397; Walker in Contribs U.S. Nat. Herb. 28 (1941) 657; Borissova in Fl. SSSR, 21 (1954) 604; Chen and Chou, Rast. pokrov r. Sulekhe (1957) 90; Grubov in Bot. mater. Gerb. Bot. inst. AN SSSR, 19 (1959) 550; Fl. Kirgiz. 9 (1960) 166; Fl. Kazakhst. 7 (1964) 464.   —*M. austriaca* Jacq. Fl. Austr. 5 (1778) 14; Krylov, Fl. Zap. Sib. 9 (1937) 2396; Grubov, Konsp. fl. MNR (1955) 239.   —*M. arvensis* L. ssp. *haplocalyx* auct. non Briq.: Hao in Engler's Bot. Jahrb. 68 (1938) 635.   —**Ic.:** Fl. SSSR, 21, Plate 32, fig. 3; Fl. Kazakhst. 7, Plate 53, fig. 3.

Described from Europe. Type in London (Linn.).

Coastal meadows, on wet and marshy banks of rivers and lakes, meanders, sometimes in plantations, roadside ditches.

IA. **Mongolia:** *Mong. Alt.* (between border and Khobdo, 1870—Kalning; Khobdosk. state farm in lower Buyantu-Gol river, 1941—Kondratenko). *Cis-Hing.* (Khalkhin-Gol river valley 13 km south-east of Khamar-Daba, willow thickets along meanders, Aug. 11, 1949—Yun.; Siguitu district, Yakshi station, No. 2832, 1954—Wang). *Cen. Khalkha* (Khurkhi river valley, July 22, 1894—Kashk.). *East. Mong.* (Tuchen town ruins, Aug. 2; Ulan-Morin river, in swamp, Aug. 23—1884, Pot.; Khailar on Argun' bank, July 18, 1909—Ivashkevich; Dariganga, Ongon-Elis sand, bank of Boro-Bulak spring, Sept. 13, 1931—Pob.; Khailar town, near solonetz reservoir, No. 2976, 1954—Wang; Shilin-Khoto town, chee grass-wheat grass steppe, 1959—Ivan.; "Sartchy, dans les foessés, July David"—Franch., l.c.). *Bas. Lakes, Ordos* (Huang He valley, way to Tsaidamin -Nor lake, Aug. 2; same site, way to Kharganty temple. Aug. 18—1871,

Przew.; 50 km south of Dzhasak town, meadow with lake amidst sand-dunes, Aug. 17, 1957—Petr.). *Alash. Gobi* ("Ho Lan Shan, No. 1099, Ching"—Walker, l.c.).

IB. Kashgar: *Nor.* (along irrigation ditches near Aksu town, Aug. 9; between Aksu and Kuchei, near Anka-Aryk village on Muzart-Dar'e, Aug. 13; along irrigation ditches near Bugur town, Aug. 20—1929, Pop.), *West.* (Yarkendsk oasis, along irrigation ditches and loessial ploughed fields, July 4, 1889—Rob.; "Kachgar, Oct. 1906, Vaillant"—Danguy, l.c.). *South.* (Sampula oasis, Sept. 9, 1885—Przew.; "on Jarkand plains beyond Sanju, 1200–3000" m. — Henderson and Hume, l.c.). *East.* (south. margin of Khamiisk oasis north of Bugas, Aug. 20, 1895—Rob.; in Toksun, No. 1009, Aug. 18, 1957—Kuan; near Bagrashkul' lake, on lowland, No. 6178, July 26, 1958—Lee and Chu; "Karachar, Sept. 15, 1907, Vaillant"—Danguy, l.c.).

IIA. Junggar: *Tarb.* (along Laty-Tumandy road, Aug. 7, 1876—Pot.; "steppe entre l'Ouchte et l'Irtich, No. 1246, Aug. 22, 1895, Chaff."—Danguy, l.c.). *Jung. Alt., Tien Shan, Jung. Gobi, Dzhark., Balkh.-Alak.*

IIIA. Qinghai: *Nanshan* (South Tetungsk mountain range, Aug. 8, 1880—Przew.; Pabatson' valley, on Tetung river, along banks of irrigation ditches in humus soil, July 24, 1908—Czet.; "Sining-fu, Schang-wu-chuang, um 2900 m, No. 787, Aug. 3"—Hao, l.c.; "Shang Hsin Chuang, No. 684, Ching"—Walker, l.c.).

General distribution: Aralo-Casp., Fore Balkh., Jung.-Tarb., Nor. Tien Shan; Europe, Caucasus, Mid. Asia, West. Sib., East. Sib., Nor. Mong. (Mong.-Daur)., China (North, Cent., East, South-West, South), Himalayas (Kashmir), Korean peninsula, Japan.

2. **M. asiatica** Boriss. in Bot. mater. Gerb. Bot. inst. AN SSSR, 16 (1954) 280; Borissova in Fl. SSSR, 21 (1954) 614; Grubov in Bot. mater. Gerb. Bot. inst. AN SSSR, 19 (1959) 550; Fl. Kazakhst. 7 (1964) 466.  —*M. sylvestris* auct non L.: Strachey, Catal. (1906) 140; Simpson in J. Linn. Soc. London (Bot.) 41 (1912–13) 436; Persson in Bot. notiser (1938) 301; Fl. Kirgiz. 9 (1960) 167.  —*M. longifolia* auct. non Huds.: Krylov, Fl. Zap. Sib. 9 (1937) 2398, p.p.  —Ic.: Fl. Kazakhst. 7, Plate 53, fig. 4.

Described from East. Kazakhstan (Saur mountain range). Type in Leningrad.

River valleys, flooded meadows, brooks, wet sections of mountain slopes, often as weed in irrigated farms; foothill deserts to midbelt of mountains.

IB. Kashgar: *West.* (Kshuiku, weed near irrigation ditch, Aug. 9, 1913—Knorring; "Opal, ca. 1440 m, Aug. 15, 1934"—Persson, l.c.). *East.* (south. extremity of Khami oasis south of Bugas, near river, Aug. 19; near water, Sept. 16—1895, Rob.; nor. of Turfan, alongside water, No. 5493, June 1; along Pichan-Nanku road, alongside water, June 15; north-east of Toksun, near water, June 19—1958, Lee and Chu).

IIA. Junggar: *Cis-Alt.* ("cultivated parts of valleys near irrigation canals, Shara-Sume, Price"—Simpson, l.c.). *Jung. Alt., Tien Shan, Jung. Gobi* (between Sandzhi and Dzhimissar, bank of irrigation canal, Aug. 17, 1898—Klem.; Shikhetsza-Nyutsyuantsza, in wet sites, No. 1376, July 15; near Fukan, near water, No. 4216, Sept. 17; near Gan'khetsz, No. 5101, Sept. 23—1957, Kuan). *Dzhark., Balkh.-Alak.*

IIIB. Tibet: *South:* ("Niti, 3450 m, Strachey et Winterbottom"—Strachey, l.c.).

General distribution: Fore Balkh., Jung.-Tarb., Nor. and Cent. Tien Shan; West. Sib. (Altay), Mid. Asia, Himalayas (west., Kashmir).

Note. *M. asiatica* is a highly polymorphic plant. Pubescence of plant as a whole and size of leaves as well as calyx and corolla vary. Closely related to *M. longifolia* (L.) Huds., differing mainly in very lax pubescence of entire plant and distinclty 2-coloured leaves.

Plants intermediate between *M. arvensis* L. and *M. asiatica* Boriss. have been collected in plantations and farm surroundings in Nor. Kashgar as well as in Fukan region in Jung. Gobi, also in farms. Similar races have been described many a time in the literature, usually treated as hybrids between these species [*M. interrupta* Boriss. in Bot. mater. Gerb. Bot. inst. AN SSSR, 16 (1954) 285].

## 39. Elsholtzia Willd.
in Roemer et Usteri, Mag. Bot. (IV) 11 (1790) 3.

1. Shrub 60 cm to 2 m tall; bracts lanceolate ...................................................
   ..................................................... 4. **E. fruticosa** (D. Don) Rehder.
+ Annual plant very short; bracts orbicular or broadly obovate ..... 2.
2. False spike regular; bracts subobtuse or very shortly acuminate at apex; corolla nearly regular ................................................................. 3.
+ False spike unilateral bracts long-acuminate; corolla more or less irregular ........................................... 1. **E. ciliata** (Thunb.) Hylander.
3. Corolla yellow ............................................... 3. **E. eriostachya** Benth.
+ Corolla pink or lilac.......................................................................... 4.
4. Bracts flabellate, densely imbricate; as a result, false spike cone-shaped; corolla puberulent; nuts glabrous ...............................................
   ................................................................. 5. **E. strobilifera** Benth.
+ Bracts broadly obovate, imbricate; corolla with dense long hairs; as a result, false spike tomentose-villous; nuts tuberculate in upper half ................................................................. 2. **E. densa** Benth.

**1. E. ciliata** (Thunb.) Hylander in Bot. notiser (1941) 129; Hara, Fl. East. Himalaya (1966) 273.   —*E. cristata* Willd. in Roemer et Usteri, Mag. Bot. (IV) 11 (1790) 5; Franch. Pl. David. 1 (1884) 234; Forbes and Hemsley, Index Fl. Sin. 2 (1902) 277; Walker in Contribs U.S. Nat. Herb. 28 (1941) 656.   — *E. patrinii* (Lepech.) Garcke, Fl. Halle, 2 (1856) 213; Krylov, Fl. Zap. Sib. 9 (1937) 2399; Kitag. Lin. Fl. Mansh. (1939) 379; Volkova in Fl. SSSR, 21 (1954) 635; Fl. Kazakhst. 7 (1964) 471.   —*Sideritis ciliata* Thunb. Fl. Japan. (1784) 245.   —*Mentha patrinii* Lepech. in Nova Acta Ac. Sci. Petrop. 1 (1787) 336.   —*Hyssopus ocymifolius* Lam. in Encycl. méth. 3 (1789.) 187.   —Ic. Willd l.c. tab. 1 (sub nom. *E. cristata*); Lepech. l.c. tab. 8 (sub nom. *M. patrinii*).

Described from Japan. Type in Upsala.

Banks of rivers and brooks, gorge floors, near springs, sparse forests, often as weed around houses.

IIIA. **Qinghai:** *Nanshan* (Chortentan temple, along gorges in forest and near springs, on humus, on humus, 2100–2400 m, Sept. 1901—Lad.).

IIIB. **Tibet:** *Weitzan* (Yantszytszyan basin, vicinity of Chzherku morastery, on sandy-rocky banks of brooks and springs, 3400 m, Aug. 14; Mekong river basin, along Chokchu river, on rocky soil and humus in juniper forest., 3600 m, Aug. 31—1900, Lad.).

**General distribution:** Europe (introduced), West. Sib. (south.), East. Sib. (west.), Far East, China (Dunbei, North), Himalayas (east.), Korean peninsula, Japan.

2. **E. densa** Benth. Labiat. gen. et sp. (1835) 714; Diels in Futterer, Durch Asien, 3 (1903) 19; Pampanini, Fl. Carac. (1930) 186; Rehder and Kobuski in J. Arn. Arb. 14 (1933) 31; Persson in Bot. notiser (1938) 301; Hand.-Mazz. in Acta Horti Gotoburg. 13 (1939) 357; Walker in Contribs U.S. Nat. Herb. 28 (1941) 656; Volkova in Fl. SSSR, 21 (1954) 636; Grubov, Konsp. fl. MNR (1955) 239; Murata in Acta Phytotax. et Geobot. 16, 1 (1955) 13; Fl. Kirgiz. 9 (1960) 169; Ikonnikov, Opred. rast. Pamira (1963) 215. —*E. calycocarpa* Diels in Fl. Centr. China (1901) 560; Hao in Engler's Bot. Jahrb. 68 (1938) 635. — *E. janthina* (Maxim.) Dunn in Notes Bot. Gard. Edinburgh, 6 (1915) 152. — *E. manshurica* (Kitag.) Kitag. Lin. Fl. Mansh. (1939) 379. —*Pogostemon janthinus* (Maxim.) Kanitz in Ber. Math.-Naturwiss. Ungarn. 3 (1886) 11; Kanitz, A. növénytani (1891) 46. —*Dysophylla janthina* Maxim. in Herb. —*Paulseniella pamirensis* Briq. in Bot. tidskr. 28 (1908) 246. — *Platyelasma densa* (Benth.) Kitag. in Rep. First Sci. Exped. Manchoukuo, 4, 2 (1935) 25. —*P. calycocarpa* (Diels) Kitag. l.c. 25. —*P. manshurica* Kitag. l.c. 25. —**Ic.:** Jacquem. Voy. Ind. tab. 131; Kitag, l.c. (1935) tab. 7 (sub nom. *P. manshurica*).

Described from India. Type in Paris.

Banks of rivers and gorge floors, meadows and steppe slopes of mountains and in rich humus soils in mountain forests, often as weed in plantations and roadsides, old pasture corrals.

IA. Mongolia: *Gobi-Alt.* (Artsa Bogdo, canyon bottoms 1620 m, No. 570, 1925—Chaney). *Alash. Gobi* ("Ho Lan Shan, in a dense stand, on edges of cultivated field, common, No. 1142, Ching"—Walker, l.c.).

IB. Kashgar: *West.* (Pasrabat pass, Aug. 2, 1909—Divn.; west of Kashgar, Bostan-Terek locality, crops, July 10, 1929—Pop.; "Jerzil, 2800 m, No. 144, July 22, 1930; Kentalek, 2700 m, No. 194, July 6, 1931; Bostan-terek, 2400 m, No. 578, July 20, 1934"—Persson, l.c.).

IIA. Junggar: *Tien Shan* (Savan district, Shichan, on roadside, No. 1241, July 7; 1 km north-east of Nyutsyuan'tsza, No. 696, July 19; south of Shichan, on roadside, No. 809, July 22; 6 km south-west of Shichan, 1500 m, No. 2304, July 26—1957, Kuan).

IIIA. Qinghai: *Nanshan* (South Tetungsk mountain range, July 25, 1872; along Yusun-Khatyma river, 2700–3000 m, in alp. region, July 24; South Tetungsk mountain range, Aug. 7—1880, Przew.; south. bank of Kuku-Nor, July 24, 1890—Gr.-Grzh.; Kuku-Nor lake, west. bank, dry Bukhata bed, 3000 m, Sept. 8, 1894—Rob.; Dulan-khit temple, 3300 m, on humus in spruce forest, Aug. 8; on sloping banks of Dan'ger-khe brook, humus mixed with pebble, around Dan'ger-tin town, 2400 m, Aug. 27—1901, Lad.; Loukhu-shan' mountain range, July 8; same site, hemp crop, July 17; Kuku-Nor lake, near river bank, on humus soil, 3600 m, Aug. 29—1908, Czet.; east. extremity of Nanshan, 25 km south of Gulan town, gentle slopes of low hills, 2435 m, hill steppe with shrubs, Aug. 12; Mon'yuna', Ganshiga river valley, left tributary of Peishikhe river, 3350–3720 m—1957, Petr.; "Kokonor: in der Nähe der Stadt Tenkar, um 3000 m [No. 1300 A—Sept. 14]; in den Tälern des Gebirges Hoto-Selgen, um 3800 m, No. 1274, Sept. 12"—Hao, l.c.). *Amdo* ("auf dem Gebirge Ja-he-mari, bei 4000 m, No. 1217, Sept. 9" — Hao, l.c.).

IIIB. Tibet: *Weitzan* (Yantszytszyan basin, Donra gorge, along Khichu river, 3900 m, in old pasture corrals, occasionally on hill slopes, July 16, 1900—Lad.). *South.* ("circa Lhasa, mt. Phali, Aug. 20; mt. Chokupori, Aug. 21; Pelon, Aug. 24—1914, Kawaguchi"—Murata, l.c.).

IIIC. Pamir (Tashkurgan, in meadow, July 25, 1913—Knorring).
General distribution: Cent. Tien Shan, East. Pam.; Mid. Asia (Pamiro-Alay), China (Dunbei, North, North-West, South-West), Himalayas (west., Kashmir).

3. **E. eriostachya** Benth. Labiat. gen. et sp. (1833) 163; Pampanini, Fl. Carac. (1930) 186; Kashyap in Proc. Indian Sci. Congr. 19 (1932) 48; Mukerjee in Rec. Bot. Survey India, 14, 1 (1940) 91; Walker in Contribs U.S. Nat. Herb. 28 (1941) 656; Murata in Acta Phytotax. et Geobot. 16, 1 (1955) 13. —*E. pusilla* Benth. Labiat. gen et sp. (1835) 714. —*E. eriostachya* Benth. var. *pusilla* Hook. f. in Fl. Brit. Ind. 4 (1885) 645; Strachey, Catal. (1906) 140. — *Aphanochilus eriostachyus* Benth. in Wall. Pl. Asia Rar. 1 (1830) 29. — *Platyelasma eriostachya* (Benth.) Kitagawa in Rep. First Sci. Exped. Manchoukuo, 4, 2 (1935) 25.

Described from India. Type in London (K). Isotype in Leningrad.

IIIB. Tibet: *Chang Tang* ("piede del Chiungàng-la, versante occ."—Pampanini, l.c.). *South.* ("Topidhunga, 4500 m, Aug., Strachey and Winter bottom"—Strachey, l.c.; "circa Shigatse, Mhali village, Aug. 1, 1914, Kawaguchii"—Murata, l.c.; "Gyantse, 3900 m, July 1925"—Kashyap, l.c.).
General distribution: Himalayas (west., east., Kashmir).

4. **E. fruticosa** (D. Don) Rehder in Pl. Wilson. 3 (1916) 381; Hand.-Mazz. Symb. Sin. 7 (1936) 934; Kitamura et Murata in Faun. et Fl. Nepal Himal. (1955) 211; Vautier in Candollea, 17 (1959) 46; Hara, Fl. East. Himalaya (1966) 274. —*E. polystachya* Benth. Labiat. gen. et sp. (1833) 161; Hook, f. Fl. Brit. Ind. 4 (1885) 643; Dunn in Notes Bot. Gard. Edinburgh, 6 (1915) 150. —*E. tristis* Lévl. et Vaniot in Feddes repert, 8 (1910) 425. —*E. dielsii* Lévl. in Feddes repert. 9 (1911) 441. —*E. souliei* Lévl. in Feddes repert. 9 (1911) 248. —*Perilla fruticosa* D. Don, Prodr. Fl. Nepal (1825) 115. — *Aphanochilus polystachyus* Benth. in Wall. Pl. Asia Rar. 1 (1830) 27. —*A. fruticosus* (D. Don) Kudo in Mém. Fac. Sci. and Agr. Taihoku Univ. 2 (1929) 61. —Ic.: Wall. Pl. Asia Rar. 1, tab. 33 (sub nom. *A. polystachyus*).

Described from East. Himalayas (Nepal). Type in London (K?).
Arid rocky slopes.

IIIB. Tibet: *Weitzan* (vicinity of Chzherku monastery, 3400 m, along hedges in ploughed fields an on dry rocky slopes of hills, Aug. 11, 1900—Lad.).
General distribution: China (Cent., South-West), Himalayas, India (Assam).

5. **E. strobilifera** Benth. Labiat. gen. et sp. (1833) 163; Marquand in J. Lin. Soc. London (Bot.) 48 (1929) 216; Kashyap in Proc. Indian Sci. Congr. 19 (1932) 48. —*E. exigua* Hand.-Mazz. in Symb. Sin. 7 (1936) 936.

Described from East. Himalayas. Type in London (K.).
In high-mountain areas.

IIIB. Tibet: *South.* ("Atsa Tso, 4200–4500 m, Aug. 26, 1924, Kingdon Ward"—Marquand, l.c.; "Gyantse, 3900 m, July 1929"—Kashyap, l.c.).
General distribution: Himalayas (west., east.), India.

### 40. Isodon (Schrad.) Kudo

in Mém. Fac. Sci. and Agr. Taihoku Univ. 2 (1929) 118. —*Plectranthus*
L'Herit. sect. Isodon Schrad. in Benth. Labiat. gen. et sp. (1832) 40.

**1. I. pharicus** (Prain) Murata in Acta Phytotax. et Geobot. 16, 1 (1955)
15. —*Plectranthus pharicus* Prain in J. As. Soc. Bengal, 59 (1890) 297;
Mukerjee in Rec. Bot. Survey India, 14, 1 (1940) 46.

Described from East. Himalayas (Phari). Type in London (K).

IIIB. Tibet: *South.* ("Circa Lhasa, Pulunka temple, Aug. 20; Panchogan, Aug. 23; Kyanko village, Sept. 16; circa Shigatse, mt. Ponbo Liuche, July 28—1914, Kawaguchi"—Murata, l.c.).
General distribution: Himalayas (east.).

### 41. Ocimum L.

Sp. pl. (1753) 597.

**1. O. basilicum** L. Sp. pl. (1753) 597; Forbes and Hemsley, Index Fl. Sin.
2 (1902) 266; Danguy in Bull. Mus. nat. hist. natur. 17 (1911) 345; Volkova in
Fl. SSSR, 21 (1954) 641; Fl. Kirgiz. 9 (1960) 170; Fl. Kazakhst. 7 (1964) 471.

Described from India. Type in London (Linn.).

Cultivated as seasoning agent; often grows as weed.

IB. Kashgar: *Nor.* ("Jardins de Koutchar, July 23, 1907, Vaillant"—Danguy, l.c.). *East.* (Turfan, Sept. 29, 1879—A. Reg.; Lyukchunsk, trough, along irrigation ditches, Oct. 8, 1895—Rob.).
IIA. Junggar: *Dzhark.* (Kul'dzha, cultivated, July 1877—A. Reg.).
General distribution: Nor. Tien Shan (cultivated); Europe, Fore Asia, Caucasus, Mid. Asia (cultivated), Far East, China (North, South-West, South, Indo-Mal., Japan.

## Family 106. SOLANACEAE Juss.[1]

1. Shrubs with numerous slender spines and red or black berries. Flowers in axillary clusters ............................................. 1. **Lycium** L.
+ Herbs or subshrubs with creeping stem, without spines. Flowers single or in inflorescence ....................................................... 2.
2. Rosetted plant, stemless or with shortened stem, with large, thick and often branched root ................................................................. 3.
+ Plant with distinct stem. ........................................................................ 4.
3. Fruit a capsule, loosely placed in highly accrescent cyst-like inflated calyx, coarsely reticulately nerved .............. 3. **Przewalskia** Maxim.
+ Fruit a berry; calyx accrescent in fruit but not modified in shape, closely enveloping berry at bottom and sides..... 8. **Mandragora** L.
4. Fruit a succulent berry. Corolla patelliform .................................... 5.
+ Fruit a capsule. Corolla campanulate or tubular-infundi buliform ...................................................................................................... 6.

---

[1] By V.I. Grubov.

5. Flowers single, white; calyx strongly accrescent and inflated cyst-like in fruit, totally enveloping orange-red berry, impressed at base, orange. Herbaceous plant with erect stem. ........................................
.................................................... 6. **Physalis** L. (*Ph. alkekengi* L.).

+ Flowers in racemose or umbellate inflorescence, white or violet; calyx not accrescent or modified, not enveloping red or black berry. Herbs or subshrubs with creeping stem .................... 7. **Solanum** L.

6. Capsule large, 4–6 cm long, with stiff tubercles and dehising by 4 valves. Flowers single in forks of stems and branches; corolla large, 6–10 cm long, tubular-infundibuliform, with broad plicate limb, white .................................. 9. **Datura** L. (*D. stramonium* L.).

+ Capsule extremely small, glabrous and opening by operculum; flowers in inflorescence or single, axillary. ............................................. 7.

7. Flowers single; corolla broadly campanulate with broad 5-lobed limb; calyx highly accrescent in fruit and turning stiff, reticulate nerved ................................. 2. **Scopolia** Jacq. (*S. tangutica* Maxim.).

+ Flowers in inflorescence; corolla infundibuliform ........................ 8.

8. Flowers in terminal leafless corymbose inflorescence, not modified in fruit. Perennial with thick taproot ........ 4. **Physochlaina** G. Don.

+ Flowers aggregated at ends of stems and branches in secund foliated inflorescence, highly elongated in fruit turning racemose or spicate. Annual or biennial with slender root. .. 5. **Hyoscyamus** L.

## 1. **Lycium** L.
Sp. pl. (1753) 191.

1. Spines numerous, short and slender, glabrous, leafless. Leaves linear or linear-lobate, more rarely spatulate, succulent and thick, subcylindrical, in clusters on shortened shoots. Berries blackish-violet with violet sap. .................................... 5. **L. ruthenicum** Murr.

+ Spines few, long and thick, foliated (or with leaf scars), rarely leafless. Leaves flat, slender, linear-lanceolate to oblong-obovate. Berries red with yellow sap ................................................................ 2.

2. Corolla tube longer than limb; stamens at base with few hairs only inside or with sparse circular pubescence not forming compact cover
................................................................................................................ 3.

+ Corolla tube not longer than limb, almost not exserted from calyx; stamens with compact cover of hairs above base .......................... 5.

3. Corolla infundibuliform with gradually enlarging tube; stamens with ring of hairs above base. Leaves lanceolate, broadest above middle, obtuse or acute, glaucescent beneath blackish on drying; branches dark grey. ..................................................... 1. **L. barbarum** L.

+ Corolla clavate with cylindrical tube; stamens pubescent above base only inside. Leaves light green, monochromatic, not blackish on drying; branches light-coloured, yellowish ........................................ 4.

4. Leaves green, linear-lanceolate to lanceolate and broadly lanceolate; up to ovate-lanceolate rarely on vigorously growing shoots; acute, rarely obtuse. Berry ovate or oblong-ovate, truncate or emarginate at tip. Corolla violet-pink; stamens not longer than corolla ...................................................................... 6. **L. truncatum** Wang.

+ Leaves glaucescent, oblong-spatulate and lanceolate to oblong-obovate, broadest above middle, shortly acuminate, more rarely acute or obtuse. Berry globose. Corolla violet-blue; stamens longer than corolla ...................................... 2. **L. dasystemum** Pojark.

5. Leaves lanceolate or linearl-lanceolate, broadest below middle, acute. Berry with callous cusp at tip. Stamens distinctly longer than corolla ................................................ 4. **L. potaninii** Pojark.

+ Leaves lanceolate or broadly lanceolate, broadest above middle, shortly acuminate or obtuse or even rounded at tip, more rarely acute. Berry without callus cusp. Stamens not longer than corolla ................................................................ 3. **L. flexicaule** Pojark.

1. **L. barbarum** L. Sp. pl. (1753) 192 excl. syn.; ej. Syst. natur. ed. 10, 2 (1759) 936; Pojark. in Bot. mater. Gerb. Bot. inst. AN SSSR, 13 (1950) 262; id. in Fl. SSSR, 22 (1955) 82. —*L. halimifolium* Mill. Gard. Dict. ed. 8 (1768) No. 6; ? Kitag. Lin. Fl. Mansh. (1939) 389. —*L. turcomanicum* auct. non Turcz.: ?Kanitz in Szechenyi, Wissensch. Ergebn. 2 (1898) 721. —*L. ruthenicum* auct. non Murr.: ? Diels in Filchner, Wissensch. Ergebn. 10, 2 (1908) 263.—*L. chinense* auct. non Mill.: ? Danguy in Bull Mus. nat. hist. natur. 17 (1911) 343; ?Rehder in J. Arn. Arb. 9 (1928) 113. —**Ic.:** Duham. Traité arbr. et arbust. (1755) tab. 121, fig. dextra; ed. 2 (1801) tab. 31 (sub nom. *L. turbinatum* Poir.); Schneid. Ill. Handb. Laubholzk. 2, fig. 396, a-f.

Described from China from a plant grown in Europe. Type lost. First figure cited by Duham. adopted as type.

Steppe loessial slopes of mountains and cliffs, river banks and shoals.

IA. **Mongolia:** *Khesi* (nor. of Lan'chzhou town, Beitashan' town, dry slopes of loessial knolls, June 24; 67 km nor. of Lan'chzhou town, wormwood-forb-saltwort subdesert, June 29—1957, Petr.).

IB. **Kashgar:** *West.* (Kashgar [evidently cultivated], 1892—Rhins.).

IIIA. **Qinghai:** *Nanshan* ("Pinfang-hsien, June 18; Sining-fu, Aug. 18; in valle Tatung-ho, Aug. 14, 1879"—Kanitz. l.c.; "Sining-fu, June 1904"—Diels, l.c.; "Houei-Ning, 1000 m, Aug. 6, 1908, Vaillant"—Danguy, l.c.; "in dry gorge between Hsining and Tankar, 2750 m, Rock"—Rehder, l.c.). *Amdo* (on Huang He river near Guidun [Guide] town, 2150 m, on silty soil, July 1, 1880—Przew.).

IIIB. **Tibet:** *South.* (Lhasa, under shrub, Aug. 8, 1878—Dung-boo [K]; Kyi Chu valley 15 miles east of Lhasa, Aug. 1904—Walton [K]).

General distribution: China (North-West); extensively distributed in crops in Europe, Mediterranean, North America.

2. **L. dasystemum** Pojark. in Bot. mater. Gerb. Bot. inst. AN SSSR, 13 (1950) 268; Pojark. in Fl. SSSR, 22 (1955) 84; Fl. Kirgiz. 9 (1960) 189; Fl. Kazakhst. 8 (1965) 16. —*L. turcomanicum* auct. non Turcz.: ? Wang in Contribs Inst. Bot. Nat. Ac. Peiping, 2, 4 (1934) 101. —**Ic.:** Pojark. l.c. (1950) 269, fig. 7; Fl. Kazakhst. 8, Plate 1, fig. 6.

Described from South. Kazakhstan (near Baiga-kum railway station on Syr-Darya). Type in Leningrad.

Dry steppe and desert slopes of mountains and knolls, coastal sand and pebble beds, tugais, more often on solonetz soils.

**IB. Kashgar:** *Nor.* ("Uchturfan, flowers; beds—Krasnov" Pojark. l.c. 1950).

**IIA. Junggar:** *Tien Shan* (Ili valley—Kunges, right bank, 42 km east of bridge on Kash, on rocky slopes of extremely barren knolls, Aug. 29, 1957—Yun.). *Jung. Gobi* (west.: hills in lower Borotala river, Takiansi village, Aug. 24, 1878—A. Reg.). *Dzhark.* (5th sumun on left bank of Ili south-west of Kul'dzha, May 29; Ili bank south-west of Kul'dzha, May 30; Suidun, July 16—1877, A. Reg.).

General distribution: Aralo-Casp. (south-east), Fore Balkh. (south.), Jung.-Tarb. (south.), Cent. Tien Shan; Mid. Asia (plains and nor. of Pam.-Al.).

Note. The distribution range of this species is given in the cited work of A.I. Pojarkova (1950, p. 242, fig. 2, 2).

3. **L. flexicaule** Pojark. in Bot. mater. Gerb. Bot. inst. AN SSSR, 13 (1950) 255; Pojark. in Fl. SSSR, 22 (1955) 81; Fl. Kirgiz. 9 (1960) 186; Fl. Kazakhst. 8 (1965) 15. —**Ic.:** Pojark. l.c. (1950) 255, fig. 5.

Described from Nor. Tien Shan (Issyk-Kul' lake). Type in Leningrad.

Dry steppe and desertified slopes of knolls and mountains, coastal cliffs in saline soils.

**IIA. Junggar:** *Dzhark.* (Suidun settlement, July 16, 1877—A. Reg.).

General distribution: Fore Balkh. (Sary-Chagan), Nor. Tien Shan.

Note. The distribution range of this species is given in the cited work of A.I. Pojarkova (1950, p. 245, fig. 3, 1).

4. **L. potaninii** Pojark. in Bot. mater. Gerb. Bot. inst. AN SSSR, 13 (1950) 265; Grubov, Konsp. fl. MNR (1955) 239. —? *L. chinense* auct non Mill.: Franch. Pl. David. 1 (1884) 220; Walker in Contribs U.S. Nat. Herb. 28 (1941) 568. —*L. barbarum* var. *chinensis* Ait.: Chen and Chou, Rast. pokrov r. Sulekhe (1957) 82. —**Ic.:** Pojark. l.c. 265, fig. 6.

Described from East. Mongolia. Type in Leningrad.

Along banks of rivers on solonetz soils, solonchaks and saline sand, desert slopes of knolls and foothills, in gravelly deserts.

**IA. Mongolia:** *East Mong.* (south.: vicinity of Khukh-Khoto town near Siustuchzhao monastery, July 15, 1884—Pot., typus!; vicinity of Guikhuachen [Khukh-Khoto] town, July 15; Chaodzhyunfyn knoll [about 12 km south of Khukh-Khoto], at foot of knoll on dry clayey soil, July 18—1884, Pot.; "autour de Sartchy, No. 2718 fl. Juin 1864, fr. Aou. 1866, David"—Franch, l.c.). *Alash. Gobi* (Khalkha, near Bortszon well, on pebble bed, rare, Aug. 19, 1873—Przew.; Gobi [left bank of Edzin-Gol, first half of July], 1886—Pot.; 15 km south of Denkou town, first terrace of Huang He river, solonchak, July 19, 1957—Petr.; "Hsin Cheng, north of

Ninghsia, on exposed, hard clay cliffs"—Walker, l.c.). *Ordos* (Huang He river valley [near Shara-dzu temple], Aug. 22, 1871—Przew.; on right bank of Huang He below Khekou town, on sandy soil, Aug. 4; Bain-Tukhum area, on sandy soil, Sept. 23—1884, Pot.). *Khesi* ("Sulekhe river"—Chen and Chou, l.c.).

IIA. Junggar: *Jung. Gobi* (south.: environs of Dzhimissara, on clayey soil around brook, Aug. 18, 1898—Klem.; Dzimusair district, San'tai village, No. 627 July 23, 1956—Ching).

General distribution: China (North and North-West, nor. Gansu).

Note. The distribution range of this species is given in the cited work of A.I. Pojarkova (1950, p. 245, fig. 3, 2).

**5. L. ruthenicum** Murr. in Comm. Soc. Sci. Götting. 2 (1779) 9; ? Diels in Futterer, Durch Asien, 3 (1903) 19; Danguy in Bull. Mus. nat. hist. natur. 17 (1911) 343; 20 (1914) 79; Ostenfeld in Hedin, S. Tibet, 6, 3 (1922) 45; Pampanini, Fl. Carac. (1930) 186; Wang in Contribs Inst. Bot. Nat. Ac. Peiping, 2, 4 (1934) 102; Persson in Bot. notiser (1938) 301; Hao in Engler's Bot. Jahrb. 68 (1938) 636; Grubov, Konsp. fl. MNR (1955) 239; Pojark. in Fl. SSSR, 22 (1955) 80; Chen and Chou, Rast. pokrov r. Sulekhe (1957) 90; Grubov in Bot. mater. Gerb. Bot. inst. AN SSSR, 19 (1959) 550; Fl. Kirgiz. 9 (1960) 185; Fl. Kazakhst. 8 (1965) 14; Hanelt und Davažamc in Feddes repert. 70 (1965) 55. —*L. tataricum* Pall. Fl. Ross. 1, 1 (1784) 78 excl. var. β; Henderson and Hume, Lahore to Jarkand (1873) 329. —**Ic.:** Fl. Kazakhst. 8, Plate 1, fig. 7.

Described from garden specimen grown from seeds sent by P. Pallas, evidently from Kazakhstan. Type lost.

Puffed and clayey solonchaks, toirims, saline sand along banks of lakes and rivers and in sondoks (small sand collections in terminal sections of large gorges) dry river-beds and deltas of gorges, tugais, solonetz banks of rivers, lakes and irrigation canals, around springs, abandoned farms, saline talus.

IA. Mongolia: *Bas. Lakes* (nor. bank of Khirgis-Nur lake near Termin arshan, precipitous rocky slope Aug. 21, 1944—Yun.; Shargain-Gobi, not far from Shargain-Tsagan-Nur lake, solonchaks, Sept. 9, 1930—Pob.; near Sundul tu-Baishing, takyr sections with saxaul, July 15, 1947—Yun.; 2 km West of Sundultu—Baishing, on flanks and floor of dry irrigation canal, Sept. 4, 1948—Grub.). *Val. Lakes* (near beach ridge two miles from lake Orok Nor, at 1150 m, No. 300, 1925—Chaney). *Gobi-Alt.* (Legin-Gol river valley, sand, early Aug. 1922—Pisarev; same site, Sept. 1, 1927—M. Simukova). *West. Gobi* (Altay somon, Bain-Undur desert, Aug. 6, 1947—E. Dagva; Ekhin-Gol area, puffed solonchak, Aug. 21, 1948—Grub.; Bon-Tooroin-Khuduk area, in clayey spread of gorge among tamarisk shrubs, June 12, 1949—Yun.). *Alash. Gobi* (up to Sogo-Nur lake meridian in east). *Khesi* (up to Min'tsin' town in east).

IB. Kashgar: all regions.

IC: Qaidam: *plains* (Urtu-Bulak spring, 2800 m, usually on sandy-clayey solonchaks, forms shrubs, Sept. 10; along Nomokhun-Gol river, 2800 m, Sept. 12— 1884, Przew.; Golmo river valley, environs of Golmo settlement, Oct. 12, 1959—Petr.).

IIA. Junggar: *Jung. Gobi.* (nor., near Ulyungur lake, Aug. 12, 1876—Pot.; north-west, Dzhair mountain range, Tuz-agny ravine, solonets, Aug. 2, 1951—Mois.; west.; south., east.).

IIIC. Pamir (Tashkurgan, on rocky terrace, July 25, 1913—Knorring; Kunlun, Mia river [Tisakrik], 2300–2800 m, July 16, 1941—Serp.

General distribution: Aralo-Casp., Fore Balkh.; Fore Asia, Caucasus (Transcaucasus) Mid. Asia, Himalayas (Kashmir).

6. **L. truncatum** Wang in Contribs Inst. Bot. Nat. Ac. Peiping, 2, 4 (1934) 103; Pojark. in Bot. mater. Gerb. Bot. inst. AN SSSR, 13 (1950) 277; Grubov, Konsp. fl. MNR (1955) 240.—?*L. turcomanicum* auct. non Turcz.: Deasy, In Tibet and Chin. Turk. (1901) 403. —**Ic.:** Wang l.c. tab. III.

Described from North China (around Taiyuan town). Type in Peking (Beijing).

Dry and usually saline loessial, clayey and rocky slopes of knolls and mountains, gorges and ravines, loessial cliffs, solonchaks and solonetz coastal sand and pebble beds, along banks of rivers and lakes, tugais, banks and beds of gorges, as well as along banks of irrigation canals, farm hedges and abandoned farms; up to 3100 m in hills.

**IA. Mongolia:** *East. Mong.* (Chaodzhyunfyn knoll [12 km from Kuku-Khoto], on dry clayey soil in foothills, July 18; right bank of Huang He river below Hekou settlement, on sandy soil, Aug. 4—1884, Pot.). *East. Gobi* (nor. extremity of Galbyn-Gobi, Undain-Gol near Tabun-obo, valley between mud cones, rubble rim of gorge, Sept. 29, 1940—Yun.). *Alash. Gobi* (Khalkha [camp near Bortszon collective], on pebble bed, occasional. Aug. 19, 1873—Przew.; Dynyuan' in town [Bain-Khoto], on edges of irrigation ditches, June 2, 1908—Czet.; Nin'sya town vicinity—Napalkov). *Khesi* (many reports from all over the region).

**IB. Kashgar:** *South.* (Keriya, along Mal'dzha river, on cliffs, common, 2200 m, May 15, 1885—Przew.; nor. slope of Russky mountain range, Bostan-Tograk river gorge, 2700 m, clayey slopes and along banks of rivers, June 11; nor. slope of Atyntag, Muna-Bulak area, 2100–2400 m, clayey-solonetz moist soil, shrubs up to 12 ft. tall, Aug. 6—1890, Rob.; "Julgan Bulak, 2300 m; Toaln Khoja, 2500 m [Karasaya region], 1898"—Deasy, l.c.).

**IC. Qaidam:** *plain* (south.: near Barun-tszasaka khyrma, Sept. 5, 1879; on Nomokhun-Gol river, 7–10 ft tall bush, Aug. 14; in gorge along Nomokhun-Gol river, about 3000 m, Aug. 15—1884, Przew.; around Barun-tszasaka khyrma, 2600 m, solonchak June 17, 1900; same site, bush up to 2 arsh. tall [1 arshine = 0.71 m], June 11, 1901; Nomokhun-taran gorge, 3000 m, clayey-rocky soil, July 30, 1901—Lad.). *Mont.* (Kurlyk-Nor lake, 2700 m, in irrigation ditches in ploughed fields, 6–7 ft tall bush, June 1, 1895— Rob.; same site, banks of lakes, 2600 m, June 25, 1901—Lad.).

**IIA. Junggar:** *Jung. Gobi* (south.: environs of Sondzhi [near Guchen], Aug. 15, 1898—Klem.; vicinity of Urumchi town, Sept. 13, 1929—Pop.; along Urumchi-Khunshan'kou road, on slope, No. 528, June 20, 1957—Kuan; environs of Fukana town, No. 2001, Sept. 21, 1957—Shen-Tyan'; Fukan town, No. 4383, Sept. 21, 1957—Kuan).

**IIIA. Qinghai:** *Amdo* (on Churmyn river 2700–2900 m, in silty soil, May 10; along Huang He river near Churmyn river estuary, May 19; along Boga-Gorgi river, 2600–2700 m, in lesert on silty soil, 1–1.5 ft tall bush, May 22—1880, Przew.).

General distribution: China (North, North-West).

Note. The distribution range of this species is given in the cited work of A.I. Pojarkova (1950, p. 242, fig. 2, 3).

## 2. Scopolia Jacq.

Observ. bot. icon. 1 (1764) 32, Lab. 20. —*Anisodus* Link et Otto, Icon. pl. select. (1828) 77.

1. **S. tangutica** Maxim. in Bull Ac. Sci. St.-Pétersb. 27 (1881) 508; Forbes and Hemsley, Index Fl. Sin. 2 (1902) 176; Diels in Futterer, Durch Asien, 3 (1903) 19; Danguy in Bull. Mus. nat. hist. natur. 17 (1911) 343; Hao in Engler's

Bot. Jahrb. 68 (1938) 636. —*Anisodus tanguticus* (Maxim.) Pascher in Feddes repert. 7 (1909) 167; Walker in Contribs U.S. Nat. Herb. 28 (1941) 658.

Described from Qinghai. Type in Leningrad. Plate IV, fig. 1. Map 2.

Wet sites under shade in fertile soils, on river banks, in bottomland deciduous forest; often as ruderal around houses and on rubbish.

IIIA. Qinghai: *Nanshan* ([Yarlyk-Gol river valley] between extreme northern and North Tetungsk mountain ranges, meadow steppe, rarely also on waste dumps around buildings, common, flowers, July 4, 1872—Przew., syntypus!; between Pinfan town and Chaku village, very rare, July 25, 1875—Pias.; Sininskie Alps, wet sites under shade along Myn'dan'-sha and Guidui-sha rivers, June 14, 1890—Gr.-Grzh.; outskirts of Donkir, on river bank in poplar groves in humus soil, Aug. 5, 1908—Czet.; "Pass zwischen Liangtschou und Pingfan-hsien im Ost-Nanschan"—Diels, l.c.: "Kansou, bords de la rivière de Jong-Ngan, alt. 3200 m, July 9, 1908, Vaillant"—Danguy, l.c.; "Liufujai, lachiungkou, on exposed, moist foothills or along roadsides, July 25, 1923, Ching"—Walker, l.c.). *Amdo* (in upper Huang He [above Churmyn river estuary], flowers, June [19-20], 1880—Przew., syntypus!; "Dahoba auf dem hohen Plateau, um 4000 m"—Hao, l.c.).

IIIB. Tibet: Weitzan (left bank of Dychu river [Mur-usu], 3950 m, on slopes, common, June 24; same site, June 15-July 3—1884, Przew.; Yantszytszyan river basin, environs of Kabchzhi-Kamba village, 3700 m, on rubbish heaps near buildings, July 20, 1900—Lad.). *South.* (Gyangtse, No. 85, July–Sept. 1904—Walton [K]; Mt. Everest Exped., Chodzong, sandy grassy plain, 4600 m, No. 34, June 6; Plumg Chu Valley, stony ground, 4100 m, No. 345, July 9—1922, Monon [K]; Mt. Everest Exped., Plain near Tashid-zong, 4300 m, near cultivation, cattle fodder, No. 135, July 6, 1933—Shebbeare [K]; Mt. Everest Exped., Shekardzong, meadow, 4300 m, No. 77, June 24, 1938—Lloyd [K]).

General distribution: China (North-West, east. Gansu).

Note. Corolla colour varies from reddish-brown to dark-violet.

## 3. Przewalskia Maxim.

in Bull. Ac. Sci. St.-Pétersb. 27 (1881) 507.

1. Calyx cyathiform in fruit 2.5–4.5 cm long, wide-open with large acute teeth; stiff with coarse network of nerves and projecting ribs. Leaves lanceolate, broadest in middle, mostly short-acuminate or acute, intensely undulate or crispate, dense, dark green marginally; lateral veins abruptly eccentric, almost at right angle in midportion. ............................ 1 P. **shebbearei** (C.E.C. Fischer) Grub.

+ Calyx ovate in fruit, 6–11 cm long, almost closed, with small obtuse teeth, slender network of nerves. Leaves oblong-obovate, broadest above middle, with rounded or obtuse tip, slightly undulate or even, slender, light green marginally; lateral veins emerging at less than 45° _____ 2. P. **tangutica** Maxim.

1. P. **shebbearei** (C.E.C. Fischer) Grub. comb. nova. —*Mandragora shebbearei* C.E.C. Fischer in Kew Bull. 6 (1934) 260.

Described from South. Tibet. Type in London (K). Plate IV, fig. 3. Map 2.

Open mountain slopes, 4500–4900 m alt.

**IIIB. Tibet:** *South.* (Khambajong, Sept. 1903—Prain [K]; Tinkye-la, on the pass, 4800 m, large rosette of lanceolate leaves with bladderlike fruits below them, No. 102, July 16, 1933, Shebeeare—typus! [K]; slopes of Chumalhari, ca. 4800 m, No. 2374, July 15, 1939—Gould [K]).
General distribution: endemic.

2. **P. tangutica** Maxim. l.c. 508; Forbes and Hemsley, Index Fl. Sin. 2 (1902) 177; Hao in Engler's Bot. Jahrb. 68 (1938) 636. —*P. roborowskii* Przew. ex Batal. in Acta Horti Petrop. 13 (1894) 380. —*Scopolia* sp. Hemsley and Pearson in Peterm. Mitteil. 28 (1900) 374; Hemsley, Fl. Tibet (1902) 192.

Described from Qinghai (Amdo). Type in Leningrad. Plate IV, fig. 1; V, 4. Map 2.

Open mountain slopes, gorge floors and on clayey or silty, often cobbled soils on banks of rivers and lakes, 3500–5000 m alt.

**IIIA. Qinghai:** *Nanshan* (site not indicated, Aug. 8, 1890—Martin; Yamatyn-umru hills, 3650–3950 m, July 13, 1894—Rob.). *Amdo* (Syansibei mountain range, silty soil on gorge floor, 3350–3650 m, common, May 29, 1880—Przew., typus!; "Kukunor, Dahoba, um 4000 m"— Hao, l.c.).

**IIIB. Tibet:** *Weitzan* (Burkhan-Budda mountain range [Dynsy-Obo area], Oct. 1, 1879— Przew.; [upper Huang He, Odon'-Tala basin], on loose rubble with loess and sand, 4250–5100 m, June 5, 1884—Przew.; upper part of south. slope of Mur-Usu and Huang He water divide, 4250–5100 m, June 8, 1884—Przew. sub nom. *P. roborowskii* Przew.; N. Tibet, Lagern 29–30 [4863 m, 35° 36' N. lat., 92°24' E. long.], Sept. 20, 1896—Hedin [K]; nor. slope of Burkhan-Budda mountain range, Nomokhun river gorge, 3950–4250 m, on humus and loose clay, May 21; Alyk-Nor lake and Alyk-Norin-Gol river, 3650– 3950 m, everywhere along banks of lakes and rivers, May 30; Russkoe lake [Orin-Nor], Haung He river banks, 4100 m, June 15—1900; Yantzytszyan river basin, near Yugindo river, 4050 m, on exposed loose clayey clocalities on southern slopes of hills, May 13, 1901—Lad.).
General distribution: endemic.

Note. The second species of this genus, *P. roborowskii* Przew. ex Batal., described by A.F. Batalin (l.c.) from N.M. Przewalsky's herbarium notes can at best be treated as a variety [P. tangutica Maxim var. roborowskii (Batal.) Grub. comb. nova]. The original author himself has acknowledged this. He distinguished his species from *P. tangutica* Maxim. only in violet colour of corolla and faintly pubescent calyx with acute teeth; the two are similar in all other respects. Jugding from the collectors' notes, the colour of a living corolla may vary as in *Scopolia tangutica* Maxim.: light yellow, yellow, brown (yellowish-brown inside corolla) and lurid with violet streaks. Study of available material showed that in yellow-coloured specimens, the calyx may be obtuse or acuminate or with acute teeth while pubescence varies from dense to scattered. Thus, distinctive features of this species pointed out by A. F. Batalin are no more than intraspecific variation of *P. tangutica* Maxim.

Only fruits and leaf remnants of Hedin's cited specimen are available but is accompanied by a satisfactory sketch of the overall plant made (judging from date on sketch) a year after its collection. The calyx of this specimen (with fruits) differs from those in other available specimens of this species and resembles preceding species—very large teeth open at tip, stiffer and not as large (6 cm long instead of 7–11 cm in others). Judging from their remnants and depiction, leaves correspond more to this species in shape, consistence and venation.

Genus *Przewalskia* Maxim. is closely related to genus *Scopolia* Jacq. and evidently derived from it. Accrescent cystiform calyx lignefying and adhering to tip of fruits in *S. tangutica* Maxim. very closely resembles *Przewalskia shebbearei* (C.E.C. Fischer) Grub. Seeds of all 3 species—*Scopolia tangutica, Przewalskia tangutica* and *P. shebbearei*—are so similar in shape, size, colour and surface structure that it is hardly possible to differentiate them.

## 4. Physochlaina G. Don

Gen. Hist. 4 (1838) 470.  —*Physochlaena* Miers in Annals and Mag. nat. hist. ser. 2, 5 (1850) 469.  —*Belenia* Decne in Jacqem. Voy. Inde, 4 (1844) Bot. 113.

1.  Corolla bluish-violet; calyx highly but unevenly accrescent in fruit, becoming membranous-cystifarm; teeth surrounding narrow throat weakly developed, obtuse. Pubescence not glandular, with long articulate hairs. ................................ 1. **Ph. physaloides** (L.) G. Don.

+  Corolla greenish-yellow; calyx uniformly accresccent in fruit and coarse while preserving tubular-campanulate shape with acute upright teeth. Pubescence glandular ...............................................
.......................................... 2. **Ph. praealta** (Decne) Miers.

1. **Ph. physaloides** (L.). G. Don, Gen. Hist. 4 (1838) 470; Forbes and Hemsley, Index Fl. Sin. 2 (1902) 176; Danguy in Bull. Mus. nat. hist. natur. 20 (1914) 79; Kitag. Lin. Fl. Mansh. (1939) 390; Krylov, Fl. Zap. Sib. 10 (1939) 2405; Semenova in Fl. SSSR, 22 (1955) 104; Grubov, Konsp. fl. MNR (1955) 240; Fl. Kazakhst. 8 (1965) 18; Hanelt und Davažamc in Feddes repert. 70 (1965) 56.  —*Ph. dahurica* Miers in Annals and Mag. nat. hist. ser. 2, 5 (1850) 471, em. Pasch. in Feddes repert. 7 (1909) 166.  —*Ph. lanosa* Pasch. l.c. 167.  —*Hyoscyamus physaloides* L. Amoen. Acad. 7 (1769) 474.  —Ic.: L. l.c. tab. 6, fig. 1; Fl. SSSR, 22, Plate IV, fig. 1; Fl. Kazakhst. 8, Plate 2, fig. 2.

Described from Siberia from garden specimens. Type in London (Linn.).

Among rocks and stones on mountain slopes, knolls and outliers, along dry ravines and gorges, floors of gullies and creek valleys, among shrubs.

IA. Mongolia: *Khobd.* (Ulan-Daban pass, among stones, June 22, 1879—Pot.; in Kharkhira hills along (Netsugun river, tributary of Namyur river, July 20, 1903—Gr. Grzh.). *Mong. Alt., Cent. Khalkha* ([collective] Bumbotu [on Darkhan-Dzam road], 1841—Kirilov; around Ikhe-Tukhum-Nor lake, Temeni-Ama ravine, June 1926—Zam.; Sukhe-Bator a jmaq, west. slope of Darkhan-Khan-Ula, small ravine along steppe slope, April 27, 1944—Yun.). *East. Mong., Gobi-Alt., East. Gobi* (Delger-Khangai mountain range, creek valley along nor. slope of mud cone, July 23, 1924—Pakhomov). *Ordos* (Ordos australis, 1876—Verlinden).

General distribution: Aralo-Casp. (east.), Fore Balkh., Jung.-Tarb.; West. Sib. (south.), East. Sib. (south.), Far East (Zeya estuary), Nor. Mong. (Fore Hubs., Hang., Mong.-Daur), China (North).

Note. Closely related species *Ph. albiflora* Grub., a single report so far in east. Hangay from Nor. Mongolia, differs from this species in white corolla and leaves aggregated under inflorescence (see Grubov, l.c.).

2. **Ph. praealta** (Decne) Miers in Annals and Mag. nat. hist. ser. 2, 5 (1850) 473; Clarke in Hook. f. Fl. Brit. Ind. 4 (1883) 244; Hemsley, Fl. Tibet (1902) 192; Hand.-Mazz. in Oesterr. bot. Z. 79 (1930) 37; Pampanini, Fl. Carac. (1930) 187; Kashyap in Proc. Indian Sci. Congr. 19 (1932) 47.  —*Ph. grandiflora* Hook. in Curtis's Bot. Mag. 77 (1851) tab. 4600.  —*Belenia praealta* Decne in Jacquem. Voy. Inde, 4 (1844) Bot. 114.  —*Scopolia praealta* Dun. in DC. Prodr. 13, 1 (1852) 554; Henderson and Hume, Lahore to Yarkand (1873)

329. —Ic.: Jacquem. Voy. Inde, Atlas, tab. 120; Curtis's Bot. Mag. 77, tab. 4600.

Described from Himalayas (Kashmir). Type in Paris.

Among stones and under rocks under shade, meadow sections, waste dumps around houses, 3600–4700 m alt.

IIIB. Tibet: *Chang Tang* (far west.—"Zufluss des Noh-zo, 4400 m, Aug. 30, 1906, Zugmayer"—Hand.-Mazz. l.c.). *South.* (plains of Tibet [Tisum], 4700 m—Strachey and Winterbottom; Mt. Everest Exped., Tingri, 4600 m, sandy soils on hillside, No. 190, July 4, 1924—Kingston [K]; Mt. Everest Exped., between Chodzong and Tashidzong, 4600 m, meadow, No. 51, June 20, 1938—Lloyd [K]; "Manasarowar, 4600 m, 1922; Taklakot, 3950 m, July 1926" —Kashyap, l.c.).

General distribution: Himalayas.

Note. Henderson and Hume's (l.c.) list reports this species from Jarkend. In fact, labels of authentic specimens of these collectors indicate Ladakh as the source. Bellew's specimen with printed label "Kashgar" (Kew herbarium) has a handwritten field label dated June 10, 1873, and thus was collected not in Kashgar but in Shayok basin of Kashmir.

## 5. Hyoscyamus L.
Sp. pl. (1753) 179.

1. Calyx urceolate in fruit with broadly deltoid short teeth; corolla twice longer than calyx, 20–45 mm long, lurid, with network of violet nerves. Large biennial ........................................ 1. **H. niger** L.

+ Calyx tubular-infundibuliform in fruit with conical base and divaricate lanceolate spiny acute teeth; corolla not longer than ca-lyx, 10–14 mm long, yellow, monochromatic. Slender annual ........ ................................................................................ 2. **H. pusillus** L.

1. **H. niger** L. Sp. pl. (1753) 179; Franch. Pl. David 1 (1884) 221; Kanitz in Szechenyi, Wissensch. Ergebn. 2 (1898) 721; Forbes and Hemsley, Index Fl. Sin. 2 (1902) 177; Diels in Futterer, Durch Asien, 3 (1903) 20; Danguy in Bull. Mus. nat. hist. natur. 17 (1911) 343; Pampanini, Fl. Carac. (1930) 187; Persson in Bot. notiser 4 (1938) 301; Krylov, Fl. Zap. Sib. 10 (1939) 2403; Ching in Bull. Fan Menor. Inst. Bial. (Bot.) 10 (1941) 262; Walker in Contribs U.S. Nat. Herb. 28 (1941) 658; Grubov, Konsp. fl. MNR (1955) 240; Pojark. in Fl. SSSR, 22 (1955) 93; Grubov in Bot. mater. Gerb. Bot. inst. AN SSSR, 19 (1959) 550; Fl. Kirgiz. 9 (1960) 190; Fl. Kazakhst. 8 (1965) 17; Hanelt und Davažamc in Feddes repert. 70 (1965) 56.   —Ic.: Reichb. Ic. Fl. Germ. XX, tab. 1623; Fl. Kazakhst. 8, Plate 2, fig. 1.

Described from Europe. Type in London (Linn.).

Fields, fallow lands, kitchen gardens, banks of irrigation ditches and canals, roadsides, wasteland around houses, manured sites and rubbish dumps of old pasture corrals, along banks of rivers and on dry pebble beds as ruderal and weed; up to 2700 m in hills.

IA. Mongolia: *Mong. Alt., Cis-Hing., Cent. Khalkha, East. Mong., Gobi-Alt., Alash. Gobi* (Alashan mountain range), *Ordos.*

118

IB. Kashgar: *Nor.* (Aksu). *East.* (Khami).
IIA. Junggar: *Jung. Alt., Tien Shan, Jung. Gobi* (south.), *Dzhark.*
IIIA. Qinghai: *Nanshan.*

General distribution: Aralo-Casp. (nor.), Fore Balkh., Jung.-Tarb., Nor. and Cent. Tien Shan; Europe, Mediterr., Balk.-Asia Minor, Fore Asia, Caucasus, Mid. Asia (mont.), West, Sib., East. Sib. (south.), Far East (south, rare), Nor. Mong. (Hent., Hang., Mong.-Daur.), China (Dunbei, North, North-West, South-West—Kam), Himalayas (west., Kashmir), Nor. Amer. (introduced), Austral. (introduced).

Note. The very closely related annual weed *H. bohemicus* F.W. Schmidt (= *H. agrestis* Kit.) has not been reported from Central Asia outside or within the USSR and finds no mention in literature. This species differs from biennial *H. niger* L. in simple unbranched tender woody root without distinct root neck, simple stem and late (late June) flowering.

2. **H. pusillus** L. Sp. pl. (1753) 180; Danguy in Bull. Mus. nat. hist. natur. 20 (1914) 79; Pampanini, Fl. Carac. (1930) 187; Krylov, Fl. Zap. Sib. 10 (1939) 2404; Grubov, Konsp. fl. MNR (1955) 240; Pojark. in Fl. SSSR, 22 (1955) 98; Fl. Kirgiz. 9 (1960) 191; Ikonnikov, Opred. rast. Pamira (1963) 216; Fl. Kazakhst. 8 (1965) 17; Hanelt und Davažamc in Feddes repert. 70 (1965) 56. —Ic.: Fl. Kazakhst. 8, Plate 2, fig. 3.

Described from Iran. Type in London (Linn.).

Rocky and rubble steppes and desert slopes of knolls and mountains coastal pebble beds and along gorge floors, banks of brooks and rivers, as well as semi-weed in hedges and along irrigation ditches in crops and fallow land.

IA. **Mongolia:** *Khobd.* (between border and Khobdo town, 1870—Kalning). *Mong. Alt.* (steppe valley of Dzhirgalante river, tributary of Bulugun, Sept. 16, 1930—Bar.; Bombotu-Khairkhan mountain range, on wet bank of El'river, Oct. 10, 1930—Pob.; western extremity of Beger-Nur basin along road to Khalyun, gorge in Gobi formations, along gorge, Aug. 23, 1943—Yun.). *Gobi-Alt.* (Legin-Gol valley near somon camp, on borders and along canal in irrigated barley crop, July 25, 1948, Grub.—extreme east. find). *West. Gobi* (Ekhin-Gol on bank of brook, Aug. 21, 1948—Grub.; same site, June 1962—Hanelt und Davažamc, l.c.).
IB. **Kashgar:** *East.* (Toksun-Turfan and Khami regions).
IIA. **Junggar:** *Jung. Alt.* (Urtaksary in Barotala river basin, Aug. 1878—A. reg.). *Tien Shan* (midcourse of Tekes, along bank, 1350–1500 m, Aug. 13, 1877—A. Reg.). *Jung. Gobi* (nor., nor.-west., south., east.) *Dzhark.* (Kul'dzha region and Ili valley).

General distribution: Aralo-Casp., Fore Balkh., Nor. and Cent. Tien Shan, East. Pam.; Europe (extreme south-east. Europ. USSR), Mediterr. (south.), Asia Minor, Fore Asia, Caucasus (Transcaucasus), Mid. Asia (plain), Himalayas (Kashmir).

### 6. Physalis L.
Sp. pl. (1753) 182.

1. **Ph. alkekengi** L. l.c.; Boiss. Fl. or. 4 (1879) 287; Franch. Pl. David. 1 (1884) 220; Forbes and Hemsley, Index Fl. Sin. 2 (1902) 173; Pampanini, Fl. Carac. (1930) 187; Pojark. in Fl. SSSR, 22 (1955) 64; Fl. Kirgiz. 9 (1960) 182; Fl. Kazakhst. 8 (1965) 12. —*Ph. glabripes* Pojark. in Bot. mater. Gerb. Bot. inst. AN SSSR, 16 (1954) 325; Pojark. in Fl. SSSR, 22 (1955) 65. —*Ph.*

*praetermissa* Pojark. l.c. (1954) 322; Pojark. l.c. (1955) 67; Fl. Kirgiz. 9 (1960) 183; Fl. Kazakhst. 8 (1965) 13. —**Ic.**: Reichb. Ic. fl. Germ. XX, tab. 1630; Fl. SSSR, 22, Plate III, fig. 2 (sub nom. *Ph. praetermissa*); Fl. Kazakhst. 8, Plate 1, fig. 4.

Described from Europe (Italy). Type in London.

Waste dumps around houses, near fencings and in wasteland, roadsides, along irrigation ditches and river banks, farm borders, as ruderal and semi-weed in gardens and kitchen gardens; widely cultivated by Chinese as medicinal and ornamental plant.

**IA. Mongolia:** *Alash. Gobi* (Dyn'yuanin oasis [Bayan-Khoto], in humus soil on fringes of irrigation ditches, June 22, 1908—Czet.). *Ordos* (Linchzhou town, grown in kitchen gardens, Oct. 3, 1884—Pot.). *Khesi* (Suchzhou [Tszyutsyuan'], Sept. 1890—Marten).

**IIA. Junggar:** *Jung. Gobi* (south.—nor. slopes of Merzbacher mountain range, 1952—Mois.). *Dzhark.* (environs of Kul'dzha town, June 19; in Kul'dzha, Aug. 22—1875, Larionov; Chimpansi village around Kul'dzha, May 8; 5th sumun on left bank of Ili south-east of Kul'dzha, May 29; Suidun settlement, July 16—1877, A. Reg.).

**General distribution:** Aralo-Casp., Fore Balkh., Jung.-Tarb., Nor. Tien Shan; Europe, Mediterr., Balk.-Asia Minor, Fore Asia, Caucasus, Mid. Asia, Far East (extreme south), China (Dunbei North, North-West, Cent., East, South-West), Korean peninsula, Japan, Nor. Amer. (introduced).

**Note.** *Ph. glabripes* Pojark. and *Ph. praetermissa* Pojark. differ from *Ph. alkekengi* L. in minor and unstable characteristics and have no definite distribution range while differences between both the new species are even less significant, a fact the author himself acknowledges [Pojarkova, l.c. (1955) 67]. Important distinct differences between *Ph. alkekengi* L. and the two new species lie only in pubescence of the pedicel and calyx in fruit: in typical European *Ph. alkekengi* L., they are pubescent while, in the latter—Chinese representatives of this species—they are generally glabrous or very faintly pubescent. Races reported by A.I. Pojarkova, as also the known var. *franchetii* Hort., are probably cultigens, the result of natural selection in the course of prolonged cultivation of this species in China for 2500 years. *Ph. alkekengi* is extensively cultivated in China and Japan as an ornamental and medicinal plant. All reports of it in China are confined to inhabited areas and quite possibly came from the Near East by the ancient commercial routes [see Hara and Kurosawa; *Physalis alkekengi* and its variation in East Asia. J. Jap. Bot. 27 (1952) 247–253]. In our opinion, therefore, the cited glabrous or faintly pubescent Chinese races should only be treated as a variety—*Ph. alkekengi* L. var. glabripes (Pojark.) Grub. comb. nova (*Ph. glabripes* Pojark. l.c. = *Ph. praetermissa* Pojark. l.c.).

### 7. Solanum L.

Sp. pl. (1753) 184.

1. Subshrubs. Flowers violet, about 15 mm in diameter .................. 2.
+ Annual. Flowers White, 6–9 mm in diameter .............................. 3
2. Leaves pinnatipartite or divided into 5–11 lobes; sometimes, especially lower and uppermost ones, as well as leaves of branches only lobed or undivided, very rarely most are undivided; stems erect or ascending. Inflorescence pyramidal-paniculate; anthers free; berry ovate ........................................... 4. **S. septemlobum** Bunge.

+ Leaves undivided, oblong-cordate; stems climbing. Inflorescence corymbose-paniculate; anthers connate; berry globose ..................
.............................................................. 1. S. depilatum Kitag.
3. Mature berries black. Stem slender, sparsely branched, glabrous and with flattened-cylindrical branches; leaves tender, bright green, glabrous, with few projecting veins, entire or with 2–3 gently inclined teeth in lower half of each side. ........................................................
.................................................................................... 2. S. nigrum L.
+ Mature berries brownish-red. Stem divaricately branched; tetrahedral, hispid and finely glandular branches; leaves compact, dull green and stiffly pubescent, with strongly projecting veins, sinuate-dentate, with 2–4 pairs of teeth ............................ 3. S. olgae Pojark.

1. S. depilatum Kitag. Lin. Fl. Mansh. (1939) 390; Pojark. in Fl. SSSR, 22 (1955) 17; Fl. Kirgiz. 9 (1960) 173; Krylov, Fl. Zap. Sib. 12 (1964) 3444; Fl. Kazakhst. 8 (1965) 6. —S. dulcamara auct. non L.: Forbes and Hemsley, Index Fl. Sin. 2 (1902) 169 p.p. —S. persicum auct. non Willd.: Krylov, Fl. Zap. Sib. 10 (1939) 2406 p. max. p. —Ic.: Fl. Kazakhst. 8, Plate 1, fig. 1.

Described from nor. Manchuria. Type in Tokyo (?).

Banks of rivers and brooks, near springs, coastal meadows and among shrubs, marshes around springs, under shade of rocks, along banks of irrigation ditches and in wasteland.

IIA. Junggar: *Jung. Alt.* (upper Borotala river, 1800 m, Aug. 1878; Archaty on Borotala river, 1500–1800 m, Sept. 5, 1880—A. Reg.). *Jung. Gobi* (nor., west. and south., in east up to Guchen). ? *Zaisan, Dzhark., Balkh.-Alak.* (bank of Churchutsu river, Aug. 10 [22], 1840—Schrenk; vicinity of Dachen [Chuguchak] town, on marsh, No. 2830, Aug. 10, 1957—Kuan).

General distribution: Fore Balkh., Jung.-Tarb., Nor. and Cent. Tien Shan; Europe (east. European part of USSR), Mid. Asia (hilly), West. Sib. (up to 61°N. lat.), East. Sib. (up to 66°N. lat.), Far East (south), China (Dunbei).

2. S. nigrum L. Sp. pl. (1753) 186 quoad var. α. *vulgare;* Franch. Pl. David. 1 (1884) 220; Forbes and Hemsley, Index Fl. Sin. 2 (1902) 171; ? Danguy in Bull. Mus. nat. hist. natur. 17 (1911) 343; Pampanini, Fl. Carac. (1930) 187; Kitag. Lin. Fl. Mansh. (1939) 391; Krylov, Fl. Zap. Sib. 10 (1939) 2407; Walker in Contribs U.S. Nat. Herb. 28 (1941) 658; Pojark. in Fl. SSSR, 22 (1955) 25; Fl. Kirgiz. 9 (1960) 175; Fl. Kazakhst. 8 (1965) 7. —Ic.: Reichb. Ic. bot. tab. 953; Fl. Kazakhst. 8, Plate 1, fig. 3.

Described from Europe. Type in London (Linn.).

Gardens, kitchen gardens, wasteland, around hedges, roadsides, waste dumps, banks of irrigation ditches and around fields, banks of rivers as ruderal and weed.

IA. Mongolia: *Alash. Gobi* (Chzhunvei, garden of experimental station at Sabotou, July 26, 1957—Petr.).

IB. Kashgar: *Nor.* ("province de Koutchar [Kucha], terrains humides, Juin 1907, Vaillant"—Danguy, l.c.). *West.* (Kashgar town, humid zone, No. 7538, Sept. 23, 1958—Lee and

Chu; "Kachgar, Oct. 1906, Vaillant"—Danguy, l.c.; "common in Jarkand, from 1200 to 2700 m"—Henderson and Hume, l.c.). *East.* (Turfan town, Sept. 22, 1879—A. Reg.).

IIA. Junggar: *Tien Shan* (upper Ili valley, in gardens, occasional, Aug. 28, 1876—Przew.). *Jung. Gobi* (Shikhetsza settlement, wasteland, No. 1372, July 13, 1957—Kuan). *Dzhark..* (Kul'dzha ton, June 28, 1875—Larionov; Ili river bank west of Kul'dzha, July 1877—A. Reg.).

IIIA. Qinghai: *Nanshan* ("Lien Cheng, along moist edges of cultivated fields" Walker, l.c.).

General distribution: Aralo-Casp. (nor.), Fore Balkh., Nor. and Cent. Tien Shan; Europe, Mediterr., Balk.-Asia Minor, Fore Asia, Caucasus, Mid. Asia, West. Sib. (south.), China, India, Korean peninsula, Japan.

3. **S. olgae** Pojark. in Bot. mater. Gerb. Bot. inst. AN SSSR, 17 (1955) 333; Pojark. in Fl. SSSR, 22 (1955) 30; Fl. Kirgiz. 9 (1960) 75; Fl. Kazakhst. 8 (1965) 8. —*S. nigrum* auct. non L.: Henderson and Hume, Lahore to jarkand (1873) 329. —?*S. alatum* auct. non Moench: Persson in Bot. notiser, 4 (1938) 301. —**Ic.:** Fl. SSSR, 22, Plate 1, fig. 3.

Described from Mid. Asia (Gissar mountain range). Type in Leningrad.

Along farm hedges and borders and irrigation ditches, in gardens and kitchen gardens, fallow and wasteland, roadsides as weed, also along river banks.

IB. Kashgar: *Nor.* (Kuerchu settlement west of Kurel' town, No. 8648, Aug. 28, 1958—Lee and Chu). *West.* (Jarkand, 1870, Henderson; Kashgar, over 1200 m, 1912—Cresswell [K]; Jarkand oasis, in gardens, ploughed fields and irrigation canals, 200 m, June 4, 1889—Rob.; Kashgar oasis near Khan-aryk village, in fields July 22, 1929—Pop.; 45–46 km nor. of Jarkand along old road to Maralbash, oasis, weed on roadsides, Sept. 30, 1958—Yun.). *East.* (nor. rim of Khami desert, Chanlyufi well, along irrigation ditches in ploughed fields, 600 m, Aug. 16; south. Tien Shan foothills [nor. of Khami], Edir-Gol river, in ploughed fields, Aug. 22—1895, Rob.; west of Myaoergou settlement north-west of Khami, No. 141, Aug. 6, 1956—Ching; Turfan district, Putougou settlement, along farm border, No. 5528, April, 1958—Lee and Chu). *South.* (Khotan, Avak village in Ashi-Darya river valley, 1600 m, in garden, Aug. 1, 1885—Przew.).

IIA. Junggar: *Tien Shan* (Yamatu, on Ili bank, near water, No. 3029, Aug. 4, 1957—Kuan). *Dzhark.* (Kul'dzha town, May; Ili river bank [near Kul'dzha], July—1877, A. Reg.).

IIIB. Tibet: *South.* ("Sangpo [Brahmapootra] Valley, July 1904"—Walton [K]).

General distribution: Aralo-Casp. (south.), Fore Balkh. (south.), Jung.-Tarb., Nor. Tien Shan; Fore Asia, Mid. Asia, China (North-West—Gansu).

Note. Difficult to distinguish from *S. nigrum* L.

4. **S. septemlobum** Bunge, Enum. Pl. China bor. (1832) 46; Franch. Pl. David 1 (1884) 220; Kanitz in Széchenyi, Wissensch. Ergebn. 2 (1898) 721; Forbes and Hemsley, Index Fl. Sin. 2 (1902) 172; Diels in Filchner, Wissensch. Ergebn. 10, 2 (1908) 263; Danguy in Bull. Mus. nat. hist. natur. 17 (1911) 343; Rehder in J. Arn. Arb. 14 (1933) 32; Kitag. Lin. fl. Mansh. (1939) 391; Walker in Contribs U.S. Nat. Herb. 28 (1941) 658; Grubov, Konsp. fl. MNR (1955) 240; Pojark. in Fl. SSSR, 22 (1955) 11.

Described from North China (around Peking [Beijing]). Type in Paris. Isotype in Leningrad.

Banks of rivers and around springs, among shrubs, on sand, under cliffs and in rocks, as well as ruderal and semi-weed on wasteland, around buildings, in wreckage, gardens and kitchen gardens.

IA. Mongolia: *East. Mong.* (Kerulen river 80 km beyond San-beise, Aug. 15, 1928—Tug.; around Kalgan, 1870, No. 113—Lom.; Muni-Ula, on rubbish among wreckage, July 19, 1871—Przew.). *East. Gobi* (Buterin-Obo area on Khara-Airik—Sain-Usu road, granitic hillocky area, among clumps, Aug. 27, 1940—Yun.). *Alash. Gobi* (Alashan hills, in grand duke's garden, June 17, 1872—Przew.; around Chzhunvei town, garden of experimental station at Sabotou, July 26, 1957—Petr.; "Holanshan, Hsijehkou, on moist rich grassland and in dry places"—Walker, l.c.). *Ordos* (Ulan-Morin river valley, Aug. 22; on sand-dunes north of Narin-Gol river, Sept. 10—1884, Pot.).

IIIA. Qinghai: *Nanshan* (South Tetungsk mountain range [near Chortynton temple], in lower belt, under rocks, June 29, 1873*; same site, in forest, Aug. 6*; North Tetungsk mountain range [east of Chortynton temple], in forest 2400–2700 m, Aug. 14*—1880, Przew.; near Chortynton temple, 2100 m, on hill slopes, waste dumps, along banks of brooks, Sept. 8*; same site, on waste dumps near buildings, near cliffs, along banks of rivers and in thickets, first 10 days of Sept.*—1901, Lad.; in upper Huang He river, near spring, 2100 m [Khagomi area], among shrubs, June 9*; same site [below Khagomi area], 2100 m, on clayey soil, June 10—1880, Przew.; along Itel'-Gol river [San'chuan'skaya valley], April 1885*—Pot.; "Sining-fu, 1904, Filchner"—Diels, l.c.; "Kan-Tsao-Tien, 2000 m, environs de Lantcheou, Aug. 3, 1908, Vaillant"—Danguy, l.c.). *Amdo* (in upper Huang He [below Churmyn river estuary], 2500–2700 m, on sandy and clayey cliffs, May 21, 1880*—Przew.; Dzhakhar mountains, Mudzhik river, July 6, 1890*—Gr. Grzh.; Ba valley in shade of willow forest, 3000 m, No. 14273, June 1926—Rock [K]*).

General distribution: East. Sib. (Dauria, Achinsk steppe), China (Dunbei, North, North-West, Cent.—nor. part).

Note. In Qinghai, *North and North-West China, along with the typical race with pinnatisect leaves, plants with wholly undivided oblong-ovate leaves are found while the rest are barely divided or only with 1–2 short basal lobes (asterisked). The latter plants additionally are invariably much less pubescent than typical plants. In no other respect do they differ from the typical race and are related to it through numerous intergrades. They owe their origin to very favourable habitat conditions, mainly soil and humidity (as could be judged from labels). We treat them as variety S. septemlobum Bunge var. *subintegrifolium* Grub. var. nova (folia plerumque integra oblongo-ovata, caeteris paucilobata. Typus: Nan-Schan, ad fl. Tetung prope templum Tschertynton, 2100 m, s.m., Sept. 8, 1901—Ladygin, in Herb. Inst. Bot. Ac. Sci. URSS in Leningrad conservatur).

## 8. Mandragora L.

Sp. pl. (1753) 181; Benth. et Hook. Gen. pl. 2 (1876) 900.

1. Calyx 15–20 mm long, incised half length into broad deltoid-ovate teeth; corolla broadly campanulate, only slightly longer than calyx, purple ................................................................1. M. caulescens Clarke.
+ Calyx 5–9 mm long, incised more than half length into ovate and oblong-ovate lobes; corolla tubular-campanulate, 1/3 longer than calyx, yellow................................................................ 2. M. tibetica Grub.

1. M. caulescens Clarke in Hook. f. Fl. Brit. Ind. 4 (1883) 242; Forbes and Hemsley, Index Fl. Sin. 2 (1902) 175; Hand.-Mazz. Symb. Sin. 7 (1936) 828.

Described from Himalayas (Sikkim). Type in London (K). Plate IV, fig. 2. Map 2.

Open rubble and meadow slopes of mountains, 3600–4500 m alt.

IIIB. Tibet: *South*. (Hang Kar Ebhu Valley and up to Hang Kar La Pass, No. 113, end-May 1904—Walsh [K]).

General distribution: China (South-West—Kam), Himalayas (east.).

Note. The single report to date for Cent. Asia, falls in the extreme western part of the distribution range of this east. Himalayan species.

2. **M. tibetica** Grub. sp. nova. Radix crassa, carnosa, palaris, 10–20 mm in diam., interdum inferne in ramos 2–3 palmatim divisa, superne transversim rugosa, contractilis; caules subterranei floriferi 1–4, ad 7 cm longi, debiles, squamis membranaceis distantibus tecti, simplices vel rarius summo apice bifidi, folia in fasciculos epigeos congesta, lanceolata ad ovata, 5–20 mm longa, in petiolum planum membranaceo-alatum laminae aequilongum angustata, margine dense et breviter ciliata. Flores solitarii, axillares, in caule 1–4, pedunculati, pedunculis longis, ad 25 mm longis, firmis, vix nutantes; calyx late campanulatus, 5–9 mm longus, infra medium in lobos 5 oblongos obtusos dense ciliatos divisus, asymmetricus, lobis tribus superioribus inferioribus vix majoribus; corolla flava, tubuloso-infundibuliformis, calycem triente superans, limbo quinque laciniato, laciniis rotundatis, inter lacinias plicata, extus dense glandulosa; stamina 5, in quadrante tubuli inferiore affixa, lobis alternantia, filamentis tenuibus, antheris e fauce haud exsertis; stylus tenuis, stigmate plus minusve capitato, staminibus vix longiore. Fructus ignotus.

Typus: Tibet, systema fl. Jangtze, amniculus Machmuchtschu 4100–4300 m, in declivibus montium argilloso-lapidosis siccis denudatis solutis, cum *Przewalskia*, No. 61, May 21, 1901, Ladygin. In Herb. Inst. Bot. Acad. Sci. URSS (Leningrad) conservatur.

Habitu, florum dispositione necnon foliorum pubescentia species nostra *M. caulescenti* Clarke affinis est sed calycis et corollae forma atque florum magnitudine ab ea differt. Plate V, fig. 2 Map 2.

IIIB. Tibet: *Weitzan* (plateau north-east of Tibet 3950–4400 m, May 1884—Przew.; Yantszytszyan river basin, Makhmukhchu brook, 4100–4300 m, on dry exposed loose clayey-rocky mountain slopes, together with *Przewalskia*, flowers yellow, highly fragrant, leaves concealed within soil, with only flowers projecting, No. 61, May 21, 1901—Lad., typus!).

General distribution: endemic.

Note. V.F. Ladygin's collections form 3 valuable herbarium sheets with some complete plant specimens under each and thus anatomical investigations of plants and chemical analysis of roots were possible. The former, done by G.A. Komar[1] in the laboratory of plant anatomy and morphology, revealed the presence of bicollateral vascular bundles characteristic of Solanaceae. The ovary consisted of 2 carpels with false walls, 3–4 lobes, central placentation and many anatropous ovules.

Chemical analysis was carried out by G.N. Yaroshevskaya[1] in the laboratory of plant chemistry and the presence of alkaloids detected.

---

[1]The assistance of both there scientists is gratefully acknowledged.

## 9. Datura L.
Sp. pl. (1753) 179.

1. **D. stramonium** L. l.c.; Franch. Pl. David. 1 (1884) 221; Forbes and Hemsley, Index Fl. Sin. 2 (1902) 176; Danguy in Bull. Mus. nat. hist. natur. 17 (1911) 343; Krylov, Fl. Zap. Sib. 10 (1939) 2402; Pojark. in Fl. SSSR, 22 (1955) 109; Fl. Kirgiz. 9 (1960) 198; Fl. Kazakhst. 8 (1965) 23. —Ic.: Reichb. Ic. fl. Germ. XX, tab. 1624, fig. 1; Fl. Kazakhst. 8, Plate 2, fig. 4; Fl. Kirgiz. 9, Plate 21, fig. 1.

Described from Europe. Type in London (Linn.).

Wasteland and trash dumps around houses, near hedges, roadsides and gutters, farm fencings and borders; as ruderal in kitchen gardens; up to 800 m in hills.

IA. **Mongolia:** *Alash. Gobi* (Alashan mountain range, garbage, common, Sept. 8, 1880 —Przew.).

IB. **Kashgar:** *Nor.* ("prov. de Koutchar, July 23, 1907, Vaillant"—Danguy, l.c.) *West.* (Jarkend, Kashgar, Yangi-Gissar). *East.* (Khami, Bugas area, 500 m, around houses, Aug. 18, 1895— (Rob.).

IIA. **Junggar:** *Tien Shan* (on Kash river, 600 m, May 18, 1878—A. Reg.; in Chapchal hills, on roadsides, No. 3088, Aug. 5, 1957—Kuan). *Dzhark.* (Kul'dzha town, 1874—Larionov).

General distribution: Aralo-Casp., Fore Blakh., Jung.-Tarb., Nor. and Cent. Tien Shan; Europe, Mediterr., Balk.-Asia Minor, Fore Asia, Caucasus, Mid. Asia (rare), China (Dunbei, North, North-West, Cent., East, South-West, South), Himalayas, Korean peninsula, Japan, Nor. Amer., South Amer., Africa.

## Family 107. SCROPHULARIACEAE Juss.[1]

1. Flowers with 5 stamens; calyx 5-partite; corolla 5-lobed, rotate, with short tube and flat limb, yellow or violet. ............. 1. Verbascum L.
+ Flowers with 2 or 4 stamens; calyx with 3, 4 or 5 lobes or teeth; corolla with 3-, 4- or 5-partite limb or bilabiate ............................ 2.
2. Flowers with 2 stamens; corolla blue, dark blue, pink, somewhat lilac or white. ................................................................................. 3.
+ Flowers with 4 stamens, 2 sometimes underdeveloped; corolla yellow, violet, pink, red, white, rarely different ................................. 4.
3. Calyx herbaceous, usually deeply partite into 4, more rarely 5 lobes; corolla with short tube or tubular-campanulate; limb of corolla with 4 (more rarely 5) lobes, rotate or bilabiate. Stem distinctly developed, erect, rarely prostrate. Annuals and perennials .....................
................................................................................. 8. Veronica L.
+ Calyx scarious, tubular, 3-5-toothed or 3-5-partite. Corolla with short or long cylindrical tube with 3–4 (5)-lobed limb, sometimes

---

[1]By L.I. Ivanina.

irregularly bilabiate. Stems shortened, ascending. Perennials ........
................................................................. 9. **Lagotis** Gaertn.

4. (2). Corolla with basal spur, bilabiate; base of lower trilobed lip inflated, more or less concealing mouth, upper lip with 2 incisions;
calyx 5-partite; stamens 4, 2 longer than others ...... 2. **Linaria** Mill.

+ Corolla without spur ......................................................................... 5.

5. Corolla campanulate or orbicular- or tubular-urceolate .............. 6.

+ Corolla with more or less long tube and bilabiate or rarely 5-lobed,
with subregular limb. ...................................................................... 8.

6. Plant stemless, very small; all leaves in radical rosette. Corolla short-
campanulate, small, light pink, with subregular 5-lobed limb. ......
................................................. 7. **Limosella** L. (*L. aquatica* L.).

+ Plant with foliated stems, very large .............................................. 7.

7. Corolla 3–4 cm long, purple, campanulate, slightly bilabiate; sta-
mens 4, staminodes lacking ...................................................................
.......................... 11. **Rehmannia** Libosch. [*R. glutinosa* (Gaertn.) DC.]

+ Corolla not longer than 1.7 cm, brown, reddish-brown or yellow,
more or less inflated, orbicular- or tubular-urceolate with irregular
short bilabiate limb; stamens 4, staminode in form of fleshy scale
generally present under upper lip, sometimes lacking ....................
......................................................................... 3. **Scrophularia** L.

8 (5). Corolla narrowly infundibuliform with 5-lobed limb, pink or lilac;
lobes of limb nearly equal, round, split almost to base. Plant up to 1
m tall, stems straight or branched; flowers small. ...........................
.............. 12. **Leptorhabdos** Schrenk [*L. parviflora* (Benth.) Benth.].

+ Corolla with more or less long tube and bilabiate limb ................. 9.

9. Upper lip flat, straight or reclinate ............................................. 10.

+ Upper lip galeate, sometimes extended into beak, convex or longitudinally folded and laterally pressed ........................................... 13.

10. Stems divaricately branched from base; leaves small, linear shedding early so that plants appear leafless. Corolla dark violet with
infundibuliform tube ......................**4. Dodartia** L. (*D. orientalis* L.).

+ Stems unbranched or sparsely branched; leaves persistent. Corolla
ivory-white, lilac or blue, with more or less long tube. .............. 11.

11. Flowers single, on slender long (1.5–2 cm long) pedicels in leaf axils,
almost all along stem; corolla with yellowish tube and white limb.
Plant with well-developed stem .... 6. **Gratiola** L. (*G. officinalis* L.).

+ Flowers aggregated in small numbers at tip of stem or in short terminal cluster, sessile or on short pedicels; corolla lilac or blue. Plant
low, stem poorly developed or plant almost stemless ................. 12.

12. Calyx 5-toothed; corolla lilac; corolla tube nearly as long as calyx.
Root slender; leaves subcoriaceous, obovate-oblong, entire ............
.......... 5. **Lancea** Hook. f. et Thoms. (*L. tibetica* Hook. f. et Thoms.).

+ Calyx 5-partite almost to base into linear-lanceolate lobes; corolla blue; corolla tube narrow and long, more than twice length of calyx. Root thickened; leaves fleshy, ovate-orbicular, with large serrate teeth. ........... 10. **Oreosolen** Hook. f. (*O. unguiculatus* Hemsl.).

13 (9). Calyx with 5 (6) lobes and additional lobes between former (as though 10-toothed); corolla 3–4 cm long, yellow, with infundibuliform tube, nearly as long as bilabiate limb; lower lip with 2 hollow umbos; flowers not many, 1 each in axils of middle leaves .............
..................................................................................... 19. **Cymbaria** L.

+ Calyx without additional lobes between teeth; corolla 0.4–3 cm long, varying in colour (white, yellow, pink, red or lilac); flowers generally many, in terminal inflorescence or in axils of upper and middle leaves ................................................................................................ 14.

14. Upper lip of corolla galeate, sometimes extended more or less into a long **beak**................................................................................ 15.

+ Upper lip of corolla with 2 divaricate lobes or undivided, convex or concave ............................................................................... 16.

15. Calyx in flated, 4-toothed, laterally compressed, with reticulate nervation, strongly accrescent after flowering, as long as circular flattened capsule. Plant annual with opposite, sessile, oblong dentate leaves subcordate at base. ............ 17. **Rhinanthus** L. [*R. songaricus* (Sterneck) B. Fedtsch.].

+ Calyx not inflated (sometimes weakly inflated), generally 5- toothed, rarely 2–4 toothed (corolla in such case with very long tube or inverted). Plants perennial, rarely annual, with whorled, alternate, rarely opposite or only radical, pinnate or undivided leaves ..........
................................................................. 18. **Pedicularis** L.

16 (14). Leaves entire. Corolla 2–3 cm long, pale yellow, with long tube, almost twice as long as limb; upper lip concave, longitudinally folded. Perennial plant ............................................................
.................... 13. **Castilleja** Mutis ex L. f. [*C. pallida* (f.) Spreng.].

+ Leaves dentate. Corolla less than 15 mm long, white, pale lilac or dark pink, with tube almost as long or shorter; than limb; upper lip convex, not longitudinally folded. Annual plant ........................ 17.

17. Upper lip bilobed, with replicated lobes, lower lip with deeply emarginate lobes. Leaves ovate, broadly ovate or subrhombic.
.......................................................................... 14. **Euphrasia** L.

+ Upper lip emarginate or undivided with margins not replicated, lobes of lower lip entire. Leaves linear or broadly lanceolate .... 18.

18. Flowers on long (a few times longer than calyx) filiform pedicels, in diffuse paniculate inflorescence; corolla white, 5–5.5 mm long; anthers glabrous. ..... 15. **Omphalothrix** Maxim. (*O. longipes* Maxim.).

+ Flowers on short (half of calyx) pedicels, on secund spicate inflorescences; corolla dark pink or reddish, 8–10 mm long; anther lobes
pubescent .............. 16. **Odontites** Ludw. [*O. serotina* (Lam.) Dum.].

## 1. Verbascum L.

Sp. pl. (1753) 177; Murb. Monogr. Verbascum (1933).

1. Flowers invariably singly on rachis, on long (10–30 mm long)
pedicels; inflorescence sparse raceme ............................................. 2.
+ Flowers in clusters of 2–7 on rachis on short (1–5 mm long) pedicels;
inflorescence branched, paniculate or very compact, spicate ....... 3.
2. Flowers violet. Nearly all leaves radical, cordate or oblong-ovate,
with hairs scattered on both surfaces. Capsule 4.5–6 mm long,
obovoid, subacute, subglabrous ...................... 3. **V. phoeniceum** L.
+ Flowers yellow or dark brown. Stem foliated; leaves glabrous, sometimes hispid, radical leaves sessile, oblong, crenate-dentate, sometimes pinnatifid at base, caulous leaves subsessile, upper ones oblong-lanceolate, more or less amplexicaule. Capsule 5–7 mm long,
globose, obtuse, glandular ...................................... 1. **V. blattaria** L.
3. Inflorescence compact spicate raceme pedicels thik, short, more or
less adnate to common rachis; anthers of 2 anterior stamens
decurrent on filament. All plants covered with grey, more rarely
light yellow tomentose pubescence; radical leaves petiolate, oblong;
caulous leaves decurrent, oblong-ovate ................... 5. **V. thapsus** L.
+ Inflorescence paniculate; pedicels not adnate to rachis; anthers of
all stamens reniform ........................................................................ 4.
4. Plant with ash-coloured or whitish pubescence, especially in inflorescence; leaves with dense tomentose pubescence on both surfaces.
Flowers 25–30 (40) mm in dian.; filaments with white papilliform
hairs; capsule obovate or broadly ellipsoidal, nearly as long as ca-
lyx .............................................................. 4. **V. songaricum** Schrenk.
+ Plant weakly pubescent; leaves green on upper surface and shortly
pubescent beneath. Flowers 20–35 mm in diam.; filaments with vio-
let-coloured papilliform hairs; capsule oblong-ellipsoidal or oblong,
1/3 longer than calyx ........................................... 2. **V. orientale** M.B.

1. **V. blattaria** L. Sp. pl. (1753) 178; Murb. Monogr. Verbascum (1933)
560; B. Fedtsch. in Fl. SSSR, 22 (1955) 167; Fl. Kirgiz. 10 (1962) 146; Fl.
Kazakhst. 8 (1965) 30.   —**Ic.:** Reichb. Ic. fl. Germ. XX, tab. 12.

Described from South. Europe. Type in London (Linn.).

River banks, wet meadows and as weed in gardens.

IIA. Junggar: *Tien Shan* (lower Kunges, steppe belt, in wet meadow, on clayey soil, 1050
m, June 21, 1877—Przew.; Kul'dzha, May 15; Piluchi, May 17; garden in Kul'dzha vicinity,
June 12; Talki gorge, July 19—1877; upper Kunges, forest belt, wet meadow, 1300 m—[1878], A.

Reg.; 15 km east of Gunlyu to Tszin'tsyuan', on roadside, No. 1084, Aug. 10—1957, Kuan).
*Dzhark.* (Suidun settlement, July 16, 1877—A. Reg.; from Kul'dzha to Chapchal, on shoal, No. 3031, Aug. 5; between Kul'dzha and Suidun, on roadside, No. 3904, Aug. 31—1957, Kuan). *Balkh.-Alak.* (2 km west of Durbul'dzhin, sophora-wormwood fallow land, Aug. 7, 1957— Yun. et al.; Dachen [Chuguchak] vicinity, in marshy sites, No. 2826, Aug. 10, 1957—Kuan). *Jung. Alt.* (Toli district, near water, No. 2985, Aug. 15—1957, Kuan).

General distribution: Aralo-Casp., Fore Balkh., Jung.-Tarb., Nor. Tien Shan; Europe, Mediterr., Balk.-Asia Minor, Fore Asia, Caucasus, Mid. Asia, West. Sib.

2. **V. orientale** M.B. Fl. tauro-cauc. 1 (1808) 160; B. Fedtsch. in Fl. SSSR, 22 (1955) 147; Fl. Kazakhst, 8 (1965) 29. —*V. chaixii* Ledeb. Fl. Ross. 3 (1849) 202. —*V. chaixii* var. *orientale* Murb. Monogr. Verbascum (1933) 413. —Ic.: Fl. Kazakhst. 8, Plate III, fig. 4.

Described from East. Europe ("South Russia and Ukraine"). Type in Leningrad.

Mountain slopes in steppe and forest belts.

IIA. Junggar: *Tien Shan* (Piluchi 2100–2400 m, July 28, 1878—A. Reg.; Borokhoro mountain range, south. slope, 6 km south of N. Ortai settlement, lower part of forest belt, scrub steppe with asp and dried apricot groves, Aug. 19, 1957—Yun.).

General distribution: Aralo-Casp., Fore Balkh., Jung.-Tarb.; Europe, Balk.-Asia Minor, Caucasus, Mid. Asia, West. Sib.

3. **V. phoeniceum** L. Sp. pl. (1753) 178; Murb. Monogr. Verbascum (1933) 582; Kryl. Fl. Zap. Sib. 10 (1939) 2414; B. Fedtsch. in Fl. SSSR, 22 (1955) 168; Fl. Kirgiz. 10 (1962) 149; Fl. Kazakhst. 8 (1965) 30. —Ic.: Reichb. Ic. fl. Germ. XX, tab. 31, 1; Fl. Kazakhst. 8, Plate IV, fig. 2.

Described from Cent. Europe. Type in London (Linn.).

Steppe slopes, pastune-lands, fallow land and irrigation canals.

IIA. Junggar: *Cis-Alt.* (15 km north-west of Shara-Sume, scrub steppe, July 7; 20 km north-west of Shara-Sume, scrub steppe, July 7—1959, Yun. et al.). *Jung. Alt.* (Urtaksary west of Sairam lake, Aug. 13, 1877; same site, July 19; around Sairam lake, July 23—1878, Fet.). *Tien Shan* (Sary-Bulak, north-west of Kul'dzha, 1200 m, April, 24, 1878; Piluchi gorge, 900– 1200 m, April 23; same site, April 24—1879, A. Reg.; around Manasa, July 17, 1908—Merzbacher; 30 km west of Chzhaosu, Tszin'tsyuan', in intermontane basin, No. 3271, Aug. 19, 1957— Kuan).

General distribution: Aralo-Casp., Jung.-Tarb., Nor. and Cent. Tien Shan; Europe, Balk.- Asia Minor, Fore Asia, Caucasus, West. Sib.

4. **V. songaricum** Schrenk in Fisch. et Mey. Enum. pl. nov. 1 (1841) 26; Murb. Monogr. Verbascum (1933) 244; B. Fedtsch. in Fl. SSSR, 22 (1955) 133; Fl. Kirgiz. 10 (1962) 145; Fl. Kazakhst. 8 (1965) 28. —*V. polystachyum* Kar. et Kir. in Bull. Soc. natur. Moscou, 14 (1841) 716. —*V. candelabrum* Kar. et Kir. l.c. 717. —Ic. Fl. Kirgiz. 10, Plate 20; Fl. Kazakhst. 8, Plate III, fig. 2.

Described from Junggar Alatau. Type in Leningrad.

Steppe and rocky slopes of mountains, in river valleys.

IIA. Junggar: *Tarb.* (Dachen [Chuguchak]—Tarbagatai, 1300 m, No. 1708, Aug. 17, 1957—Kuan). *Tien Shan* (Talki river gorge, July 19, 1877—A. Reg.). *Balkh.-Alak.* (Dachen, in desert, No. 2882, Aug. 12, 1957—Kuan).

General distribution: Aralo-Casp., Jung.-Tarb., Nor. Tien Shan; Fore Asia (Iran), Mid. Asia.

Note. References in the original description of the species and on the label of the type that this species was collected "in Karatau foothills, June 11, 1840", as pointed out by B.A. Fedtschenko [Fl. SSSR, 22 (1955) 134], should be regarded as erroneous since Schrenk on that day was in Junggar Alatau, on the upper Karatal river.

5. **V. thapsus** L. Sp. pl. (1753) 177; Franch. in Bull. Soc. Bot. Franch, 47 (1900) 11; Pampanini, Fl. Carac. (1930) 188; Murb. Monogr. Verbascum (1933) 120; Pai in Contribs Inst. Bot. Nat. Ac. Peiping, 2, 7 (1933) 182; Kryl. Fl. Zap. Sib. 10 (1939) 2410; B. Fedtsch. in Fl. SSSR, 22 (1955) 128; Fl. Kirgiz. 10 (1962) 144; Fl. Kazakhst. 8 (1965) 27. —*V. schraderi* G. Mey. Chlor. Hannov. (1836) 326. —Ic.: Reichb. Ic. fl. Germ. XXII, tab. 16; Fl. Kazakhst. 8, Plate III, fig. 2.

Described from Europe. Type in London (Linn.).

Banks of rivers and brooks, meadows, talus and rock exposures, 1000–3000 m.

IIA. Junggar: *Jung. Alt.* (Toli district, No. 1136, Aug. 7, 1957—Kuan). *Tien Shan* (Kunges river, meadow, 1000 m, June 19, 1877—Przew.; Piluchi, June 19; Talki river gorge, July 18; Dzhagastai, 1600–2300 m, Aug. 9—1877; Kunges river [1878]; below Aryslyn, 2000 m, June 20; Khap chagai-Aryslyn, 2000–2300 m, July 7; Aryslyn river gorge, 2800–3000 m, July 10—1879; Tekes river gorge, 1880—A. Reg.; environs of Tekes river [July 4, 1886]—Krasnov; "Ouroumtai, under Bogdo Ula, alt. 1070 m, No. 3167, Aug. 14, 1931—Liou".—Pai, l.c.).

General distribution: Fore Balkh., Jung.-Tarb., Nor. Tien Shan; Europe, Mediterr., Balk.-Asia Minor, Fore Asia, Caucasus, Mid. Asia, West. Sib. (Altay), East. Sib., China (South-West), Himalayas (west., Kashmir); introduced in Japan and Nor. Amer.

## 2. Linaria Mill.[1]
### Gard. Dict. (1768) No. 14.

1. Inflorescence rachis, calyx and pedicels with dense and villous pubescence of simple and glandular hairs; calyx lobes 5–6 mm long, linear-lanceolate ................................................ 3. **L. buriatica** Turcz.

+ Inflorescence rachis, calyx and pedicels glabrous or with short glandular hairs; calyx lobes 1.5–4 mm long, ovate, oblong-ovate or lanceolate ........................................................................................ 2.

2. Seeds discoid, sharply tuberculate at centre. Corolla yellow; lower lip with bright orange spot; spur broadly conical (2–3 mm broad at base); flowers aggregated into dense more or less long raceme; calyx lobes acute ................................. 1. **L. acutiloba** Fisch. ex Reichb.

---

[1]Drawing up an accurate picture of the species composition of genus *Linaria* in Central Asia is difficult at present because material on *Linaria* species described from contiguous Central Asian regions is extremely fragmentary and partly not amenable to reliable identification.

+ Seeds discoid, glabrous at centre. Corolla violet, yellow or yellow-ish–violet; lower lip with yellow or orange spot; spur slender (generally 1 mm, rarely 1.5–2 mm broad at base); flowers aggregated into lax and long, rarely compact and short raceme or 1–3 at branch ends; calyx lobes obtuse, rounded or acuminate ............................ 3.

3. Corolla yellow ............................................................................................ 4.

+ Corolla violet or violet–yellow (upper and lower lips violet but tube and spurs yellow) ..................................................................................... 5.

4. Corolla 8–12 mm long (excluding spur), pale yellow, with 2 bright yellow bands on umbo of lower lip; corolla tube gradually narrowing at base. Plant 10–20 cm tall ............................ 2. **L. altaica** Fisch.

+ Corolla 13–15 mm long (excluding spur), yellow with orange spots in throat; corolla tube abruptly narrowing at base. Plant (10) 20–30 cm tall. ............................................ 5. **L. pedicellata** Kuprian.

5. Corolla with yellow tube and spur and brownish–violet lips. Stems single or 2–3; lower caulous leaves broadly lanceolate or lanceolate-linear, 2–3 cm long, (1.5) 3–5 mm broad, with 3 veins, narrowing towards base, shortly acuminate at tip. Capsule ellipsoidal, 8–10 mm long .............................................................. 4. **L. hepatica** Bunge.

+ Corolla with light or dark violet tube and dark violet lips and spur. Stems numerous or rarely single; caulous leaves lanceolate, linear-lanceolate or linear, not narrowing towards base, gradually acuminate at tip. Capsule globose, 4–6 (8) mm in diam. ............................ 6.

6. Stems numerous, conspicuously branched in middle and upper portions (branches emerge almost at right angle to stem or erect), vegetative branches lacking at base of stems. Flowers in small numbers aggregated at ends of stems; corolla dark violet with yellow spot in throat ............................ 6. **L. ramosa** (Kar. et Kir.) Kuprian.

+ Stems few, rarely single, poorly branched at base or in upper portion of stem (branches upright), vegetative branches present at base of stems. Flowers aggregated at branch ends in more or less lax long inflorescence; lower and upper lips bright violet, tube and spur pale violet with orange spot in throat .......... 7. **L. transiliensis** Kuprian.

1. **L. acutiloba** Fisch. ex Reichb. Ic. pl. crit. 5 (1827) 14; Kuprian. in Tr. Bot. inst. AN SSSR, ser. 1, 9 (1950) 47; Grubov, Konsp. fl. MNR (1955) 240; Kuprian. in Fl. SSSR, 22 (1955) 202; Sergievskaja in Kryl. Fl. Zap. Sib. 12, 2 (1964) 3445; Fl. Kazakhst. 8 (1965) 35. —*L. vulgaris* auct. non Mill.: Maxim. in Mem. Ac. Sci. St.-Pétersb. Sav. Etrang. 9 (1859) 484; Franch. Pl. David. 1 (1884) 221; Forbes and Hemsley, Index Fl. Sin. 2 (1902) 178; Sapozhn. Mong. Alt. (1911) 379; Danguy in Bull. Mus. Nat. hist. natur. 17, 7 (1911) 554; Pai in Contribs Inst. Bot. Nat. Ac. Peiping, 2, 7 (1934) 182, p.p.; Kitag. Lin. Fl. Mansh. (1939) 393, p.p.; Li in Bot. Bull. Ac. Sinica (Taipai) 3 (1961) 205,

p.p. —*L. vulgaris* var. *latifolia* Kryl. in Fl. Zap. Sib. 10 (1939) 2418. —**Ic.:** Reichb. l.c. fig. 611.

Described from Dauria from plants grown from seeds sent to Reichenbach by Fischer. Type in Leningrad.

In wild rye, feather grass, forb-tansy, scrub-meadow and other steppes, on dry rocky, clayey-rocky and rubble slopes, rock crevices among granite outcrops, talus and cliffs, sandy-clayey soil on dry river-beds, sand, in thickets, occasionally in larch forests and forb meadows from plains to alpine belt, as well as weed among crops and in fallow land.

**IA. Mongolia:** *Khobd.* (on border to Khobdo town road, between 2nd and 3rd posts, June 18; same site, June 20—1870, Kalning). *Mong. Alt.* (V. Kobdossk lake, July 19, 1909—San.; Khara-Dzarga mountain range, Shatyn-Gol river, rocky gorge, Aug. 27, 1930—Pob.; Dzun-Bulak area 32 km south-west of Yusun-Bulak along road to Shargain-Gobi, among barley crops, July 15, 1947—Yun.) *Cis-Hing.* (environs of Trekhrech'e, No. 1356, July 14, Argun' district, Maritka river, 650–750 m, No. 1911, Aug. 13—1951, Wang and Li) *Cen. Khalkha, East. Mong., Gobi-Alt.* (Gurban-Bongo, Gurban-Saikhan). *East. Gobi* (without date or location—1831, Bunge and 1841, Kirilov). *Ordos* (Saksygyr river, July 24; Guan'-dzhatagai area, Sept. 6—1884, Pot.; Ordos australis [1877]—Verlinden).

**IIA. Junggar:** *Cis-Altay* (Shara-Sume, on slope, No. 1300, Aug. 3; in Fuyun' town region, 1200 m, No. 1808, Aug. 13; Shara-Sume, 1400 m, No. 2490, Aug. 27—1956, Ching; 20 km north-west of Shara-Sume, scrub steppe, July 7, 1959—Yun. et al.). *Jung. Alt.* (Urtaksary river, Borotala region, July 19, 1878—Fet.; upper Borotala, 1800–2100 m, Aug. 8, 1878—A. Reg.). *Tien Shan* (environs of Sairam-Nur lake, 2100–2750 m). *Balkh.-Alak.* (Churchutsu river near Chuguchak town, Aug. 9, 1840—Schrenk).

**General distribution:** Fore Balkh., Jung.-Tarb.; Arct. (Europ.), West. Sib., East. Sib., Nor. Mongolia, China (Altay, Dunbei).

**Note.** Very closely related to *L. vulgaris* Mill. which not so far detected in Cent. Asia. *L. acutiloba* is a fairly polymorphic species with very broad ecological range. Most typical plants (leaves about 3–6 cm long and 5–8 mm broad, with 3 veins, stem simple or branched in inflorescence, 20–40 cm tall, calyx lobes 1.2–1.5 mm broad) are confined to hilly slopes of Jung. Alatau and Tien Shan and often found in meadows and river valleys of Mongolia and Cent. Khalkha region and on slopes of Mong. and Gobi Altay. Narrow-leaved race of this species— f. *angustifolia* Serg.—predominates in steppes, on dry rocky slopes, sandy-pebble soils in Cent. Khalkha, in river valleys in Tien Shan and especially in East. Mongolia. This race is characterised by narrow (2–4 mm broad) leaves invariably with 1 vein, 20–50 cm tall stems, generally branched in inflorescence, calyx lobes 0.9–1.5 mm broad. Plants collected by V.I. Grubov and A.A. Yunatov in Gobi Altay (Ikhe-Bogdo town) differ significantly from typical plants and we separate them into a distinct variety, var. *pygmea* Ivanina. —Caules 7–20 cm (nec 20–50 cm) alti; folia 1.5–2 cm (nec 2–8 cm) longa; inflorescentia pauciflora, floribus 1–7 (nec 8–20), brevis, 3–6 cm (nec 10–20 cm) longa. The small plant size and few-flowered inflorescence (sometimes with deformed flowers) are evidently explained by the fact that these specimens are found at the upper limit of the distribution range of species in the hills.

In most plants growing in herb meadows and sand in East. Mongolia and Ordos, stems branch in inflorescence, sometimes right from base, very densely foliated, calyx lobes with marginal glandular pubescence. Stray specimens from these same Mongolian regions (Khailar town vicinity, meadow on sand flat, No. 619, Oct. 6, 1951—A.R. Lee (1959)) have additionally subwhorled leaves and, as a result, resemble *L. melampyroides* Kuprian (differ in orange, not violet, spot on lower lip of corolla, broader lobes of calyx, erect stems, etc.).

2. **L. altaica** Fisch. in Ledeb. Fl. alt. 2 (1830) 448; Kuprian. in Tr. Bot. inst. AN SSSR, ser. 1, 2 (1936) 29; Kryl. Fl. Zap. Sib. 10 (1939) 2420, p.p.; Grubov, Konsp. fl. MNR (1955) 241; Kuprian, in Fl. SSSR, 22 (1955) 207; Fl. Kazakhst. 8 (1965) 39. —*L. odora* α *major* Kryl. Fl. alt. 4 (1907) 927. —*L. odora* auct. non Fisch. Sapozhn. Mong. Alt. (1911) 379. —*L. uralensis* Kotov in Bot. zh. AN UkSSR, 3, 3–4 (1946) 26. —*L. dmitrievae* Semiotr. in Fl. Kazakhst. 8 (1965) 40, 421. —Ic.: Kuprian. l.c. (1936) fig. 3; Fl. Kazakhst. 8, Plate V, fig. 6.

Described from Altay. Type in Leningrad.

Rocky steppe and desert-steppe valley slopes, mud cones and low mountains.

IA. Mongolia: *Mong. Alt.* (north-west, Oigur river valley, Tsagan-Gol river, July 16, 1909—Sap.). *Bas. Lakes* (Buyantu river, plantations, Aug. 28, 1930—Bar.).

IIA. Junggar: *Jung. Alt.* (Dzhair mountain range, 25 km west of Aktam, along Urumchi-Chuguchak road, desert steppe, on knolls No. 1092, Aug. 3; same site, 1–2 km north of Otu settlement, steppe belt, on bank of valley of small steppe brook, No. 1138, Aug. 4—1957, Yun. et al.; same site, low hills, No. 939, Aug. 4; 25 km north of Toli settlement, in ditch, No. 1451, Aug. 9—1957, Kuan).

General distribution: *Fore Balkh.* (Ulutau); *West. Sib.* (South. Urals, Mugodzhary and Altay).

Note. *L. altaica* shows considerable variation (over its range as a whole as well as in Cent. Asia) and some specimens are difficult to distinguish from *L. debilis* Kuprian. Plants gathered from Mong. Altay differ from typical plants in pubescence on calyx (similar to *L. debilis* in this respect) while specimens from the environs of Toli are prominent with very long spur (but altogether glabrous calyx, as in typical *L. altaica*, predominates).

3. **L. buriatica** Turcz. Cat. Baikal (1837) 14, nom.; id. in Bull. Soc. natur. Moscou, 24 (1851) 302; Kuprian. in Sov. bot. 4 (1936) 117; Kitag. Lin. Fl. Mansh. (1939) 393; Grubov, Konsp. fl. MNR (1955) 241; Kuprian. in Fl. SSSR, 22 (1955) 196. —Ic.: Kuprian, l.c. (1936) fig. 3.

Described from East. Siberia (Ol'khon island on Baikal). Type in Leningrad.

Rocky and rubble steppe mountain slopes, sandy steppes, coastal sand, sand floors of gorges, among shrubs.

IA. Mongolia: *Khobd.* (mountains north-east of Ureg-Nur lake, June 22, 1879—Pot.). *Cent. Khalkha, East. Mong.* (between Kulusutaev post and Dalai-Nur lake, 1870—Lom.; west. bank of Buir-Nur lake, on sandy soil, June 15; Kulun-Buir-Nurskaya plain, June 25—1899, Pot. and Sold., Manchuria station, steppe, June 5, 1902—Litw.; same site, 1915—E. Nechaeva; Dariganga, east of Ongon-Elis sand, Khoshun-Khuduk well, sandy steppe on spurs, Sept. 20, 1931—Pob.).

General distribution: East. Sib. (south), Nor. Mong., China (North).

4. **L. hepatica** Bunge in Ledeb. Ic. pl. fl. ross. 1 (1829) 22 et Fl. alt. 2 (1830) 445; Kryl. Fl. Zap. Sib. 10 (1939) 2421; Grubov, Konsp. fl. MNR (1955) 241; Kuprian. in Fl. SSSR, 22 (1955) 205; Fl. Kazakhst. 8 (1965) 36. —*L. macroura* γ. *hepatica* (Bunge) Benth. in DC. Prodr. 10 (1846) 273. —Ic.: Ledeb. l.c. tab. 91; Fl. Kazakhst. 8, Plate V, fig. 2.

Described from East. Kazakhstan. Type in Paris (?).

Rubble and rocky steppe mountain slopes, sandy-pebble floors of gorges and shoals.

IA. Mongolia: *Mong. Alt.* (Adzhi-Bogdo mountain range, Dzusylyn gorge, June 29; Tatal river, on wet soil, July 8—1877; Pot.; south. slope of Mong. Altay, Kostun area, June 9, 1903—Gr.-Grzh.; Bulgan somon, Ulyasutuin-Gol area, 15 km from estuary, floor of lateral dry creek valley, sand bed of gorge, July 21, 1947—Yun.).

IIA. Junggar: *Zaisan* (Besh-Kudyk collective, rocky-sandy steppe, on flat top of upland, June 18, 1914—Schischk.).

General distribution: Fore Balkh. (Zaisan), Jung.-Tarb.; West. Sib. (Altay).

Note. Specimens from Zaisan basin (including those from Kazakhstan, Balkany cape on Zaisan lake) differ from typical *L. hepatica* in narrow (linear-filiform) leaves, approaching *L. pedicellata* Kuprian. in this characteristic. However, the corolla of Zaisan plants is bicoloured (with violet lips and yellow tube and spur) as in *L. hepatica*. Zaisan plants are evidently hybrids between these two species.

5. **L. pedicellata** Kuprian. in Bot. mater. Gerb. Bot. inst. AN SSSR, 11 (1949) 161 and in Tr. Bot. inst. AN SSSR, ser. 1, 9 (1950) 49; Kuprian. in Fl. SSSR, 22 (1955) 211; Fl. Kazakhst. 8 (1965) 42.    —Ic.: Fl. SSSR, 22, Plate IX, fig. 3.

Described from East. Kazakhstan (Balkhash region). Type in Leningrad.

Sand, rocky-sandy steppes and as weed on fallow land and plantations.

IA. Mongolia: *Mong. Alt.* (Mankhan somon, Bodkhon-Gol area, 30–35 km south of Tugrik-Sume, weeds on fallow land and crops, Aug. 11, 1945—Yun.). *Bas. Lakes* (around Bichigin-Nuru cliff, on rocky talus, July 18; near entrance to Ulan-Sair creek valley [Darbin-Nuru], July 22—1896, Klem.).

IIA. Junggar: *Jung. Gobi* (Baitak-Bogdo mountain range, nor. slope, Ulyaste-Gola gorge, 3–4 km from entrance, on flanks and floodplain terrace, Sept. 12, 1948—Grub.). *Zaisan* (Mai-Kapchagai hills, rocky slopes, June 6; between Kaba river and Besh-Kudyk well, sandy-rocky steppe, June 17; Besh-Kudyk well, rocky-sandy steppe, on flat top of upland, June 18—1914, Schischk.).

General distribution: Fore Balkh., Jung.-Tarb. (Chingil'dy environs), Nor. Tien Shan (Chu-Ili hills).

Note. Herbarium material for this species is scant and polymorphic. Specimens collected by E. Klements in Basin of Lakes (July 22, 1896) differ significantly from typical ones (low stems, rather short leaves, short pedicel and narrow corolla) and resemble *L. sessilis* Kuprian. described from Pamir. Hybrids of *L. pedicellata* and *L. hepatica* (or *L. transiliensis*) are evidently found in Zaisan basin.

6. **L. ramosa** (Kar. et Kir.) Kuprian. in Fl. SSSR, 22 (1955) 207; Fl. Kazakhst. 8 (1965) 39.    —*L. praecox* β. *ramosa* Kar. et Kir. in Bull. Soc. natur. Moscou. 15 (1842) 145.

Described from Junggar. Type in Moscow. Isotype in Leningrad.

Sandy and sandy-rubble deserts.

IIA. Junggar: *Jung. Gobi* (lower Borotola 450–600 m, Aug. 22, 1878—A. Reg.; along Kul'dzha-Urumchi road, 517th km, rubble desert, 750–900 m, No. 2161, Aug. 29, 1957— Kuan).

General distribution: Fore Balkh., Jung.-Tarb., Nor. Tien Shan.

7. **L. transiliensis** Kuprian. in Tr. Bot. inst. AN SSSR, ser. 1, 9 (1950) 69; Kuprian. in Fl. SSSR, 22 (1955) 206; Fl. Kirgiz. 10 (1962) 155; Fl. Kazakhst. 8 (1965) 38. —*L. odora* δ. *violacea* Ledeb. Fl. Ross. 3 (1849) 208. —Ic.: Fl. SSSR, 22, Plate IX, fig. 7; Fl. Kirgiz. 10, Plate 21; Fl. Kazakhst. 8, Plate V, fig. 3.

Described from Tien Shan (Transili Alatau). Type in Leningrad.

Rubble and rocky slopes of mountains in steppes in forest and steppe belts, 600–2450 m alt.

IB. Kashgar: *East.* (Turfan, 850 m, Sept. 1879—A. Reg.).

IIA. Junggar: *Jung. Alt.* (Urtaksary river, July 19, 1878—Fet.; Khorgos river, April 22; confluence of Khorgos and Alimtu, April 22—1877; midcourse of Khorgos, 1200–1500 m, May 15, 1878—A. Reg.). *Tien Shan* (Muzart foothill, Aug. 15, 1877; environs of Kul'dzha, 1200, May 21; Shary-Su river, 2100–2450 m, May 26; Khanakhai, 1200–1500 m, June 16—1878; Pilyuchi river, April 22; Taldy river, Iren-Khabirga mountain range, 1500–2450 m, May 15; upper Taldy, May 15—1879, A. Reg.; Kungei valley, May 1, 1908—Merzbacher). *Dzhark.* (Bayandai village on outskirts of Kul'dzha town, 600–1200 m, May 6, 1878—A. Reg.; 1 km south of mines in Dzhagastai along Ili, Aug. 7, 1957—Kuan).

General distribution: Fore Balkh., Jung.-Tarb., Nor. and Cent. Tien Shan, East. Pam.; Mid. Asia (Pamiro-Alay, West. Tien Shan).

Note. Plants from Jung. Alatau differ from typical ones in small size of corolla and calyx with glandular hairs (former somewhat resemble Altay species *L. bungei* Kuprian.) while specimens collected in steppe of Ili basin differ in low stems, few-flowered inflorescence and calyx with glandular hairs.

## 3. Scrophularia L.

Sp. pl. (1753) 619; Stiefelh. in Engler's Bot. Jahrb. 44 (1910) 406.

1. Corolla yellow or yellowish-white, 9–17 mm long. .......................2.
+ Corolla dark purple, brown or green, 4–7 mm long .....................4.
2. Calyx 6–7 mm long, deeply 5-fid, with tube about 1/2 its length and unequal (subbilabiate) limb. Plant 3–10 (15) cm tall; leaf blade 1–2.5 cm long, 0.7–1.5 cm broad, with faintly visible secondary veins beneath ........................................................ 10. **S. przewalskii** Batalin.
+ Calyx 4–5 mm long, 5-partite, with nearly equal lobes. Plant 15–40 (65) cm tall; leaf blade 2.5–7 (15) cm long, 1.5–5 cm broad, with distinct secondary veins beneath .........................................................3.
3. Plant glabrous; leaves runcinate or sharply toothed; bracts 10–16 mm long, ovate, longer than calyx ........... 1. **S. alaschanica** Batalin.
+ Plant with glandular and simple hairs; leaves unevenly dentate; bracts linear, 4–5 mm long, shorter than calyx. ....... **S. altaica** Murr.

4. Leaves round-cordate or rounded at base; under surface of lower leaves with distinct anastomosing secondary veins ...................... 5.

+ Leaves more or less cuneately narrowed, gradually passing into petiole, more rarely truncate; under surface of lower leaves with non-anastomosing secondary veins.................................................... 6.

5. Stem 4-angled, angles of stem and petiole winged. Calyx with rounded lobes, broadly scarious along margin; corolla greenish-red. Plant glabrous. ...................................... 2. **S. alata** Gilib.

+ Stem subglabrous, wingless. Calyx with oblong or oblong-lanceolate obtuse immarginate (or with very narrow, scarious margin) lobes; corolla green. Plant covered with glandular and simple hairs.
........................................ 5. **S. heucheriiflora** Schrenk ex Fisch. et Mey.

6. Leaves pinnatipartite or pinnatisect ..............................................
.................................................7. **S. kiriloviana** Schischk. ex Gorschk.

+ Leaves undivided or not deeply pinnatilobed, lower ones sometimes lyrate ............................................................................... 7.

7. Stamens included in corolla; sepals with more or less broad membranous margin. ................................................................................ 8.

+ Stamens exserted from corolla; sepals with narrow white or green membranous margin. ....................................................................... 9.

8. Flowers aggregated into sparse paniculate inflorescence; staminodes clavate. Plant with canescent pubescence, bushy, broom-like; stems numerous, slightly woody at base, 10–20 cm tall. ........4. **S. dentata** Royle ex Benth.

+ Flowers aggregated into very dense oblong-ovate or oblong inflorescence; staminodes lacking or 0.2 mm long, oblong-linear. Plant glabrous except for glandular-pubescent inflorescence, with some herbaceous stems, 7–12 cm tall. ...................... 11. **S. regelii** Ivanina.

9′ (7). Staminodes orbicular-tetragonal, 0.8–0.9 mm long, 1–1.2 mm broad, gradually narrowing at base. Plant 50–80 cm tall, stems thick, 5–7 mm in diam.; leaves 6–8 cm long and 1.5–4 cm broad ....................
.............................................................. 9. **S. potaninii** Ivanina.

+ Staminodes lanceolate, oblong, ovate or obovate, 0.5–1 mm long, 0.2–0.9 mm broad. Plant 10–40 cm tall, stems 2–4 mm in diam.; leaves 1–6 cm long, 0.2–3 cm broad. ................................................. 10.

10. Leaves greyish, 0.8–1.5 cm long, 0.2–0.8 cm broad, obtuse, leaf blade shortly glandular-hairy .................... 3. **S. canescens** Bong.

+ Leaves green, 1.5–4 (6) cm long, 0.7–3 cm broad, subacute, leaf blade glabrous or diffusely glandular-hairy............................................ 11.

11. Leaves linear or lanceolate, 1.5–3 cm long, 0.2–0.5 cm broad, entire or remotely dentate, lower sometimes coarsely serrate stems remotely foliated. Inflorescence 1–7 cm long, flowers aggregated singly or in pairs; staminodes oblong-linear, 0.6–0.8 mm long, 0.2–0.4

mm broad, acute. Plant 8–25 cm tall. ...............................................
................................................ 8. **S. pamirica** (B. Fedtsch.) Ivanina.

+ Leaves oblong-elliptical or ovate-lanceolate, 2–6 cm long, 1–2.5 cm broad, coarsely dentate or lower caulous leaves lyrate. Inflorescence 7–30 cm long, flowers aggregated in groups of 2 to 5, more rarely singly or inpairs; staminode oblong or oblong-ovate, 0.5–1.0 mm long, 0.2–0.8 mm broad. Plant 20–70 cm tall. ... 6. **S. incisa** Weinm.

1. **S. alaschanica** Batalin in Acta Horti Petrop. 13 (1894) 380; Stiefelh. in Engler's Bot. Jahrb. 44 (1910) 461; Pai in Contribs Inst. Bot. Nat. Ac. Peiping, 2, 7 (1934) 183; Hao in Engler's Bot. Jahrb. 68 (1938) 638; Ching in Bull. Fan. Mem. Inst. Biol. (Bot.) 10, 5 (1941) 262; Walker in Contribs U.S. Nat. Herb. 28 (1941) 660; Li in Lioydia, 16, 3 (1953) 169.—*S. delavayi* auct. non Franch.: Walker in Contribs U.S. Nat. Herb. 28 (1941) 660.

Described from Mongolia. Type in Leningrad. Plate VI. fig. 2.

Foot of rocks under shade, 1350–2400 m alt.

IA. **Mongolia:** *Alash. Gobi* (in midportion of Alashan mountain range, west. slope, June 23, 1873—Przew., typus!; "Ha La Hu Kou, a valley north-west side of the Ho Lan Shan range, its mouth 30 li from Wang Yeh Fu; Ho Lan Shan, on moist, gravelly valley bottoms, Ching"—Walker, l.c.).

IIIA. **Qinghai:** *Amdo* ("Mingke [Ming-ge-schan], alt. 3900 m, No. 1007, Aug. 25, 1930, Hopkinson"—Pai, l.c.).

General distribution: endemic.

2. **S. alata** Gilib. Fl. lith. (1781) 127; Gorschkova in Fl. SSSR, 22 (1955) 270; Fl. Kirgiz. 10 (1962) 158; Fl. Kazakhst. 8 (1965) 54. —*S. aquatica* auct. non L.: Kryl. Fl. Zap. Sib. 10 (1939) 2427. —Ic.: Fl. Kazakhst. 8, Plate VI, fig. 6.

Described from Europe (Grodno town). Type in Kiev.

Wet and marshy meadows, around lakes and ditches, 1600–2000 m s.m.

IIA. **Junggar:** *Cis-Alt.* (south. spur of Mong. Altay, along edge of irrigation ditch, 1700 m, No. 2592, Aug. 29, 1956—Ching). Tarb. (environs of Dachen [Chuguchak], in marshy site, No. 2833, Aug. 10, 1957—Kuan). *Tien Shan* (Talki river gorge, July 18, 1877; Urtaksary river, 1800 m, Aug. 4, 1878; Borgaty river, 1600–2000 m, July 4 and July 8—1879, A. Reg.; Urumchi region, pass between Tien Shan and Bogdo-Ula hill, near Saëpu lake, in marshes, Sept. 5, 1929—Pop.; Dzhagastai, on edge of irrigation ditch, No. 3913, Aug. 8; Fukan district, near Tyan'chi lake, near water, No. 4379, Sept. 20—1957, Kuan).

General distribution: Aralo-Casp., Fore Balkh., Jung.-Tarb., Nor. and Cent. Tien Shan; Europe, Mediterr., Balk.-Asia Minor, Fore Asia, Mid. Asia, West. Sib. (Altay), East. Sib.

**S. altaica** Murr. in Comment. Soc. Sc. Götting. (1781) 35; Kryl. Fl. Zap. Sib. 10 (1939) 2424; Gorschkova in Fl. SSSR, 22 (1955) 261; fl. Kazakhst. 10 (1965) 53. —Ic.: Fl. SSSR, 22, Plate XI, fig. 4.

Described from Altay. Type in Berlin (?).

On rocks under shade.

Found in Hangay region bordering valley of lakes in Nor. Mongolia (Bulgansk Ajmak, Under-Ulan somon, Chulutuin-Khurym area, rocky slope, June 27, 1956—D. Banzragch).

IA. Mongolia: *Mong. Alt.* (occurrence likely).
General distribution: West. Sib. (Altay), East. Sib. (Ang.-Sayan), Nor. Mong. (Hang.).

3. **S. canescens** Bong. in Bull. Ac. Sci. St.-Pétersb. 8 (1841) 340; Franch. Pl. David. 1 (1884) 221; Pai in Contribs Inst. Bot. Nat. Ac. Peiping, 2, 7 (1934) 184; Kryl. Fl. Zap. Sib. 10 (1939) 2429; Gorschkova in Fl. SSSR, 22 (1955) 298; Fl. Kazakhst. 8 (1965) 58. —*S. cretacea* Fisch. var. *glabrata* (Franch.) Stiefelh. in Engler's Bot. Jahrb. 44 (1910) 477; Li in Lloydia, 16, 3 (1953) 179. —Ic.: Bong. in Mém. Ac. Sci. St.-Pétersb. ser. VI, 2, tab, 12; Fl. SSSR, 22, Plate XIII, fig. 1.

Described from East. Kazakhstan (Zaisan lake). Type in Leningrad.

Sandy and pebble banks, solonetz meadows, along valleys and dry river-beds.

IA. Mongolia: *East. Mong.* (Inshan': Ourato, bords du torrent de Kuentileen dans le sable, June 1866—David.).

IIA. Junggar: *Tien Shan* ("near Ouroumtai, No. 2820, July 25, 1931, Liou"—Pai, l.c.).

General distribution: Aralo-Casp., Fore Balkh., Jung.-Tarb., East. Tien Shan; Mid. Asia, China (North).

Note. David's specimens from Kuenti-Lien river sand in the Inshan' system (East. Mongolia) differ from typical Kazakhstan specimens (var. *canescens*) in near-total absence of farinaceous-greyish pubescence and much smaller leaves. Franchet (l.c.) treated these plants as a distinct variety (var. *glabrata* Franch.). However, Stiefelhagen (l.c.) and later Li (l.c.) placed them under *S. cretacea* Fisch. distributed only in the Volga basin. We cannot concur with this status since *S. cretacea* growing on lime and chalk slopes (not on valley sand) are densely covered with white glandular pubescence, leaves coarsely dentate (not at all entire) and upper lip 3 (not 1.5–2) times longer than lower. Subglabrous plants with small obtuse leaves (up to 0.7 cm long, 0.6 cm broad) are found within the main range of *S. canescens*. Specimens of this variety exhibit some similarity also with *S. incisa* Weinm. in similar staminodes, calyx lobes and colour of corolla, but differ distinctly in small size of leaves and corolla.

4. **S. dentata** Royle ex Benth. Scroph. Ind. (1835) 19; Hemsley; Fl. Tibet (1902) 192; Ostenfeld and Paulsen in Hedin, S. Tibet, 6, 3 (1922) 45; Pampanini, Fl. Carac. (1930) 188; ej. Agg. Fl. Carac. (1934) 171.

Described from Himalayas (Kumaon). Type in Leningrad.

Slopes in alpine belt, 4000–5000 m alt.

IIIB. Tibet: *Chang Tang* ("Without locality, Deasy and Pike"—Hemsley, l.c.). *South.* ("S.W. Tibet, between Camp 194 [30° 04' N. lat., 83°01' E. long.], Gjiangtju-Kaman, 4461 m and Camp 195 [30°14' N. lat., 82°57' E. long.], Thärck, 4657 m valley of Upper Tsangpo, July 6, 1907, Hedin"—Ostenfeld and Paulsen, l.c.).

General distribution: Himalayas (west., Kashmir).

5. **S. heucheriiflora** Schrenk ex Fisch. et Mey. Enum. pl. nov. 1 (1841) 25; Kryl. Fl. Zap. Sib. 10 (1939) 2425; Gorschkova in Fl. SSSR, 22 (1955) 260;

Fl. Kirgiz. 10 (1962) 158; Fl. Kazakhst. 8 (1965) 52. —Ic.: Fl. SSSR, 22, Plate XI, fig. 1.

Described from East. Kazakhstan (Ayadyr). Type in Leningrad.

Gorges and along mountain rivers valleys, rocky slopes in forest belt.

IIA. Junggar: *Tien Shan* (Sary-Bulak river, 1300-2000 m, April; Suidun river, 1200–1800 m, April; Almaty river near Kul'dzha, 1200–1500 m, May—1878, A. Reg.).

General distribution: Fore Balkh., Jung.-Tarb., Nor. and Cent. Tien Shan; Mid. Asia (Pamiro-Alay), West. Sib.

6. **S. incisa** Weinm. Bot. Gart. Univ. Dorp. (1810) 136; Maxim. in Bull. Soc. natur. Moscou, 54 (1879) 34; Stiefelh. in Engler's Bot. Jahrb. 44 (1910) 478; Sapozhn. Mong. Alt. (1911) 380; Danguy in Bull. Mus. nat. hist. natur. 20 (1914) 80; Rehder and Kobuski in J. Arn. Arb. 14 (1933) 32; Pai in Contribs Inst. Bot. Nat. Ac. Peiping, 2, 7 (1934) 184; Hao in Engler's Bot. Jahrb. 68 (1938) 636; Kryl. Fl. Zap. Sib. 10 (1939) 2428; Kitag. Lin. Fl. Mansh. (1939) 397; Walker in Contribs U.S. Nat. Herb. 28 (1941) 660; Li in Lloydia, 16, 3 (1953) 179; Grubov, Konsp. fl. MNR (1955) 241; Gorschkova in Fl. SSSR, 22 (1955) 306; Fl. Kazakhst. 8 (1965) 61; Fl. Kirgiz. Dop. 1 (1967) 106. —*S. patriniana* Wyder, Essi Mon. Scroph. (1828) 39. —*S. orientalis* auct. non L.: Maxim. in Mém. Ac. Sci. St.-Pétersb. Sav. Etrang. 9 (1859) 484. —*S. integrifolia* auct. non Pavlov: Fl. Kirgiz. 10 (1962) 159, p.p. —Ic.: Ledeb. Ic. pl. fl. ross. 2, tab. 156.

Described from specimens grown in Tartu botanical garden from Siberian seeds. Type lost. Plate VII, fig. 2.

Rocky and stony steppe slopes, along flanks and gorges of dry creek valleys, on coastal and river-bed pebbles, around gorges, in sand, sandy-rocky and clayey dry steppes.

IA. Mongolia: *Khobd.* (Burtu area, Kharkhira hill group, July 17, 1903—Gr. Grzh.). *Mong. Alt., Cis-Hing.* (near Yakshi railway station, rocks, June 11 and Aug. 18—1902, Litw.). *Cent. Khalkha* (midcourse of Kerulen, steppe near Dalaibeise camp, 1899—Pal.). *East. Mong.* ("ancied fond du Dalai-Nor, sables, No. 1710, June 14, 1896, Chaff."—Danguy, l.c.). *Bas. Lakes* (Shuryk river valley, July 23, 1877; vicinity of Khirgis-Nur lake, July 28, 1879—Pot.; Dzabkhyn river valley, below Tsagan-Oloma, gorge near Slava foothill, July 22, 1950—Kuznetsov). *Val. Lakes* (Ongiin-Gol river, dry floor of Lamynkhor gorge, July 21, 1926—Lis.; Del'ger-Hangai somon, Ongiin-Gol river floodplain, 30 km below Khushu-Khida, solonchak meadow combined with pebble shoals, July 7, 1943—Yun.; Ongiin-Gol river valley 10 km north of Khayashine-Usu-Khuduk, floodplain along border of river-bed, Oct. 22, 1947—Grub. and Kal.). *Gobi-Alt., East. Gobi.* (Alashan-Urgu road, Turkhum-Sume area, May 17, 1909—Davydenko). *Khesi* (15 km nor. of Yunchan town, rocky slopes of Beidashan' hill, June 28, 1958—Yun.).

IB. Kashgar: *Nor.* (Uchturfan, Airi area, along brook, between stones, June 4; Taret river, Charlysha tributary, common, June 19—1909, Divn.). *West.* (Kenkol area, Togoibashi, July 30, 1913—Knorring; 30 km before Turugart [on Kashgar road], steppe belt, along gorge valley, 3100 m, July 20, 1959—Yun. et al.).

IC. Qaidam: *montane* (south. vicinity of Dulan-Khit temple, on pebble of dry river-beds, brooks and overflows, 3050 m, Aug. 10, 1901—Lad.).

IIA. Junggar: *Tien Shan* (in estuary above 'Nanshan' kou [Karlyktag], on pebble bed, June 6, 1877—Pot.; south. slope of Tien Shan, toward Khami desert, on clayey and rocky soil, May 25, 1879—Przew.; M. Yuldus basin, right bank of Khaidyk-Gol river, 10 km east of Bain-Bulak settlement, on road to Karashar, steppe belt, on pebble bed, Aug. 11, 1958—Yun. et al.; hill road from Bartu village to Khomote timber factory, 2160 m, No. 6974, Aug. 3; 2 km southwest of Bain-Bulak settlement, Khotun-Sumbul, on exposed slope, 2650 m, No. 6440, Aug. 10—1958, Lee et al.), *Jung.-Gobi.* (from Tien Shan-Laoba to Myaoergou, No. 2413, Aug. 3; from Tsitai to Beidashan', on shaded slope, 1700 m, No. 5252, Sept. 28—1957, Kuan).

IIIA. Qinghai: *Nan Shan* (on barren slope in clayey soil and humus [North Tetungsk mountain range], June 21, 1872—Przew.; Shonpyn village, Lan'chzha-Lunva river valley, on dry sandy-rocky soil, May 14, 1885—Pot.; environs of Pinfansyan village on nor. slope [Huang He valley], June 14, 1885—Pias.; Humboldt mountain range, Argalin-Ula, Sharagol'dzhin river valley, meadow, May 17, 1894—Rob.; Kuku-Nor lake, sandbank, 3650 m, Aug. 29, 1908—Czet.; Artyn-tag mountain range, rocky slopes of gorge, 2800 m; Aug. 1958; meadow on east bank of Kuku-Nor lake, 5200 m, Aug. 5, 1959—Petr.; "Shui Mo Kou, near Lien Ch'eng, on exposed moist foothills of sandy soil, No. 364, Ching"—Walker, l.c.). *Amdo* (Baga-Gorgi river, on clayey soil, May 14, 1880—Przew.;"Radja and Yellow River gorges, Rock"—Rehder and Kobuski, l.c.; "in der Nähe des Klosters Ta-schin-cze, 3600 m, No. 1031, Aug. 26, 1930, Hao"— Hao, l.c.).

IIIC. Pamir (Kulan-aryk area, between Zad settlement and Tashui, 3500–4800 m, July 29, 1942—Serp).

General distribution: Aralo-Casp., Fore Balkh., Jung.-Tarb., Nor. and Cent. Tien Shan; Mid. Asia (Pamiro-Alay), West. and East. Sib., Nor. Mong. (Hang., Mong.-Daur), China (Altay, North, North-West).

Note. Species widely distributed in Mongolia but very rare in Junggar, Kashgar and other areas; varies considerably especially in vegetative characteristics (form and size of leaves and inflorescence) and form of staminodes (oblong to ovate-orbicular). Plants found on steppe slopes of high mountains in Kashgar differ from typical plants in lyrate radical and caulous leaves and 10–20 cm tall stems. *S. incisa*, differing from typical plants in pinnately incised leaves, is found in the eastern part of East. Tien Shan.

7. **S. kiriloviana** Schischk. ex Gorschk. in Fl. SSSR, 22 (1955) 306; Fl. Kirgiz. 10 (1962) 160; Fl. Kazakhst. 8 (1965) 60.    —*S. pinnata* Kar. et Kir. in Bull. Soc. natur. Moscou, 14 (1841) 719. non Mill. (1768).    —*S. pinnata* Kar. et Kir.    *subpinnata* Fisch. et Mey. Ind. sem. Horti Petrop. 10 (1845) 35.    — *S. incisa* Weinm. var. *alpina* Kar. et Kir. in Bull. Soc. natur. Moscou, 15 (1842) 414, p.p.    —*S. incisa* Weinm. var. *pinnata* Trautv. in Bull. Soc. natur. Moscou, 39 (1866) 435, p.p.    —Ic.: Fl. Kirgiz. 10, Plate 22, fig. 2.

Described from Tarbagatai (Chegarak-asu river). Type in Leningrad. Coniferous forests and tall-grass subalpine meadows.

IIA. Junggar: *Tarb.* (nor. of Dachen [Chuguchak], on shaded slope, No. 1551, Aug. 17, 1957—Kuan). *Jung. Alt.* (Urtaksary, west of Sairam lake, July 19, 1878—Fet.; Toli district, on nor. slope of water divide, 2400 m, No. 1225, Aug.; same site, 3400 m, No. 1223, Aug. 6; same site, Albakzin town, on slope, Nos. 2610, 2723, Aug. 7; 15 km nor. of Ulastai, No. 1235, Aug. 8; 20 km east of Ven'tsyuan' [Arasana], on slope in forest, No. 4677, Aug. 27— 1957, Kuan). *Tien Shan* (Talki river gorge around Kul'dzha, July 18; Sairam-Nur lake, 2100–2750 m, July 20–24; midcourse of Muzart, 1800–2450 m, Aug. 20–21; same site, Aug. 21–22—1877; between Kegen and Khorgos rivers, 600–900 m, May 12; Khorgos river, 1600–1800 m, May 15; upper Khorgos, 2750 m, Aug. 11—1878, south of confluence of Kargol and Pass rivers near Nilki, 2450–2750 m, June 16; Mengute, 3050—3350 m, July 4—1879; Kash river valley, June—1881, A. Reg.; Badan' to Danu, on slope, 2700 m, No. 2011, July 19; Danukhe river valley, on slope, 2600 m, No. 2129, July 21—1957, Kuan; Manas river basin, Ulan-Usu river valley, 8–9 km above discharge of Koisu river into it, forest belt, on pebble bed, along floor of right bank of creek valley, July 18, 1957—Yun. et al.).

General distribution: Fore Balkh., Jung. Tarb., Nor. and Cent. Tien Shan; Mid. Asia (West. Tien Shan and Pamiro-Alay).

Note. Species closely related to *S. incisa* Weinm. and associated with it through intermediates but we treat it as an independent species because of the presence of a large number of specimens with a stable complex of distinctive characteristics (plant 40–85 cm tall, leaves pinnatisect or pinnatipartite, 6–10 cm long, 2.5–5 cm broad, with acute, linear-oblong, incised, serrate-dentate lobes; calyx with broadly scarious lobes), confined to spruce forest zone in West. East. Tien Shan and Junggar (within the USSR—Tarbagatai, east. Nor. Tien Shan). Plants with characteristics of *S. incisa* are rare in these sites. Somewhat transitional forms are distributed in east. East. Tien Shan, in Jung. Alatau and in Kashgar. We have named them *S. incisa* var. *alpina* and placed them appropriately. The presence of plants with characteristics of *S. incisa* (confined to steppe belt or on open rocky slopes or talus in forest or subalpine belts) in territories in habited by *S. kiriloviana* may be regarded as evidence of the individualism of the two species.

The genesis is not clear of the following race differing from typical plants in rather small size, pinnately lobed leaves with obtuse oblong-ovate lobes and orbicular-rhombic staminodes: *S. kiriloviana* Schisck. var. *flexuosa* Ivanina. —Planta perennis, 20–30 cm alta; caules basi sublignescentes quadrangulares. Folia opposita oblongo-ovata vel ovata, 1–4 cm longa, 0.5–2 cm lata, petiolata, petiolis 1–2 cm longis, pinnatilobata, lobis oblongo-ovatis. Flores singuliterni in inflorescentiam anguste paniculatam laxam flexuosam congesti; calyx ca 2 mm longus brunneus, laciniatus, laciniis margine late aureomembranaceis; corolla 6–6.5 mm longa; filamenta flexuosa, inclusa; staminodium orbiculari-rhombicum, 0.6–0.8 mm longum , 0.8 mm latum. Capsula ignota.

Typus: Dschungaria, Irenchabirga, fl. Taldy, 2400 m, June 24, 1879, A. Regel. In Herb. Inst. Bot. Acad. Sci. URSS (Leningrad) conservatur. Plate VII, fig. 3.

8. S. pamirica (O. Fedtsch.) Ivanina, comb. nova.—*S. incisa* var. *pamirica* O. Fedtsch. et var. *angustifolia* O. Fedtsch. in Tr. Bot. sada, 21 (1903) 391. — *S. incisa* auct. non Weinm.: Ikonnikov in Dokl. AN TadzhSSR, 20 (1957) 55; ibid., Opred. rast. Pamira (1963) 217.

Described from East. Pamir. type in Leningrad. Plate VII, fig. 4.

Rocky slopes and talus in high-mountain belt.

IIIC. Pamir (Muztagata foothill, on rocky talus, July 20, 1909—Divn.).
General distribution: East. Pam.

9. S. potaninii Ivanina, sp. nova.

Planta perennis. Caules solitarii herbacei, teretes, 50–80 cm alti, crassi (ad 7 mm in diam.). Folia alternantia, ovata vel oblongo-elliptica, inferiora et media 6–8 cm longa, ca 3 cm lata, petiolis 1.5–3 cm longis praedita, incisodentata. Flores in cymis terni-septeni, congesti, inflorescentiam paniculatam pyramidatam formantes; calyx 2–2.5 mm longus, laciniis anguste viridipaleaceis; corolla ca 5 mm longa, lobis atro-purpureis. Stamina vix exserta; staminodium quadrangulari-rotundatum, 0.8–0.9 mm longum, 1–1.2 mm latum. Capsula ignota.

Typus: Gobi, fl. Bajan-gol, brachium fl. Lonsyr, in solo lapidosa humida, June 1, 1886, Potanin. In Herb. Inst. Bot. Acad. Sci. URSS (Leningrad) conservatur.

Affinitas, Species *S. koelzii* Pennell plus minusve affinis sed foliis majoribus, 6–8 cm longis (nec 1–2 cm longis), inflorescentia lata pyramidali-paniculata (nec angusta racemosa) et pedicellis pallide (nec obscure) glanduloso-pilosis differt.

Plate VI, fig. 1.

Wet rocky river banks.

**IA. Mongolia:** *Khesi* (Bayan-Gol river, tributary of Lonsyr river, wet rocky soil, June 1, 1886—Pot., typus!).

**General distribution:** endemic.

**Note.** Species more or less closely related to *S. koelzii* Pennell but differs in large leaves, 6–8 (not 1–2) cm long, broad pyramidal-paniculate (not narrow racemose) inflorescence, pedicels with light (not dark) glandular hairs; well distinguished from *S. incisa* Weinm. in nearly flabellate staminodes and general shape.

10. **S. przewalskii** Batalin in Acta Horti Petrop. 13 (1894) 382; Stiefelh. in Engler's Bot. Jahrb. 44 (1910) 469; Marquand in J. Linn. Soc. London (Bot.) 48 (1929) 219; Li in Lloydia, 16, 3 (1953) 178.

Described from Tibet (Weitzan). Type in Leningrad.

Plate VI. fig. 3.

Rocky slopes, rocks and river-beds in alpine belt, 4000–4700 m alt.

**IA. Mongolia:** *Alash. Gobi* (Alashan mountain range, Yamata gorge, east. slope, along river-bed, on humus soil, May 5, 1908—Czet.).

**IIIA. Qinghai:** *Nanshan* (Choibsen-Khit temple, 1901—Lad.).

**IIIB. Tibet:** *Weitzan* (on south. slope of water divide of Huang He and Yantszytszyan rivers, 4300 m, May 29; hills on Bychu river, left tributary of Yantszytszyan river, 4700 m, June 3— 1884, Przew., typus!; Yantszytszyan river basin, Chzhabu-Vrun area, under pass and on slope, on wet sticky clay, 4350–4500 m, July 9, 1900; Tszergen area, on wet [freezing at night] clayey-rocky exposed rocks on nor. mountain slope, 4150 m, May 14; Lomlun area, 4400 m, May 15—1901, Lad.).

**General distribution:** endemic.

11. **S. regelii** Ivanina, sp. nova.

Planta perennis. Caules numerosi subquadrangulares, 6–10 cm alti. Folia opposita oblongo-ovata vel oblonga, 1–3 cm longa, 0.5–1 cm lata, incisa vel lobata, lobis lanceolatis unibiparibus. Flores singuli-terni in inflorescentiam densam oblongo-ovatam vel oblongam; calyx laciniatus, laciniis late et pallide viridi-paleaceis; corolla 4–5 mm longa; stamina inclusa; staminodium nullum (vel raro lanceolatum 0.2 mm longum). Capsula ignota.

Typus: irenchabirga, fl. Taldy, 2800 m, May 26, 1879, A. Regel. In Herb. Inst. Bot. Acad. Sci. URSS (Leningrad) conservatur.

Affinitas. Species *S. dentatae* et *S. incisae* plus minusve affinis a quibus tamen caulibus humilibus, inflorescentia oblonga-ovata vel oblonga densa (nec paniculata angusta laxa), laciniis calycis late (nec anguste) paleaceis necnon notis ceteris distat.

Plate VII, fig. 1.

River valleys around 2800 m alt.

IIA. Junggar: *Tien Shan* (Irenkhabirga mountain range, 2800 m, May 26, 1879—A. Reg.).
General distribtion: endemic.
Note. Species more or less closely related to *S. dentata* Royle ex Benth. and *S. incisa* Weinm. but differing in low stems, dense oblong-ovate or oblong (not narrow, sparsely paniculate) inflorescence, broadly (not narrowly) scarious calyx lobes, absence of staminodes and other features.

## 4. Dodartia L.

Sp. pl. (1753) 633.

1. **D. orientalis** L. Sp. pl. (1753) 633; Henderson and Hume, Lahore to Jarkand (1873) 254; Diels in Futterer, Durch Asien, 3 (1903) 20; Sapozhn. Mong. Alt. (1911) 380; Pai in Contribs Inst. Bot. Nat. Ac. Peiping, 2, 7 (1934) 190; Persson in Bot. notiser (1938) 301; Kryl. Fl. Zap. Sib. 10 (1939) 2430; Grubov, Konsp. fl. MNR (1955) 241; Gorschkova in Fl. SSSR, 22 (1955) 319; Chen and Chou, Rast. pokrov r. Sulekhe (1957) 90; fl. Kirgiz. 10 (1962) 164; Fl. Kazakhst. 8 (1965) 46. —*D. atrocoerulea* N. Pavl. in Vestn. AN KazSSR, 5 (1952) 91. —Ic.: Curtis's Bot. Mag. 48, tab. 2199; Fl. Kazakhst. 8, Plate VI, fig. 3.

Described from Caucasus (Ararat hill). Type in London (Linn.).

Clayey and pebble coastal alluvium, fringes of toirims, solonchak meadows and chee grass thickets, desert and desert-steppe rubble trails as well as weed in old ploughed fields and irrigation ditches, in crops and on roadsides.

IA. Mongolia: *Khobd.* (on way from border to Khobdo town, between 2nd and 3rd pickets, June 18, 1870—Kalning). *Mong. Alt.* (Bulugun river bank, beyond Dzhirgalanta estuary, July 25, 1898—Klem.; Bulugunskii region, Tsagan-Tyunge, Sept. 22, 1930—Bar.). *East. Mong.* ("P'atzupulung, near Wulashan, alt. 1100 m, No. 3177, Aug. 29, 1931, Hsia"—Pai, l.c.). *Alash. Gobi* (Dzhintasy village, July 3, 1886—Pot.; Shara-Burdu area, Edzin-Gol river, marsh overgrown with reeds, solonchaks with sparse floodplain poplar forests with tamarisk undergrowth, June 20, 1909—Lad.; Edzin-Gol valley, near upper Ontsin-Gol, Bukhan-Khub area, sand-mounds June 7, 1926—Glag.; Shitszuidzhan' village, in wheat crop on Huang-He river bank, near pass, July 12, 1957—Petr.). *Ordos* (Ordos, 1884, Bretschneider). *Khesi* (Sachzou, Aug. 1 and 4, 1875—Pias.; same site, on solonetz clayey soil, June 14, 1879—Przew.; between Shakhe and Fuiitin villages, on sandy soil, June 5, 1886—Pot.; Su-tcheou, May 30, 1890—Martin; Sachzou oasis, along irrigation ditches and borders of ploughed fields, May 6, 1894—Rob.; 12 km south-west of An'si, semi-overgrown sand at site of Tubaochin fort ruins, Sangun sand, July 25, 1958—Petr.;"zwischen Su-tschôu and Liang-tschôu, Sept. 30, 1890"—Diels, l.c.; "Sulekhe river"—Chen and Chou, l.c.).

IB. Kashgar: *Nor.* (Taushkan-Darya river, in old ploughed fields and irrigation ditches, 1500 m, June 16, 1889—Rob.; environs of Uchturfan, along hill valleys; between Dzham village and Aksu river—May 1903, Merzbacher; Uchturfan, bank of irrigation canal, May 10,

1908—Divn.; "Jakaerik, ca 1070 m, Aug. 7, 1932"—Persson, l.c.). *West.* (environs of Yangi-Gissar, on borders of wheat field, May 24, 1909—Divn.; along Kashgar-Ulugchat road, in river gorge, 2140 m, No. 384, June 17, Artush, Khalatsi, 1650 m, No. 9786, June 22—1959, Lee and Chu; 75 km west-north-west of Kashgar, along road to Kensu mine, chee grass thickets on second terrace, June 17, 1959—Yun. et al.; "this is a common bush in the desert between Sanju and Jarkand"—Henderson and Hume, l.c.; "*Jarkand*, 1350 m, No. 160, May 23, 1930; Kashgar, 1330 m, No. 496, April 24, 1934"—Persson, l.c.). *South.* (nor. foothill of Kunlun, from Kargailyk to Ladak, Tash-Bulak river, in old ploughed fields and along brooks, 1500–1600 m, July 26, 1889—Rob.). *East.* (Khami, on clayey soil, May 30, 1877—Pot.; Khami desert, on clayey saline soil, June 6, 1879—Przew.; upper Algoi, 2400 m, Sept. 11; north-west of Turfan, on Algoi river, Sept. 14; Turfan, Sept. 29—1879, A. Reg.; north-west of Khami, No. 104, July 6, 1956—Ching; in Khami region, Nos. 427, 453, May 21; No. 999, June 23—1957, Kuan; west of Khami, Utakelik, No. 5407, May 22; Turfan district, state farm "Krasnaya Zvezda", No. 5448, May 26; Toksun district, 650 m, in desert, No. 7499, June 16—1958, Lee and Chu).

IIA. *Junggar:* **Cis-Alt.** (Kiikty-Sai river valley, Ch. Irtysh river basin, Aug. 5, 1906—Sap.; Qinhe district, No. 739, Aug. 1; in Qinhe [Chingil'] region, in field, No. 1628, Aug. 11—1956, Ching; 15 km north-west of Shara-Sume, scrub steppe, July 7, 1959—Yun. et al.). **Tarb.** (Saur-Argaltyn, May 20–June 28, 1877—Pev.). **Jung. Alt.** (north-west of Dzhair mountain range trail, 24 km north-east of Toli settlement on road to Temirtam, Modun-Obo brook, on wet pebble beds, Aug. 5, 1957—Yun. et al.; in Ven'tsyuan' region, No. 1293, Aug. 31, 1957—Kuan). **Tien Shan** (desert on lower Kunges, 900–1050 m, Sept. 4, 1876—Przew.; Nanshankou picket, on rocky soil, June 7, 1877—Pot.; Kul'dzha, July 3, 1877—A. Reg.). **Jung. Gobi, Dzhark., Zaisan** ("Ul'kun-Ulasty, Urta-Ulasty"—Sapozhn. l.c.).

**General distribution:** Aralo-Casp., Fore Balkh., Nor. and Cent. Tien Shan; Europe (south), Asia Minor, Fore Asia (Iran), Caucasus, Mid. Asia.

## 5. Lancea Hook f. et Thoms.
### in J. Bot. Kew Misc. 9 (1857) 244.

1. **L. tibetica** Hook. f. et Thoms. in J. Bot. Kew Misc. 9 (1857) 244; Hook. f. Fl. Brit. Ind. 4 (1884) 260; Henderson and Hume, Lahore to Jarkand (1873) 256; Kanitz in Széchenyi, Wissen-schaff. Ergebn. 2 (1898) 722; Diels in Futterer, Durch Asien, 3 (1903) 20; Paulsen in Hedin, S. Tibet, 6, 3 (1922) 43; Pampanini, Fl. Carac. (1930) 188; Pai in Contribs Inst. Bot. Nat. Ac. Peiping, 2, 7 (1934) 190; Walker in Contribs U.S. Nat. Herb. 28 (1941) 658; Grubov, Konsp. fl. MNR (1955) 241.   —Ic.: Hook, f. et Thoms. l.c. tab. 7.

Described from Himalayas. Type in London (K). Plate VIII, fig. 1.

Wet, often solonetzic banks of rivers and lakes, sometimes in spruce forests, up to 4650 m. Map 3.

IIIA. **Qinghai:** *Nanshan* (South Tetung mountain range [Soli-soroksum hill], in valley, forest, on pebble bed, in wet places, often in groups, July 3, 1872, in Tetung river valley, lower and middle south. slope, in wet sites, often in groups, May 26, 1873; Huang He river basin [on Dzhakharchu river], June 24, 1880—Przew.; Chadzhi gorge, May 12; Sinin valley, Myndan'sha river, May 27 and June 5, 1890—Gr.-Grzh.; South Kukunor mountain range, nor. slope, near Tszagastyten-Korol' pass, 3350–3450 m, Aug. 15, 1901; near Donkyr monastery, on foot-hills, in humus, Aug. 5, 1908—Czet.; "in fruticetis collium circa viam versus Wo-so-ling, June 18; Joung-tschang-shien, June 12—1879"—Kanitz, l.c.; "Liu Fu Jai [P'ing Fan Hsien]; many growing together on exposed, moist, valley bottoms, common, Ching"—Walker, l.c.). *Amdo* (Dulan-khit temple, spruce forest, in humus, 3350 m, Aug. 9, 1901—Lad.).

IIIB. Tibet: *Changtang* ("Inner Tibet, Camp 78 [31°40' N. lat., 88°22' E. long.], shore of lake Naktsong'tso, 4636 m, Sept. 11, 1901, Hedin"—Paulsen, l.c.)". *Weitzan* (Yantszytszyan river, moist banks, common, 4000 m, June 23; along Dzhagyn-Gol river, on banks, July 23—1884, Przew.; in isthmus separating Russkoe lake from Ekspeditsii lake, on hills, in moist clay covered with fine pebbles, 4150 m, June 30, 1900—Lad.; "Tosson-nor, steppenbaden, July 12, 1909, Filchner"—Diels, l.c.).

General distribution: Nor. Mongolia (Hang.), China (North-West, South-West), Himalayas.

## 6. Gratiola L.
Sp. pl. (1753) 17.

1. **G. officinalis** L. Sp. pl. (1753) 17; Kryl. Fl. Zap. Sib. 10 (1939) 2431; Gorschkova in Fl. SSSR, 22 (1955) 322; Fl. Kazakhst. 8 (1965) 47. —Ic.: fl. Kazakhst, 8, Plate VI, fig. 4.

Described from Europe. Type in London (Linn.).

Banks of rivers and reservoirs, marshes and floodplain meadows.

IIA. Junggar: *Jung. Gobi* (Ch. Irtysh river, Dyurbel'dzhin pass, south-east of Kran river basin, Aug. 24, 1876—Pot.).

General distribution: Aralo-Casp., Fore Balkh., Jung.-Tarb., Nor. Tien Shan; Europe, Mediterr., Balk.-Asia Minor, Fore Asia, Caucasus, Mid. Asia, West. Sib., Japan (in crop), Nor. America.

## 7. Limosella L.
Sp. pl. (1753) 631.

1. **L. aquatica** L. Sp. pl. (1753) 631; Kryl. Fl. Zap. Sib. 10 (1939) 2432; Grubov, Konsp. fl. MNR (1955) 241; Gorschkova in Fl. SSSR, 22 (1955) 324; Fl. Kirgiz. 10 (1962) 165; Fl. Kazakhst. 8 (1965) 50. —*Limosella* sp. Pampanini, Fl. Carac. (1930) 188. —Ic.: Fl. Kazakhst. 8, Plate VI, fig. 5.

Described from Nor. Europe. Type in London (Linn.).

Silty and sandy lake banks, floodplains of rivers, flooded meadows, along pools and irrigation canals.

IA. Mongolia: *Val. Lakes*, (Tuin Gol, edge of pool at 1200 m, No. 330, 1925—Chaney). *East. Gobi* (Delger-Hangay somon, Khushu-khid monastery on Ongiin-Gol, river floodplain in river-bed area, Oct. 20, 1947—Grub. and Kal.).

IIA. Junggar: *Jung. Gobi* (Ch. Irtysh river, Dyurbel'dzhin pass, Aug. 24, 1876—Pot.).

General distribution: Aralo-Casp., Fore Balkh., Nor. Tien Shan; Arct., Europe, Caucasus, Mid. Asia (Pamiro-Alay), West. and East. Sib., Nor. Mongolia, China (North), Himalayas, Nor. and South Americas, Africa, Australia.

## 8. Veronica L.
Sp. pl. (1753) 9. —*Veronicastrum* Heist. ex Fabr. Enum. meth. (1759) 111.

1. Plants of aquatic or wet habitats. Capsule inflated ........................ 2.

+ Plants of dry terrestrial habitats. Capsule more or less laterally flattened or weakly inflated ................................................................... 3.

2. Leaves on short, more or less distinct petioles, with obtuse or
rounded tip, oblong-elliptical more rarely orbicular or oblong; stem
cylindrical, dense ..................................... 3. **V. beccabunga** L.

+ Leaves sessile (lowermost sometimes on very short petiole), semi-
amplexicaul, acute or acuminate, oblong or lanceolate, rarely oblong-ovate; stems nearly 4-angled fistular ....................................
........................................................ 1. **V. anagallis-aquatica** L.

3. Perennials ............................................................................................ 4.

+ Annuals .............................................................................................. 18.

4. Inflorescence axillary, opposite or flowers singly in leaf axils; main
stem terminating in vegetative shoots ........................................... 5.

+ Inflorescence terminal, racemose, paniculate-racemose, spicate or
subcapitate; stems and their branches terminating in inflorescence
......................................................................................................... 6.

5. Flowers single in axils of bracts, approximate above in racemose
inflorescence, not sharply separated from main part of stem. Lower
leaves orbicular or ovate, obtuse, entire or crenate; upper oblong-
ovate or lanceolate ................................... 15. **V. serpyllifolia** L.

+ Flowers in axillary opposite racemose inflorescence, sharply separated from main part of stem. Leaves oblong-ovate to lanceolate
and linear-lanceolate, acuminate, serrate or incised-dentate ..........
........................................................... 8. **V. krylovii** Schischk.

6. Corolla clavate, with long tube (moe than 2/3 of corolla) and short
limb. All leaves aggregated into verticels of 3–9 each ....................
................................................................... 24. **V. sibirica** L.

+ Corolla rotate, with very short tube or with tube 1/3–1/2 of corolla
and prominent limb (more than 1/2 of corolla). Leaves alternate,
opposite or sometimes whorled, 3–4 together ............................... 7.

7. Inflorescence capitate; calyx 5-partite ........................................... 8.

+ Inflorescence racemose, paniculate-racemose or spicate; calyx 4-
partite ............................................................................................... 10.

8. Plant with fibrous roots, not forming mat, 10–40 cm tall; leaves 2–
2.5 cm long; bracts linear, ciliate. Stamens 1/2 of corolla lobes. Capsule oblong-ovoid, 9–10 mm long, narrowing at tip ........................
.................................................................. 4. **V. ciliata** Fisch.

+ Plant with slender creeping rhizome, forming compact or lax mat,
5–15 cm tall; leaves 0.5–2 cm long; bracts ovate-lanceolate, pubescent or glabrous. Stamens as long or longer than corolla. Capsule
obovate, 5–6 mm long, not narrowing at tip ................................. 9.

9. Plant forming lax mat; stems interrupted, branched midlength;
leaves oblong-elliptical or ovate, subacute, serrate-dentate, diffusely

pubescent an upper surface; glabrous beneath. Calyx lobes lanceolate, hairy along margin. ..................... 12. **V. macrostemon** Bunge.

+ Plant forming compact mat; stems branched; leaves oblong, obovate or orbicular-ovate, subobtuse, obtuse-dentate or crenate, slightly pubescent on both surfaces. Calyx lobes ovate or oblong, ciliate. .. .............................................................................6. **V. densiflora** Ledeb.

10 (7). Leaves alternate, rarely some opposite ........................................ 11.

+ Leaves opposite or sometimes whorled (mostly 3, rarely 4, together) ...............................................................................................14.

11. Leaves ovate or oblong in general shape, pinnatisect into linear or lanceolate long lobes. Stamens scarcely exserted from corolla........ ............................................................................... 13. **V. pinnata** L.

+ Leaves linear or lanceolate, undivided, small- or large-toothed, serrulate or entire. Stamens twice longer than corolla limb. ....... 12.

12. Roots slender, fibrous, short, not woody. Leaves 4–20 mm broad, serrulate or crenulate ...................... 10. **V. linariifolia** Pall. ex Link.

+ Rhizome more or less thick and long, woody. Leaves 0.5–4 mm broad with interrupted, uneven, large teeth, or entire ...........................13.

13. Plant green, subglabrous or covered with short curved hairs; leaves with uneven large teeth, or entire. Calyx lobes acuminate, lanceolate or ovate-oblong, glandular-ciliate along margin ..................... ............................................................................ 9. **V. laeta** Kar. et Kir.

+ Plant with dense grey pubescence due to short curved hairs; leaves entire or with interrupted teeth. Calyx lobes acute, ovate, rarely ovate-oblong, densely pubescent. ........ 2. **V. arenosa** (Serg.) Boriss.

14 (10). Flowers sessile or subsessile (pedicels of lower flowers 1–2 mm long); bracts longer than pedicels. Leaves sessile or narrowing at base into petiole, crenulate or scarcely crenate. Plants generally 10–20 cm tall. .....................................................................................15.

+ Pedicels longer than 2 mm; bracts shorter than pedicels or as long. Leaves petiolate or rarely narrowing at base into petiole, sharply serrate, biserrate or bidentate, deeply and coarsely irregularly dentate or coarsely serrate. Plants generally 30–80 cm tall ........16.

15. Plant with dense white pubescence of slender crispate adherent hairs. Corolla with very short tube, lobes ovate or broadly ovate .. ...............................................................................7. **V. incana** L.

+ Plant green or greyish-green, pubescent with short patent hairs mixed with glandular hairs. Corolla with short tube, lobes lanceolate or oblong .................................... 14. **V. porphyriana** Pavl.

16. Roots slender, fibrous, short; leaves ovate-oblong or deltoid, more or less deeply coarsely irregularly dentate or coarsely serrate, with dense glandular pubescence, rarely diffusely pilose. Corolla about

7 mm long, white or pink, occasionally dark blue ...........................
................................................................ 5. **V. dahurica** Stev.

+ Rhizome creeping, long; leaves linear, linear-lanceolate or broadly lanceolate, sharply serrate or biserrate, glabrous or densely covered with short crispate hairs. Corolla 5–6 mm long, blue, bluish-violet, sometimes pink. ......................................................................17.

17. Leaves cuneately narrowed at base into short petiole, stiff, whorled, 3, rarely 4 in each verticil or opposite, covered with simple crispate hairs. Inflorescence paniculate-racemose with several elongated racemes aggregated at tip; all lobes of corolla ovate...................... .
......................................................... 16. **V. spuria** L.

+ Leaves not deeply cordate at base, orbicular or truncate, opposite, rarely whorled, 3–4 in each verticil, glabrous, sometimes slightly pubescent along veins beneath. Inflorescence single raceme, sometimes with few lateral racemes; 1 lobe of corolla orbicular, rest oblong. ........................................................ 11. **V. longifolia** L.

18 (3). Seeds flat .......................................... 20 **V. perpusilla** Boiss.

+ Seeds carinate, concave on one side and convex on other ..........19.

19. Caulous leaves 4, whorled at base of branching stem; leafy bracts strongly differing from caulous leaves, linear. Capsule with ovate-orbicular or orbicular lobes. ...................... 23. **V. tenuissima** Boriss.

+ Caulous leaves opposite or alternate; leafy bracts similar to caulous leaves or different, lanceolate. Capsule with oblong or ovate lobes ................................................................................................20.

20. Capsule lobes erect or insignificantly divergent at acute angle (not more than 45°), calyx lobes connate in pairs at base or almost up to middle. ...................................... 22. **V. rubrifolia** Boiss.

+ Capsule lobes diverge differently (at more than 45°) or horizontal; calyx lobes partite almost up to base or connate in pairs very shortly ................................................................................................21.

21. Caulous leaves broadly ovate to orbicular, coarsely crenate-dentate. Capsule 8–10 mm broad. Plant 15–70 (120) cm tall ...........................
.......................................................... 21. **V. persica** Poir.

+ Caulous leaves lanceolate to ovate, entire or remotely serrate-dentate. Capsule 2–5 mm broad. Plant 5–15 cm tall. ...................22.

22. Calyx lobes lanceolate or oblong, long-acuminate and acuminulate, glabrous or sparsely hairy, 1.5–2 times longer than capsule; seeds deeply transversely rugose ................... 19. **V. campylopoda** Boiss.

+ Calyx lobes ovate or broadly elliptical, short-acuminate, glandular-hairy, nearly as long as capsule; seeds glabrous or slightly rugose ................................................................................................23.

23. Corolla shorter than calyx; calyx lobes at base very shortly connate in pairs; pedicel after anthesis deflexed or erect. Capsule with

oblong lobes; seeds undulate only along margin. All leaves opposite, oblong or lanceolate, acuminate, entire or remotely serrate-dentate ............................................................................ 18. **V. biloba L.**

+ Corolla as long or longer than calyx; calyx lobes partite almost up to base; pedicel arcuately upcurved after anthesis. Capsule with ovate lobes; seeds subglabrous or indistinctly undulate. Lower leaves opposite, upper alternate, ovate to lanceolate, almost incised serrate-dentate along margin............. 17. **V. argute-serrata** Regel et Schmalh.

## Subgenus Veronica[1]

### A. Perennials

1. **V. anagallis-aquatica** L. Sp. pl. (1753) 12; Kitag. Lin. Fl. Mansh. (1939) 397; Li in Proc. Ac. Natur. Sci. Philad. 104 (1952) 217; Borissova in Fl. SSSR, 22 (1955) 469; Yamazaki in J. Fac. Sci. Univ. Tokyo, sect. III, 7, 2 (1957) 159; Fl. Kirgiz. 10 (1962) 180; Fl. Kazakhst. 8 (1965) 88. —*V. anagallis* auct.: Franch. Pl. David. 1 (1884) 224; Forbes and Hemsley, Index Fl. Sin. 2 (1902) 198; Sapozhn. Mong. Alt. (1911) 380; Danguy in Bull. Mus. nat hist. natur. 5 (1911) 14; Pampanini, Fl. Carac. (1930) 189; Pai in Contribs Inst. Bot. Nat. Ac. Peiping, 2, 7 (1934) 195; Persson in Bot. notiser (1938) 301; Kryl. Fl. Zap. Sib. 10 (1939) 2454; Walker in Contribs U.S. Nat. Herb. 28 (1941) 660; Grubov, Konsp. fl. MNR (1955) 241—Ic.: Fl. Kazakhst. 8, Plate X, fig. 7.

Described from Europe, Type in London (Linn.).

Banks of rivers, brooks and irrigation ditches, around springs, moist and marshy meadows, near and in water.

IA. Mongolia: *Mong. Alt.* (Tatalty river, right tributary of Bulugun, Aug. 25, 1898—Klem.; Khairkhan river, near water, Oct. 10, 1930—Pob.; Khan-Taishiri mountain range, south. trail, Dzun-Bulak spring, on way to Bain-somon, on quicksand, in spring sources, Sept. 3, 1948—Grub.). *Cent. Khalkha* (Kerulen river valley, near water in wet sites, July 26; Ikhe-Bulak, on bank of brook, Sept. 2—1924, Lis.; upper course of Ubur-Dzhargalante river, near Botoga town, in brook, Sept. 11, 1925—Krasch. and Zam.). *East. Mong.* (along right bank of Huang He river, below Hekou, in water, Aug. 4, 1884—Pot.; Dariganga, 22 km west of Baishintu-Sume, Bagalyutne spring, on wet bank of spring, Sept. 10; Ongon-Elisu sand, Boro-Bulak spring, on bank of spring, Sept. 13—1931, Pob.; 60 km north of Dushnya town, in Khantaichuan river valley, near emergence of spring water, Aug. 13, 1957—Petr.). *Val. Lakes* (south. bank of Orok-Nur lake, in meadow, Sept. 2, 1886—Pot.; left bank of Baidarik, on sand, July 20, 1894—Klem.; Orok-Nor, border of moist lagoon at 1100 m, No. 318, 1925—Chaney). *Gobi-Alt.* (Bain-Tukhum area and Gurban-Saikhan mountain range). *West. Gobi* (Tsagan-Bulak area, on south. foothill of Tsagan-Bogdo mountain range, solonchak-like small grass meadow, along spring,

---

[1]In view of the small number of Central Asian species, their delimitation into section is not given. The monographs of Koch (1833), Römpp (1927) and Stroh (1941) are not cited as they do not contain significant data on Central Asian species.

Aug. 1, 1943—Yun.). *Alash. Gobi* (Alashan, Dyn'yuanin oasis, bank of brook, June 27, 1908—Czet.; Noyan somon, Mukhor-shanda spring along border road to south of Tostu-Nuru mountain range, near water among dense reed thickets, Aug. 13, 1948—Grub.: "Suiynan, No. 2607, July 1931, Hsia"—Pai, l.c.). *Ordos* (Dzhasygen-Qaidam area, July 30, 1884—Pot.). *Khesi* (Sutcheau [Tszyutsyuan'], June 26, 1890—Martin; Danhe river, Satszau-Yuan'tszy area, along springs, in water, 1200 m, June 25, 1895—Rob.; Chzhan'e 40 km nor. of Yunchan town, Nin"yanlu settlement, in river valley, on wet meadow, July 1, 1958—Petr.).

IB. **Kashgar:** *Nor.* (Uchturfan, Yaman-Su river, on sasa solonchaks, June 9, 1908—Divn.). *West.* (Jarkand-Darya river, along river overflow regions, June 15, 1889—Rob.; near Yangishar town, along irrigation ditches, July 25, 1929—Pop.). *East.* (Tsagan-Su river, tributary of Khaidyk-Gol, 1500 m, Aug. 12, 1893; Adak area, Aug. 29, 1895—Rob.; Choltag mountain range, near Argai-Bulak picket, near spring, Sept. 1, 1929—Pop.).

IIA. **Junggar:** *Cis-Alt.* (Durul'chi river valley, forest slopes, July 29, 1906—Sap.). *Tien Shan* (Talki river, July 15 and 22—1877, A. Reg.; Talki, 1950 m, July 10, 1878—Fet.; "Kunges, ca. 1130 m, Aug. 23, 1932—Persson, l.c.). *Jung. Gobi* (around Guchen, July 1876—Pev.; Urungu river, Aug. 26, 1876—Pot.; Tashyu area, moist meadow, Sept. 1–10, 1895—Rob.; Bodonchiin-Gol near Bodonchiin-Baishing, near water on bank, Sept. 12; Baitak-Bogdo-Nuru, Takhiltu-Ula, Ulyastu-Gol, 5 km from estuary, in water along river bank, Sept. 17—1948, Grub.; 8–10 km south of Darbaty river [at its intersection with Karamai-Alatay road], in water and bank of spring, June 20; 3–4 km nor. of Staryi Kuitun settlement [on Shikho-Manas road], sasa zone spring outlets, forb-sedge marshy meadow and grassy marsh, June 30, July 6, July 7—1957, Yun. et al.). *Zaisan* (between Kara area and village on Kabe river, along banks of irrigation ditch, June 16; between Kaba river and Besh-Kuduk collective, near irrigation ditch, June 17— 1914, Schischk.). *Dzhark.* (Kul'dzha, 1876—Golike; on Ili river, May 16; bank of Ili river west of Kul'dzha, July 5; Dzhagastai, Aug. 6—1877, A. Reg.). *Balkh.-Alak.* (on Churchutsu river, Aug. 10, 1840—Schrenk.).

IIIA. **Qinghai:** *Nanshan* (Chortentan temple, 2100 m, Sept. 8, 1901—Lad.; "Ping Fan Hsien [Hsi Mi Jai] along shaded stream banks, common, Ching"—Walker, l.c.).

**General distribution:** Aralo-Casp., Fore Balkh., Jung.-Tarb., Nor. and Cent. Tien Shan; Europe, Mediterr., Balk.-Asia Minor, Fore Asia, Caucasus, Mid. Asia, West. and East. Sib., Nor. Mong. (Hent., Hang., Mong.-Daur.), China (North, North-West, Cent., East, South-West, South), Himalayas, Korean peninsula, Japan, Indo-Mal., Nor. and South Americas (introduced), Africa.

**Note.** Widely distributed species restricted to wet habitats. Small, low (3–10 cm tall), thin-stemmed plants are found in some regions of Mongolia (West. Gobi and Khesi) and Junggar. These plants can be regarded as annuals. However (as pointed out by A.P. Gamayunova and A.A. Dmitrieva in Fl. Kazakhstana), articulation of subsoil horizontal portion reveals their difference from annual plants.

2. **V. arenosa** (Serg.) Boriss. in Fl. SSSR, 22 (1955) 390; Fl. Kazakhst. 8 (1965) 72. —*V. laeta* var. *arenosa* Serg. in Kryl. Fl. Zap. Sib. 10 (1939) 2446. Described from East. Kazakhstan (Akkum sand in Zaisan basin). Type in Leningrad.

Sand and along steppe slopes.

IA. **Mongolia:** *Mong. Alt.* (2–3 km south of Tamchi lake, in valley, July 17; nor. trail of Bus-Khairkhan mountain range, depression in upper part of trail, July 17; Bulgan somon, upper Bayan-Gol river, steppe slope, July 23—1947, Yun.).

**General distribution:** Fore Balkh., West. Sib. (Altay).

3. **V. beccabunga** L. Sp. pl. (1753) 12; Pampanini, Fl. Carac. (1930) 189; Kryl. Fl. Zap. Sib. 10 (1939) 2455; Li in Proc. Ac. Natur. Sci. Philad. 104

(1952) 216; Borissova in Fl. SSSR, 22 (1955) 475; Fl. Kirgiz. 10 (1962) 181; Ikonnikov, Opred. rast. Pamira (1963) 218; Fl. Kazakhst. 8 (1965) 91. — Ic.: Fl. SSSR, 22, Plate XXI, fig. 2; Fl. Kazakhst. 8, Plate X, fig. 10.

Described from West. Europe. Type in London (Linn.).

In wet mountain meadows, along banks of rivers, brooks and irrigation ditches, up to 2800 m.

IA. Mongolia: *Mong. Alt.* (Tsinkir river valley, Tuguryuk plain, Aug. 15, 1930—Bar.).

IIA. Junggar: *Tien Shan* (M. Yuldus, 2250–2750 m, May 30, 1877—Przew.; Asu river valley, July 4, 1877—Fet.; Talki gorge, July 18; Sairam lake, south. bank, 2100 m, July 20; Talki brook, July 22—1877; Aruslyn, 2450 m, July 15, 1879—A. Reg.).

General distribution: Aralo-Casp., Fore Balkh., Jung.-Tarb., Nor. and Cent. Tien Shan, East. Pam.; Europe, Mediterr., Balk.-Asia Minor, Fore Asia, Caucasus, Mid. Asia, West. Sib., Far East (introduced), China (South-West), Himalayas (west., Kashmir).

4. **V. ciliata** Fisch. in Mém. Soc. natur. Moscou, 3 (1812) 56; Strachey, Catal. (1906) 128; Franch. in Bull. Soc. Bot. France, 47 (1907) 19; Pampanini, Fl. Carac. (1930) 188; Rehder and Kobuski in J. Arn. Arb. 14 (1933) 32; Pai in Contribs Inst. Bot. Nat. Ac. Peiping, 2, 7 (1934) 196; Walker in Contribs U.S. Nat. Herb. 28 (1941) 660; Li in Proc. Ac. Natur. Sci. Philad. 104 (1952) 216; Grubov, Konsp. fl. MNR (1955) 241; Borissova in Fl. SSSR, 22 (1955) 490; Yamazaki in J. Fac. Sci. Univ. Tokyo, sect. III, 7, 2 (1957) 151; Fl. Kazakhst. (1965) 95. —*V. macrocarpa* Turcz. ex Steud. Nomencl. 2 (1843) 758. —Ic.: Fl. SSSR, 22, Plate XXII, fig. 2; Fl. Kazakhst. 8, Plate X, fig. 14.

Described from Transbaikal. Type in Leningrad.

Wet alpine meadows and rocky slopes, near brooks and along banks of rivers and springs, on clayey-rocky cliffs and rocks, 2100–4300 m alt.

IB. Kashgar: *South.* (Kyuk-Egil' river valley, mountain range, moist rocky sites, 3800 m, July 11; Keriya, mountain range, in alpine belt, 3350 m, Aug. 5—1885, Przew.).

IIA. Junggar: *Tien Shan* (Muzart gorge, 3050–3500 m, Aug. 18; 3200 m, Aug. 19—1877; Shary-Su river, 2100–2450 m, Agyaz river, 2100–2450 m, June 26—1878; Kumbel' river, 3050 m, June 3; along Borborogusun river, 2750 m, June 15—1879, A. Reg.; die Mundung den Kinsu in den Kok-su, Aug. 3, 1907—Merzb.).

IIIA. Qinghai: *Nanshan* (South Tetungs mountain range, in river valleys on wet soils in alpine belt, common, July 13 and July 28—1872; Nanshan alps, on marshy soil and in alpine meadows along river banks, common, July 30, 1879; between Nanshan and Donkyru mountain ranges along Rako-Gol river, in wet alpine meadows, common 3050–3350 m, July 21, 1880; nor. slope of Humboldt mountain range [Aragalin-Ula], Ulan-Bulak, on alp. slopes, June 24, 1884—Przew.; Sulekhe river, alpine belt, on wet clay with sand, 3650–4300 m, June 2, 1894—Rob.; Sinin hills, Myndan'sha river, May 26, 1890—Gr.-Grzh.; Kuku-Nor lake, south. bank, among grasses, on humus, 3050 m, Aug. 18; around Dangertin town [Huangyuan], hill slope, on humus, 2100–2450 m, Aug. 27, 1901—Lad.; "Ta-P'an Shan, on exposed moist grassy slopes, Ching"—Walker, l.c.). *Amdo* (Dzurgin, alps around Dzhakhar, July 25, 1880—Przew.).

IIIB. Tibet: *Weitzan* (Konchyunchu river, on rocks, 4000–4300 m, occasional, July 1; dividing ridge between Talachu and Bychu rivers, in alpine meadows, 4300 m, July 4; on hill slope towards Talachu river, on rocks, occasional, July 6—1884, Przew.; Yantszytszyan river basin, Nruchu area, in juniper forest, on clay [humus], 3550 m, June 25; north-west. bank of Russkoe lake, on clay cliffs and among large rocks, June 27; Chzhabu-vrun area, on clayey-rocky slopes of hillocks, 4300 m, July 9—1900; Burkhan-Budda mountain range, Ikhe-Gol gorge, July 23, 1901—Lad.). *South.* ("Tibet, Topidhunga, 4000–4600 m, Strachey and Winterbottom"—Strachey, l.c.).

General distribution: Jung.--Tarb., Nor. Tien Shan; Mid. Asia, West. Sib. (Altay), East. Sib., Nor. Mong. (Fore Hubs, Hent., Hang.), China (North, South-West), Himalayas (west.).

Note. Chen and Chou [Rast. pokrov r. Sulekhe (1957) 90] cite for Mongolia (Sulekhe river basin) *V. szechuanica* Batalin (some what resembles *V. ciliata* in its capitate inflorescence and form and pubescence of leaves) described from North-West China and reported from South-West China. The report of this species in Mongolia is dubious. It is probably *V. ciliata* which occurs in this region.

5. **V. dahurica** Stev. in Mém. Soc. natur. Moscou, 5 (1817) 33; Borissova in Fl. SSSR, 22 (1955) 372; Yamazaki in J. Fac. Sci. Univ. Tokyo, sect. III, 7, 2 (1957) 135. —*V. grandis* Fisch. ex Spreng. Neue Entdeck. 2 (1821) 122; Franch. Pl. David. 1 (1884) 223; Pai in Contribs Inst. Bot. Nat. Ac. Peiping, 2, 7 (1934) 197; Kitag. Lin. Fl. Mansh. (1939) 398; Li in Proc. Ac. Natur. Sci. Philad. (1952) 202; Grubov, Konsp. fl. MNR (1955) 242. —*V. longifolia* var. *grandis* (Fisch.). Turcz. in Bull. Soc. natur. Moscou, 24 (1851) 312. —**Ic.:** Bibl. Bot. 194, tab. B, fig. 12.

Described from Dauria. Type in Leningrad.

Rocky and stony slopes, in steppes on sand and sandy pebble alluvium.

IA. **Mongolia:** *Cis-Hing.* (Khalkhin-Gol river, Sept. 1, 1928—Tug.; Yakshi station, rocks 1954—Wang). *East. Mong.* (Khailar town, on Sishan' hills, No. 2113, Aug. 69 [sic], 1951—A.R. Lee (1959)); same site, on sandy hill No. 2927, 1954—Wang; Khailar town, meadow and true steppe, 1959—Ivan.; "dans les montagnes de l'Oulachan, No. 2939, July 1886"—David, l.c.).

IIA. **Junggar:** *Tien Shan* (near Kapsalan pass, alpine meadow, 3650–4000 m, June 16, 1893—Rob.; Tsanma river valley, Dagdyn-Gol area, subalpine burnet meadow, Aug. 8, 1958—Yun. et al.).

General distribution: East. Sib., Far East, Nor. Mongolia (Hent., Mong.-Daur.), China (Dunbei, North, East), Korean peninsula, Japan.

6. **V. densiflora** Ledeb. Fl. alt. 1 (1829) 34; Sapozhn. Mong. Alt. (1911) 380; Kryl. Fl. Zap. Sib. 10 (1939) 2448; Grubov, Konsp. fl. MNR (1955) 242; Borissova in Fl. SSSR, 22 (1955) 486; Fl. Kazakhst. 8 (1965) 94. —**Ic.:** Fl. SSSR, 22, Plate XXII, fig. 4.

Described from Altay (Koksu river). Type in Leningrad.

Rubble and stony slopes, bald peaks, forest to apline belts.

IA. **Mongolia:** *Mong. Alt.* (south. slope of Urmogaity pass, June 24, 1903—Gr.-Grzh.; Beloi Kobdo source, alpine tundra, June 29, 1906—Sap.).

General distribution: Jung.-Tarb., Nor. Tien Shan; West and East. Sib.; Far East.

7. **V. incana** L. Sp. pl. (1753) 10; Forbes and Hemsley, Index Fl. Sin. 2 (1902) 198; Danguy in Bull. Mus. nat. hist. natur. 20 (1914) 80; Kryl. Fl. Zap. Sib. 10 (1939) 2444; Kitag. Lin. Fl. Mansh. (1939) 398; Li in Proc. Ac. Natur. Sci. Philad. 104 (1952) 201; Jernakov in Acts Pedol. Sin. 2, 4 (1954) 276;

Grubov, Konsp. fl. MNR (1955) 242; Borissova in Fl. SSSR, 22 (1955) 377; Yamazaki in J. Fac. Sci. Univ. Tokyo, Sect. III, 7, 2 (1957) 140; Fl. Kazakhst. 8 (1965) 68. —Ic.: Fl. Kazakhst. 8, Plate VIII, fig. 4.

Described from Europe. Type in London (Linn.).

Steppe rocky and stony slopes, montane-meadow and sandy steppes, sometimes in larch forests.

IA. **Mongolia:** *Mong. Alt.* (15 km east-south-east of Yusun-Bulak, Taishiri-Ula range, nor. slope, lower mountain belt, meadow steppe along creek valley floors, July 11, 1945—Yun.), *Cis-Hing.* (Abder [Abderiin-Gol] river, June 24, 1899 Pot. and Sold.) *Cent. Khakha* (Kerulen midcourse, Bain-Khan hill slope, 1899—Pall.; around Ikhe-Tukhum-Nor lake, Ulan-Delger hill, June; Ongon-Khairkhan hills, June—1926, Zam; Kholt area, steppe meadow, Aug. 23, 1926—Ik.-Gal.; Delger-Hangay—Ulan-Bator road, on slope of rocky Khairkhan mountains, Sept. 27, 1931—Ik.-Gal.; 1–2 km east of Idermeg-Khid, grass-forb steppe, chestnut-brown loamy sand, June 1945; 20–25 km north of Undur-Khan, wormwood-snakeweed-chee grass steppe, July 25, 1949—Yun.; Bain-Baratuin somon, old road to Dalan-Dzadagad, Bain-Ula hill, on granite rocks, July 12, 1948—Grub.). *East. Mong.* (between Kulusutai and Dolon-Nor lake, 1870—Lom.; Zodol somon, north-west. fringe of Zodol-Khan-Ula, steppe on basalts, May 14, 1944—Yun.; around Khailar town, meadow on sandy flat, No. 620, June 10; same site, Sishan' hill, No. 2116, Aug. 29—1951, A.R. Lee (1959)); Khailar town, sandy hill, No. 2924, 1954—Wang). *Gobi-Alt.* (Dzun-Saikhan mountains, west. part, steppe slope in upper belt of range, June 19, 1945—Yun.).

General distribution: Aralo-Casp., Fore Balkh., Jung.-Tarb.; Arct. (Asian), Europe, West. and East. Sib., Nor. Mong., China (Dunbei), Korean peninsula, Japan.

8. **V. krylovii** Schischk. in Kryl. Fl. Zap. Sib. 10 (1939) 2457; Borissova in Fl. SSSR, 22 (1955) 436; fl. Kazakhst. 8 (1965) 86. —*V. teucrium* auct. non L.: Bunge in Ledeb. Fl. alt. 1 (1829) 40; ? Danguy in Bull. Mus. nat. hist. natur. 20 (1914) 81; Kitag. Lin. Fl. Mansh. (1939) 399. —Ic.: Fl. SSSR, 22, Plate XIX, fig. 2; Fl. Kazakhst. 8, Plate X, fig. 3.

Described from Altay. Type in Leningrad.

Meadows, steppe and rocky slopes.

IA. **Mongolia:** *East. Mong.* ("vallée du Kéroulen, May 1896, Chaff."—Dancuy, l.c.).
IIA. **Junggar:** *Cis-Alt.* (Kurtu river, June 19, 1903—Gr.-Grzh.).
General distribution: Aralo-Casp., Jung.-Tarb.; Mid. Asia, West. and East. Sib.

9. **V. laeta** Kar. et Kir. in Bull. Soc. natur. Moscou, 15 (1842) 414; Danguy in Bull. Mus. nat. hist. natur. 20 (1914) 80; Kryl. Fl. Zap. Sib. 10 (1939) 2445; Grubov, Konsp. fl. MNR (1955) 242; Borissova in Fl. SSSR, 22 (1955) 389; Fl. Kirgiz. 10 (1962) 173; Fl. Kazakhst. 8 (1965) 70. —Ic.: Fl. SSSR, 22, Plate XV, fig. 1; Fl. Kazakhst. 8, Plate IX, fig. 1.

Described from Junggar Alatau (Sarkhan river). Type in Leningrad.

Steppe rocky and stony slopes, foothills and midbelt of mountains to 1800 m.

IIA. **Junggar:** *Cis-Alt.* (on south. rocky slope, Kandagatai river estuary, Sept. 14, 1876—Pot.; 20 km north-west of Shara-Sume, scrub-steppe meadow, July 7; 35 km south of Koktogoi, on road to Ertai, foothills, scrub-wheat grass steppe, on rocky south. slope, July 14—1959,

Yun. et al.; "Altai, steppes entre l'Irtish et Kobdo, terrains frais, No. 1092, Sept. 4, 1895—Chaff." Danguy, l.c.). *Jung. Alt.* (Dzhair range, Kyr hill, 8 km north-west of Otu peak, in grasslans among granites, 1200 m, July 17, 1953—Mois.). *Tien Shan* (Pilyuchi gorge, 900–1200 m, July 22; along Kash river, Aug. 18—1878; Nilki foothills, 1500–1800 m, June 9–21, 1879—A. Reg.; Ketmen' range, 8–10 km beyond Sarbushin settlement, along road to Ili town [Kul'dzha] in Kzyl-Kure, steppe belt, mountain steppe north of valley slope, Aug. 23, 1957—Yun. et al.). *Jung. Gobi* (Burchum [Chenkur], Aug. 16, 1903—Sap.; nor. slope of Baitak-Bogdo range, Ulyastu-Gol gorge, 7 km from estuary, on northern rocky and stony slopes, Sept. 18, 1948 — Grub.).

General distribution: Fore Balkh., Jung.-Tarb., Nor. Tien Shan; Mid. Asia, Nor. Mongolia (Hang.).

10. **V. linariifolia** Pall. ex Link in Jahrb. Gewachskunde, 1–3 (1820) 35; Kitag. Lin. Fl. Mansh. (1939) 398; Li in Proc. Ac. Natur. Sci. Philad. 104 (1952) 203; Grubov, Konsp. fl. MNR (1955) 242; Borissova in Fl. SSSR, 22 (1955) 386; Yamazaki in J. Fac. Sci. Univ. Tokyo, sect. III, 7, 2 (1957) 139; Dashnyam in Bot. zh. 50 (1965) 1641. —*V. angustifolia* Fisch. ex Link. Enum. pl. hort. Berol. 1 (1821) 19; Hance in J. Linn. Soc. London (Bot.) 13 (1873) 84. —*V. cartilaginea* Ledeb. Fl. alt. 1 (1829) 28. —*V. rubicunda* Ledeb. l.c. 28. —*V. paniculata* β *angustifolia* Benth. in DC. Prodr. 10 (1846) 465, p.p. —*V. spuria* var. *angustifolia* Makino in Bot. Mag. 10 (1896) 252, p.p. —*V. spuria* auct. non L.: Franch. Pl. David. 1 (1884) 223; Forbes and Hemsley, Index Fl. Sin. 2 (1902) 200; Sapozhn. Mong. Alt. (1911) 380; Pai in Contribs Inst. Bot. Nat. Ac. Peiping, 2, 7 (1934) 198, p.p. —Ic.: Fl. SSSR, 22, Plate XV, fig. 3.

Described from Dauria. Type in Berlin.

Steppe and meadow slopes of mountains and in river valleys.

IA. **Mongolia:** *Cis-Hing.* (around Trekhrech'e mountain slope, 750–850 m, No. 1367, July 14, 1954—Wang). *Cent. Khalkha* (around Orgochen-Sume monastery [47° N. lat., 104° E. long.], Aug. 23, 1925—Krasch. and Zam.). *East. Mong.* (Khuntu somon, 17–20 km east-south-east of Bain-Tsagan, chee grass-tansy steppe, Aug. 6, 1949—Yun.; Khailar town, in steppe, on sandy hill, No. 2952, 1954—Wang). *Ordos* (Ordos, 1877—Verlinden; Ulan-Morin river valley, riverine sandy shoals, Aug. 21, 1884—Pot.).

General distribution: East. Sib., Far East, Nor. Mongolia (Hent., Hang., Mong.-Daur.), China (Dunbei, North, East, Cent., South-West), Korean peninsula, Japan.

11. **V. longifolia** L. Sp. pl. (1753) 10; Maxim. in Mém. Ac. Sci. St.-Pétersb. Sav. Etrang. 9 (1859) 207; Forbes and Hemsley, Index Fl. Sin. 2 (1902) 198; Sapozhn. Mong. Alt. (1911) 380; Danguy in Bull. Mus. nat. hist. natur. 20 (1914) 80; Pai in Contribs Inst. Bot. Nat. Ac. Peiping, 2, 7 (1934) 197; Kryl. Fl. Zap. Sib. 10 (1939) 2447; Kitag. Lin. Fl. Mansh. (1939) 399; Li in Proc. Ac. Natur. Sci. Philad. 104 (1952) 202; Grubov, Konsp. fl. MNR (1955) 242; Borissova in Fl. SSSR, 22 (1955) 367; Yamazaki in J. Fac. Sci. Univ. Tokyo, sect. III, 7, 2 (1957) 135; Fl. Kirgiz. 10 (1962) 399; Fl. Kazakhst. 8 (1965) 67; Dashnyam in Bot. zh. 50 (1965) 1642. —*V. maritima* L. Sp. pl. (1753) 10. — Ic.: Fl. Kazakhst. 8, Plate VIII, fig. 2.

Described from Europe. Type in London (Linn.).

Coastal meadows, larch forests, thickets, coastal willow groves in forest and subalpine belts.

IA. **Mongolia:** *Mong. Alt.* (Aksu sources, alpine tundra, June 29, 1906—Sap.). *Cis-Hing.* (near Yakshi railway station, meadow, Aug. 19, 1902—Litw.; around Trekhrech'e, 750–850 m, hilly valley in marsh, No. 1204, July 9; same site, 700–750 m, No. 1278, July 10—1951; Genkhe station, in meadow, No. 2373, 1954—Wang).

*Cent. Khalkha, Bas. Lakes* (around Ubsa lake, Ulan-Natsyn river valley, on moist soil, July 15, 1879—Pot.).

IIA. **Junggar:** *Cis-Alt.* (Kandagatai, Sept. 15, 1876—Pot.). *Tien Shan* (Borgaty gorge, 1500–1800 m, July 4; nor. side of Kash river valley, 1500–1800 m, July 5; Yultu to Aryslyn, 2100–2450 m, July 7; Aryslyn gorge, 2450–3000 m, July 10, 15 and 19; Mengute gorge, 3050–3350 m, July—1879, A. Reg.). *Jung. Gobi* (Ch. Irtysh, Dyurbel'dzhin, in meadow, Aug. 24, 1876—Pot.; "entre l'Ouchte et l'Irtish, Aug. 23, 1895, Chaff"—Danguy, l.c.). *Balkh.-Alak.* (Churchutsu river near Chuguchak, Aug. 10, 1840—Schrenk).

**General distribution:** Aralo-Casp., Fore Balkh., Jung.-Tarb., Nor. and Cent. Tien Shan; Europe, Caucasus (Fore Caucasus), West. and East. Sib., Far East, Nor. Mongolia.

12. **V. macrostemon** Bunge in Ledeb. Fl. alt. 1 (1829) 35; Sapozhn. Mong. Alt. (1911) 380; Kryl. Fl. Zap. Sib. 10 (1939) 2447; Grubov, Konsp. fl. MNR (1955) 242; Borissova in Fl. SSSR, 22 (1955) 485; Fl. Kazakhst. 8 (1965) 93. — *F. lütkeana* auct. non Rupr.: Borissova, l.c. quoad pl. dzungar. —**Ic.:** Fl. Kazakhst. 8, Plate X, fig. 12.

Described from Altay. Type in Leningrad.

Rocks and rocky slopes, placers, among rubble-lichen tundra, 1700–3300 m alt.

IA. **Mongolia:** *Mong. Alt.* (Bzau-Kul' sources, alpine tundra and placer, July 11, 1906—Sap.).

IIA. **Junggar:** *Tien Shan* (from upper Taldy and Kash, 3050–3100 m, May 21; Taldy river, 2450–2750 m, May 26 and 27; from Tyumbedan and Kum-Daban, 2750 m, May 28 and 29; Kumbel', June 3; Dzhunkur-Daban [Chunkur-Daban], between Ulastai and Borborogusun, 3050 m, June 13—1879, A. Reg.; south of Yakou, Barchat, nor. slope, 3000–3300 m, No. 1713, Aug. 31; Nilki-Tszinkho, slope near rocks, No. 4066, Sept. 1—1957, Kuan).

**General distribution:** Jung.-Tarb., Nor. Tien Shan; Mid. Asia (mountains), West. and East. Siberia, Nor. Mongolia (Fore Hubs.), Himalayas (west.).

13. **V. pinnata** L. Mant. 1 (1767) 24; Sapozhn. Mong. Alt. (1911) 380; Danguy in Bull. Mus. nat. hist. natur. 20 (1914) 80; Kryl. Fl. Zap. Sib. 10 (1939) 2446; Grubov, Konsp. fl. MNR (1955) 242; Borissova in Fl. SSSR, 22 (1955) 391; Fl. Kazakhst. 8 (1965) 73. —**Ic.:** Fl. Kazakhst. 8, Plate IX, fig. 2.

Described from Siberia. Type in London (Linn.).

Rocky slopes, pebble beds and sand in mountain-steppe and desert-steppe belts.

IA. **Mongolia:** *Khobd.* (Kharkhira river valley, barren pebble-bed, July 9; Turgun' river, July 21—1879, Pot.; steppe and mountains south of Ureg-Nur lake, July 29—Gr.-Grzh.; near outlet of Kharkhyra river on Ubsa-Nursk plain, Sept. 4; south. bank of Ureg-Nur lake, Sept. 15 and nor. bank of lake, Oct. 15—1931, Bar.). *Mong. Alt.* (between 2nd and 3rd pickets toward Kobdo, June 19 and 20—1870, Kalning; Tsagan-Gol midvalley near Tsagan-Kobu, standing moraine and pebble beds, June 28, 1905, Sap.; Khara-Dzarga hills valley of Boro-Gol river, east. rocky slopes, Aug. 24, 1930—Pob.; 40 km nor. of Kobdo, along road to Tsagan-Nur, hummocky area, rocky slope, Aug. 7, 1945—Yun.; "Oiguriin-Gol, moraine"—Sap. l.c.). *Bas. Lakes* (40 km south of Ulyasutai, along road to Tsagan-Olom, Shurygiin-Gol area, near road bend, sand, 1947—Yun.).

General distribution: Fore Balkh., Jung.-Tarb., Nor. Tien Shan; West. and East. Siberia, Nor. Mongolia (Hang.).

14. **V. porphyriana** Pavl. in Vestn. AN KazSSR, 4 (1951) 92; Borissova in Fl. SSSR, 22 (1955) 382; Fl. Kirgiz. 10 (1962) 172; Fl. Kazakhst. 8 (1965) 69; Dashnyam in Bot. zh. 50 (1965) 1642. —*P. spicata* auct. non L.: Pai in Contribs Inst. Bot. Nat. Ac. Peiping, 2, 7 (1934) 198, p.p.; Kryl. Fl. Zap. Sib. 10 (1939) 2441, p.p.; Li in Proc. Ac. Natur. Sci. Philad. 104 (1952) 2090; Grubov, Konsp. fl. MNR (1955) 243. —*P. spicata* var. *viscosissima* Kar. et Kir. in Bull. Soc. natur. Moscou, 14 (1841) 721; Kryl. Fl. Zap. Sib. 10 (1939) 2442. —**Ic.:** Fl. Kazakhst. 8, Plate VIII, fig. 6.

Described from Transili Alatau. Type in Moscow.

Subalpine meadows and rocky steppe slopes of mountains.

IA. Mongolia: *Mong. Alt.* (Ulyasta river upper course, hill slope, July 26; Kharchatei [Khartsiktei] bank, slope above forest fringe, July 27—1898, Klem.; Indertiin-Gol river upper course, 4–5 km below somon site, subalp. steppe, July 25, 1947—Yun.).

IIA. Junggar: *Jung. Alt.* (Dagambel' mountains, Maili range, 4 km south of Dzhirmasu picket [between Kul'denen and Otu pickets], 1300 m, July 18, 1953—Mois.; Dzhair pass, along Chuguchak-Shikho road, mountain-steppe belt, ravines of mud cones, Aug. 9, 1957—Yun. et al.). *Tien Shan* (east up to Urumchi).

General distribution: Jung.-Tarb., Nor. and Cent. Tien Shan; Mid. Asia (hilly), West. Siberia.

Note. Polymorphic species closely related to *V. spicata* L. Greyish-green and green plants with ovate and, very rarely, linear-oblong leaves are found among Cent. Asian specimens.

15. **V. serpyllifolia** L. Sp. pl. (1753) 12; Forbes and Hemsley, Index Fl. Sin. 2 (1902) 199; Pai in Contribs Inst. Bot. Nat. Ac. Peiping, 2, 7 (1934) 198; Kryl. Fl. Zap. Sib. 10 (1939) 2450; Kitag. Lin. Fl. Mansh. (1939) 399; Li in Proc. Ac. Natur. Sci. Philad. 104 (1952) 205; Borissova in Fl. SSSR, 22 (1955) 365; Yamazaki in J. Fac. Sci. Univ. Tokyo, sect. III, 7, 2 (1957) 148; Fl. Kirgiz. 10 (1962) 170; Fl. Kazakhst. 8 (1965) 66. —*V. tenella* All. Fl. Pedem. 1 (1785) 75; Yamazaki, l.c. 148. —**Ic.:** Fl. Kazakhst. 8, Plate VIII, fig. 1.

Described from Europe. Type in London (Linn.).

Steppe and meadow slopes.

IIA. Junggar: *Tien Shan* (nor. slope of Narat mountain range on Tsanma river, common, 3000 m, June 8, 1877—Przew.).

General distribution: Jung.-Tarb., Nor. and Cent. Tien Shan.; Europe, Mediterr.,., Balk.-Asia Minor, Fore Asia, Caucasus, Mid. Asia, West. and East. Sib., Far East, China (Dunbei, North, North-West, Cent., South-West), Himalayas, Korean peninsula, Japan, Nor. America, Australia.

16. **V. spuria** L. Sp. pl. (1753) 10; Kryl. Fl. Zap. Sib. 10 (1939) 2440; Grubov, Konsp. fl. MNR (1955) 243; Borissova in Fl. SSSR, 22 (1955) 376; Fl.

Kirgiz. 10 (1962) 171; Fl. Kazakhst. 8, (1965) 67.   *V. paniculata* L. Sp. pl. (1762) 18.   —Ic.: Fl. Kazakhst. 8, Plate VIII, fig. 3.

Described from Europe and Siberia. Type in London (Linn.).

Steppe and meadow slopes.

IA. **Mongolia:** *Mong. Alt.* (on way to Kobdo, June 20, 1870—Kalning).

IIA. **Junggar:** *Tarb.* (Almala river valley, north-east. part of Barlyk mountain range, 1300 m, Aug. 4, 1953—Mois.).

**General distribution:** Aralo-Casp., Fore Balkh., Jung.-Tarb., Nor. and Cent. Tien Shan; Europe, West. Sib., Nor. Mongolia (Hent., Mong.-Daur.)

Note. The specimens listed above differ from typical specimens; these are probably hybrids *V. spuria* L. x *V. longifolia* L. Tarbagatai plants have subglabrous green (not greyish tomentose-pubescent) leaves while Mongolian plants have a very small 3–4 (not 5–6) mm long corolla.

## B. Annuals

17. **V. argute-serrata** Regel et Schmalh. in Acta Horti Petrop. 5 (1877) 626; Borissova in Fl. SSSR, 22 (1955) 394; Fl. Kirgiz. 10 (1962) 174; Fl. Kazakhst. 8 (1965) 74.   —*V. campylopoda* auct. non Boiss.: Römpp in Beih. Feddes repert. 50 (1928) 80.   —Ic.: Fl. Kazakhst. 8, Plate IX, fig. 4.

Described from Junggar Alatau (Karakol river valley). Type in Leningrad.

Steppe slope in foothills and up to forest belt, sometimes in alpine belt.

IIA. **Junggar:** *Tien Shan* (Biangou area, Dasingou river upper course, Sept. 25, 1929—Pop.).

**General distribution:** Jung.-Tarb., Nor. Tien Shan; Balk.-Asia Minor, Fore Asia, Caucasus, Mid. Asia.

18. **V. biloba** L. Mant. 2 (1771) 172; Franch. in Bull. Soc. Bot. France, 47 (1907) 20; Pampanini, Fl. Carac. (1930) 189; Pai in Contribs Inst. Bot. Nat. Ac. Peiping, 2, 7 (1934) 172; Kryl. Fl. Zap. Sib. 10 (1939) 2452; Li in Proc. Ac. Natur. Sci. Philad. 104 (1952) 208; Grubov, Konsp. fl. MNR (1955) 241; Borissova in Fl. SSSR (1955) 392; Fl. Kirgiz. 10 (1962) 173; Ikonnikov, Opred. rast. Pamira (1963) 218; Fl. Kazakhst. 8 (1965) 73.   —Ic.: Fl. SSSR, 22, Plate XVI, fig. 5.

Described from Asia Minor. Type in London (Linn.).

Moist sandy-pebble river banks, rocks, rubble and rocky talus in upper and middle mountain belts.

IA. **Mongolia:** *Khobd.* (Khatu-Gol [Katu] river valley, June 15, 1879—Pot.). *Mong. Alt.* (Khara-Dzarga mountain range, on rocks and along bank of Naishuren-Gol river, Sept. 2, 1930—Pob.).

IB. **Kashgar:** *Nor.* (Uchturfan, Uital river near Karol, around Bedel' pass, June 27, 1908—Divn.).

IIA. **Junggar:** *Tien Shan* (around Kul'dzha, May 1877; Bayandai, 600–1900 m, May 6, 1878; Borborogusun, 900–1200 m, April 27 and 28, 1879—A. Reg.; Manas river basin, Ulan-Usu river valley at its confluence with Dzhartas, upper forest boundary along rubble talus, slope with nor.-nor.-east. exposure, Aug. 18, 1957—Yun. et al.; "roadside of Mayashan, Bogdo-shan, No. 3482, Aug. 1931, Liou"—Pai, l.c.).

III. Qinghai: *Nanshan* (South Tetung mountain range, in wet sites, rare, July 25, 1872; south. slope on Tetung river, near water, common, July 30, 1880—Przew.). *Amdo* (Dulan-khit temple, open glade in spruce forest, 3350 m, Aug. 9, 1901—Lad.).

IIIC. Pamir (Muztagata foothill, rocky talus, July 20, 1909—Divn.).

General distribution: Jung.-Tarb., Nor. Tien Shan, East. Pam., Balk.-Asia Minor, Fore Asia, Caucasus, Mid. Asia (hilly), West. Sib. (Altay), Nor. Mong. (Hang.), China (North-West, South-West). Himalayas (west., Kashmir).

19. **V. campylopoda** Boiss. Diagn. pl. or. 1, 4 (1844) 80; Kryl. Fl. Zap. Sib. 10 (1939) 2452; Borissova in Fl. SSSR, 22 (1955) 397; Fl. Kazakhst. 8 (1965) 76. —*V. biloba* var. *dasycarpa* Trautv. in Bull Soc. natur. Moscou, 39 (1866) 440. —Ic.: Fl. Kazakhst. 8, Plate IX, fig. 7.

Described from Near East. Type in Geneva.

Dry mountain slopes and loessial hill, from foothills to alpine belt; some-times as weed in kitchen gardens and in ploughed fields.

IB. Kashgar: *West.* (Kingtau mountain range, nor. slope, 3–4 km from Kosh-Kulak settle-ment, juniper thicket, June 10, 1959—Yun. et al.).

IIA. Junggar: *Tien Shan* (Khorgos, April 22; around Kul'dzha, May—1877; Almaty gorge, 1000 m, April 1878—A. Reg.).

IIIC. Pamir (Ulug-tuz gorge, on right bank of Charlym, river June 27, 1909—Divn.).

General distribution: Aralo-Casp., Fore Balkh., Jung.-Tarb., Nor. and Cent. Tien Shan; Mediterr., Balk.-Asia Minor, Fore Asia, Caucasus, Mid. Asia, West. Sib. (Altay), Himalayas.

20. **V. perpusilla** Boiss. Diagn. pl. or. 1, 7 (1846) 43; Borissova in Fl. SSSR, 22 (1955) 425; Fl. Kirgiz. 10 (1962) 180; Fl. Kazakhst. (1965) 84. —*V. nudicaulis* auct. non Lam.: Kar. et Kir. in Bull. Soc. natur. Moscou, 15 (1842) 415. —*V. nudicaulis* var. *glabrata* Trautv. in Bull. Soc. natur. Moscou, 39 (1855) 539; Kryl. Fl. Zap. Sib. 10 (1939) 2453; Grubov, Konsp. fl. MNR (1955) 242. —*V. acinifolia* var. *karelinii* et var. *glabrata* Trautv. in Bull. Soc. natur. Moscou, 39 (1866) 439.

Described from Fore Asia. Type in Geneva. Isotype in Leningrad.

Marshy meadows and wet banks of brooks in high-alpine belt.

IA. Mongolia: *Mong. Alt.* (Bulgan somon, upper Indertiin-Gola, high-alpine belt, marshy meadow, July 24, 1947—Yun.).

General distribution: Aralo-Casp., Jung.-Tarb., Nor. Tien Shan; Europe (Cent. Urals, Crimea), Fore Asia Caucasus, Mid. Asia, West. Sib.

21. **V. persica** Poir. Dict. Encycl. meth. 8 (1808) 542; Kryl. Fl. Zap. Sib. 10 (1939) 2453; Li in Proc. Ac. Natur. Sci. Philad. 104 (1952) 208; Borissova in Fl. SSSR, 22 (1955) 411; Fl. Kazakhst. 8 (1965) 80. —*V. tournefortii* Gmel. Fl. Bad. 1 (1805) 39, non Villars. (1779), nec F.W. Schmidt (1791); Walker in Contribs U.S. Nat. Herb. 28; (1941) 660. —Ic.: Fl. SSSR, 22, Plate XVII, fig. 4.

Described from Fore Asia. Type in Paris.

Meadow and scrub slopes as well as weed on roadsides and in fields, plains to high mountains.

IIIA. **Qinghai:** *Nanshan* ("La Ch'iung Kou, on partially shaded, moist, grassy slopes, Ching"—Walker, l.c.).

General distribution: Jung.-Tarb., Nor. Tien Shan; Europe, Balk.—Asia Minor, Mediterr., Mid. Asia, Far East, China (East. Taiwan), Himalayas (west., Kashmir).

Note. Walker (l.c.) includes this species in his list of plants with a question-mark probably because the corolla in the plants collected was pink (not blue or lilac-blue, common for this species). In all other respects, however (up to 120 cm long stem, orbicular leaves), *V. persica* Poir. is readily distinguished from other annual veronica species.

22. **V. rubrifolia** Boiss. Diagn. pl. or. 1, 12 (1853) 46; Borissova in Fl. SSSR, 22 (1955) 396; Fl. Kirgiz. 10 (1962) 174; Ikonnikov, Opred. rast. Pamira (1963) 217; Fl. Kazakhst. 8 (1965) 75. —*V. ferganica* M. Pop. in Tr. Turkest. gos. univ. 4 (1922) 64. —Ic.: Fl. SSSR, 22, Plate XVIII, fig. 2; Fl. Kazakhst. 8, Plate IX, fig. 6.

Described from Fore Asia. Type in Geneva.

Rocky slopes, talus and exposures, in meadows, from foothills to alpine belt.

IIA. **Junggar:** *Tien Shan* (Bayandai, near Kul'dzha, 600–1200 m, May 6, 1878—A. Reg.).

General distribution: Aralo-Casp., Fore Balkh., Jung.-Tarb., Nor. Tien Shan; Mid. Asia, Fore Asia, West. Sib.

23. **V. tenuissima** Boriss. in Fl. SSSR, 22 (1955) 403; Fl. Kirgiz. 10 (1962) 175; Fl. Kazakhst. 8 (1965) 77. —*V. tetraphylla* auct. non Boeb. ex Georgi ("*tetraphyllos*"): Pop. in Tr. Turkest. gos. univ. 4 (1922) 65. —Ic.: Fl. SSSR, 22, Plate XVI, fig. 1.

Described from West. Tien Shan (Sary-Tau hills). Type in Tashkent (TAK). Isotype in Leningrad.

Loamy plains and foothills.

IIA. **Junggar:** *Dzhark.* (around Kul'dzha, May 1877—A. Reg.).

General distribution: Aralo-Casp., Fore Balkh., Nor. Tien Shan; Fore Asia, Mid. Asia.

Subgenus Veronicastrum (Heister) Boriss.

24. **V. sibirica** L. Sp. pl., ed. 2 ( 1762) 12; Franch. Pl. David. 1 (1884) 223; Kitag. Lin. Fl. Mansh. (1939) 399; Grubov, Konsp. fl. MNR (1955) 243; Borissova in Fl. SSSR, 22 (1955) 495. —*V. virginica* auct. non L.: Forbes and Hemsley, Index Fl. Sin. 2 (1902) 200. —*Veronicastrum sibiricum* (L.) Pennell in Monogr. Ac. Natur. Sci. Philad. 1 (1935) 321; Yamazaki in J. Fac. Sci. Univ. Tokyo, sect. III, 7, 2 (1957) 132. —Ic.: Fl. SSSR, 22, Plate XXIII, fig. 2.

Described from Siberia. Type in London (Linn.)

Meadows, among shrubs and in mountain forests.

**IA. Mongolia:** *Cis-Hing.* (Yakshi railway station, July 7, 1901—Lipsky; Numuryg-Gol river basin, Bain-Gol river valley, reed grass meadow, Aug. 8, 1949—Yun.; environs of Trekhrech'e, in mountain valley, around thickets, July 9, 1951; Gen'khe station, meadow, 1954—Wang). *East. Mong.* (Khailar town, meadow steppe, 1959—Ivan.; Muni-Ula mountain range, nor. slope, in forest, on moist soil, July 6, 1871—Przew.). *Ordos* (Ordos, 1877—Verlinden).

**General distribution:** East. Sib., Far East, Nor. Mong. (Hent., Mong.-Daur.)., China (Dunbei).

## 9. Lagotis Gaertn.

in Nov. Comment. Ac. Sci. Petrop. 14 (1770) 553.  —*Gymnandra* Pall. Reise, 3 (1776) 710.

1.  Leaves linear-lanceolate, longer than inflorescence; plant with long creeping surface shoots ......................... 1. **L. brachystachya** Maxim.
+  Leaves orbicular, elliptical or oblong-ovate, shorter than inflorescence; surface shoots lacking ............................................................ 2.
2.  Plant with leafless flowering stem; inflorescence capitate or globose. Sepals lanceolate; filaments distinctly developed and nearly as long as upper lip ........................................ 5.**L. ramalana** Batalin.
+  Plant with foliated flowering stems; inflorescence oblong or more or less long-spicate. Sepals ovate, almost free or calyx tubular, split in front, with 2 teeth on top; filaments short or anthers subsessile ................................................................................................. 3.
3.  Corolla lobes as long or longer than tube, nearly equal, lanceolate; corolla lilac ................................................... 2. **L. brevituba** Maxim.
+  Corolla lobes 1/2–2/3 of tube, unequal: upper lip oblong-elliptical or ovate, undivided or with 2 teeth at tip (roughly bilobed), lobes of lower lip linear-oblong or linear; corolla blue or canescent ..... 4.
4.  Calyx with 2 nearly free sepals; corolla blue; lower lip 3–4 partite. Plant with flattened, 5–10 cm tall stems ......3. **L. decumbens** Rupr.
+  Calyx tubular, split in front, with 2 teeth on top; corolla canescent; lower lip with 2 linear or linear oblong lobes. Plant with 10–40 cm tall stem, erect or ascending at base ........ 4. **L. integrifolia** (Willd.). Schischk.

1. **L. brachystachya** Maxim. in Bull. Ac. Sci. St.-Pétersb. 27 (1881) 525; Hemsley in J. Linn. Soc. London (Bot.) 30 (1894) 138; ej. Fl. Tibet (1902) 193; Paulsen in Hedin, S. Tibet, 6, 3 (1922) 43; Rehder and Kobuski in J. Arn. Arb. 14 (1933) 32; Li in Brittonia, 8, 1 (1954) 28.  —*Kokonoria stolonifera* Keng et Keng f. in J. Wash. Ac. Sci. 35 (1945) 375.

Described from Qinghai. Type in Leningrad. Plate VIII, fig. 2.

Sandy and clayey soil in river valleys and on mountain slopes, 2450–5100 m alt.

**IC. Qaidam:** *hilly* (25 km before Barun-Tszasak, 2450 m, May 20, 1884—Przew.).

**IIIA. Qinghai:** *Nanshan* ("near the ruined city Ch'ahancheng, about 30 miles east of Lake Kokonor, Huan-yuan-hsien, formerly known as Tan-ke-erh, Aug. 10, 1944—Keng"—Li, l.c.). *Amdo* (upper Huang He [en route to Baga-gorgi river], May 20; upper Huang He [through Syansibei mountain range], May 29—1880, Przew., lectotypus!; "Radja and Yellow River gorges, Rock"—Rehder, l.c.).

IIIB. Tibet: *Chang Tang* ("in a stream, 35°15' N. lat., 90°20' E. long. 4800 m, Aug. 5, Wellby et Malcolm"—Hemsley, l.c. [1902]; "N. Tibet, Camp 17 [35°48' N. lat., 89°06' E. long.], in a river, 5073 m, Sept. 1, 1896; E. Tibet, near Camp 44 [35°32'N. lat., 88°52' E. long.], 5127 m, Aug. 6, 1901, Hedin"—Paulsen, l.c.). *Weitzan* (Nor. Tibet [12 km before Huang He sources], 4300 m, May 27; upper Dzhagyn-Gol river, 4300 m, June 9; in water divide of Huang He and Yangtze rivers, June 12—1884, Przew.; "Hill-slope 2 miles N. of Murus River, 33°53' N. lat., 91°31' E. long., headwaters of Jangtsekiang, sandy soil, some clay, at 4500 m, June 21, 1892—Rockhill"—Hemsley, l.c. [1902]).

General distribution: China (North-West).

Note. K.I. Maximowicz's analysed and sketched specimen is selected here as a lectotype.

2. **L. brevituba** Maxim. in Bull. Ac. Sci. St.-Pétersb. 27 (1881) 524; Li in Brittonia, 8, 1 (1954) 26.

Described from Qinghai. Type in Leningrad. Plate VIII, fig. 3.

Rocky slopes, 3400–4400 m alt.

IIIA. **Qinghai:** *Nanshan* (Sodi-soroksum hill, 3400–4000 m, July 30–31, 1872—Przew., lectotypus!). *Amdo* (in upper Huang He, 1880—Przew.; "Jupar Range, Jugar-tsargen, No. 1432, Rock"—Li, l.c.).

General distribution: endemic.

Note. K.I. Maximowicz's analysed and sketched specimen is selected here as a lectotype.

3. **L. decumbens** Rupr. in Osten-Sacken und Rupr. Sertum tiansch. (1869) 64; Krasnov in Bot. zap. Peterb. univ. 2, 1 (1887–88) 19; Deasy, In Tibet and Chin. Turk. (1901) 398; Hemsley, Fl. Tibet (1902) 194; Vikulova in Fl. SSSR, 22 (1955) 500; Fl. Kirgiz, 10 (1962) 186; Ikonnikov, Opred. rast. Pamira (1963) 218. —*L. grigorjevii* Krassn. in Bot. zap. Peterb. univ. 2, 1 (1887–88) 19. —*L. glauca* ssp. *australis* Maxim. in Bull. Ac. Sci. St.-Pétersb. 27 (1881) 524. —Ic.: Fl. SSSR, 22, Plate XXIV, fig. 1.

Described from Cent. Tien Shan (Dzhaman-Daban mountain range). Type in Leningrad.

Moraines, around glaciers, banks of brooks, on rocky and rubble slopes and talus, 3400–5200 m alt.

IIA. **Junggar:** *Tien Shan* (Danu river, on shoal, 3400 m, No. 2104, July 21; along water divide between Daban and Danu rivers, on rubble slope, 3600–3900 m, Nos. 545 and 2146, July 23—1957, Kuan; Manas river basin, Danu-Daban crossing, between upper courses of Ulan-Usu and Danu rivers, nival belt, along fine rubble talus with *Thylacospermum* blankets, July 23, 1957—Yun. et al.; "prope Musart et Jir-tass"—Krasnov, l.c.).

IIIB. **Tibet:** *Chang Tang* ("Tibet, Camp 12 [34°41' N. lat., 81°24' E. long.], 5200 m, July 5, 1896; Aksu, 4800 m, 1898"—Deasy, l.c.).

General distribution: Nor. and Cent. Tien Shan, East. Pamir; Mid. Asia (Pamiro-Alay); Himalayas (west., Kashmir).

4. **L. integrifolia** (Willd.) Schischk. in Fl. SSSR, 22 (1955) 502; Fl. Kirgiz. 10 (1962) 186; Fl. Kazakhst. 8 (1965) 96. —*L. glauca* var. *pallasii* Trautv. in Bull. Soc. natur. Moscou, 39 (1866) 375; Li in Brittonia, 8, 1 (1954) 25. —*L.*

*pallasii* (Cham. et Schlecht.) Rupr. in Osten-Sacken und Rupr. Sertum tiansch.
(1869) 64. —*L. glauca* ssp. *borealis* var. *pallasii* Maxim. in Bull. Ac. Sci. St.-
Pétersb. 27 (1881) 522, p.p. —*L. glauca* auct. non Gaertn.: Hemsley, Fl.
Tibet (1902) 194; Danguy in Bull. Mus. nat. hist. natur. 20 (1914) 82; Rehder
and Kobuski in J. Arn. Arb. 14 (1933) 32; Li in Brittonia, 8, 1 (1954) 25. —
*L. altaica* (Willd.) Smirn. in Byull. Mosk. obshch. ispyt. prir. 46, 2 (1937) 97;
Kryl. Fl. Zap. Sib. 10 (1939) 2463; Grubov, Konsp. fl. MNR (1955) 243. —
*Gymnandra integrifolia* Willd. in Ges. Nat. Fr. Berl. Mag. 5 (1811) 392. —*G.
altaica* Willd. ibid. 393. —*G. elongata* Willd. ibid. 395. —*G. pallasii* Cham.
et Schlecht. in Linnaea, 2 (1827) 564, p.p. —**Ic.:** Fl. SSSR, 22, Plate XXIV,
fig. 3.

Described from Siberia. Type in Berlin.

Moist and marshy meadows, alpine tundra and moist rubble-rocky
placers, banks of brooks in alpine and subalpine belts and larch forests in
upper forest belt.

IA. **Mongolia:** *Mong. Alt.* (upper Tsagan-Gol river, anticline in Kharsalu, alpine tundra,
June 29 [TK]; Tsagan-Gol river, rocky crest around lake glacier, placers, July 6—1905; Kak-
Kul' lake, between Tsagan-Gol and Kobdo rivers, alpine tundra, June 22, 1906; Urmogaity
pass, rubble slopes, July 11, 1908; Aksu river, slopes of glacier and moraine, July 23, 1900—
Sap. [TK]; Tolbo-Kungei-Alatau mountain range, high-alpine belt, cobresia meadow, 3200 m,
Aug. 5, 1945; Adzhi-Bogdo mountain range, rubble-rocky placers in alpine belt, Aug. 6 and 7;
top of Kharagaitu-Khutul' pass, among placers in snow and on alpine grasslands, Aug. 24—
1947, Yun.; 15 km south of Yusun-Bulak, nor. slope of Khan-Taishiri mountain range, under
gentle larch forest, Sept. 1, 1948—Grub.; "Altai, entre l'Irtich et Kobdo, 2200–2500 m, Chaff."—
Danguy, l.c.). *Gobi-Alt.* (Ikhe-Bogdo mountain range, cobresia-sedge meadow on rock scree
and upland fescue-sedge steppe, June 28 and 29, 1945—Yun.; Ikhe-Bogdo mountain range,
Narin-Khurimt gorge, among rocks, 3500 m, July 29, 1948—Grub.).

IIA. **Junggar:** *Cis-Alt.* (between Fuyun [Koktogoi] and Bulgan river, 2600 m, No. 1953,
Aug. 19, 1956—Ching). *Tarb.* (Saur mountain range, south. slope Karagaitu river valley, right
of Bain-Tsagan creek valley, alpine belt, cobresia meadow, June 23, 1957—Yun. et al.; nor. of
Dachen town, Tarbagatai hills, Koktubai, on slope, 2250 m, No. 1567, Aug. 13, 1957—Kuan).
*Jung. Alt.* (Toli district, on slope, No. 2650, Aug. 6; Nilka district, 60 km nor. of Ulastai, on
slope, No. 3994, Aug. 31—1957, Kuan). *Tien Shan* (Talki river, July 22; Aktash, Aug. 11—1877;
Dzhagastai hills, 2450–2750 m, June 20; Sumbe gorge, 2750 m, July 29; Bogdo-Ula hill, 3050 m,
July—1878; Taldy river, Iren-Khabirga mountain range, 2750 m, May 26; Borgaty gorge, 2450–
2750 m, June 4; Aryslyn, on Kash river, 2750–3050 m, July 12; same site, 3050–3350 m, July
13—1879, A. Reg.; Sairam lake, July 18; Yuldus river, Sept.—1878, Fet.; 10 km nor. of Chzhaos,
No. 3004, Aug. 15 (?); Barchat, 3 km south of Yakou, 2900 m, No. 1665, Aug. 30—1957, Kuan).

IIIA. **Qinghai:** *Amdo* (? "alpine region between Radja and Jupar ranges, Rock"—Rehder,
l.c.).

General distribution: Jung.-Tarb., Nor. and Cent. Tien Shan; West. Sib. (Altay), East.
Sib., Nor. Mongolia (Fore Hubs., Hent., Hang.).

5. **L. ramalana** Batalin in Acta Horti Petrop. 14 (1895) 177; Li in Brittonia,
8, 1 (1954) 24.

Described from South-West China (Sikang province). Type in Leningrad. Plate VIII, fig. 4.

Alpine wet meadows, 3700–4300 m alt.

IIIA. Qinghai: *Amdo* ("alpine region between Radja and Jupar Range, near top of Wajo la, No. 14098, Rock"—Li, l.c.).

General distribution: China (North-West, South-West).

Note. Li (l.c.). also placed in this species the plants described as *L. praecox* W.W. Smith from Yunnan province. Specimens of the latter, however, differ distinctly from *L. ramalana* in partite calyx lobes and large plant size; for this reason, we have not included *L. pracecox* under synonyms of *L. ramalana*.

## 10. Oreosolen Hook. f.
### Fl. Brit. Ind. 4 (1884) 318.

1. **O. unguiculatus** Hemsl. in Kew Bull. (1896) 213; ej. Fl. Tibet (1902) 193; Paulsen in Hedin, S. Tibet, 6, 3 (1922) 43.   —**Ic.:** Hook. Ic. pl. 25, tab. 2467.

Described from South. Tibet. Type in London (K).

High-alpine belt, 4750–4900 m alt.

IIIB. Tibet: *Chang Tang* ("inner Tibet, nameless valley, between Camp 70 and Camp 71 32°41′ N. lat., 88°45′ E. long.], alt. 4889 m, Sept. 1, 1901, Hedin"—Paulsen, l.c.). *South.* "chiefly from Gooring Valley, 30°12′ N. lat., 90°25′ E. long., at about 4750 m, St. George, July–Aug. 1895, Littledale, typus"—Hemsl. l.c.).

General distribution: endemic.

## 11. Rehmannia Libosch.
### ex Fisch. et Mey. Ind. sem. Horti Petrop. 1 (1835) 36; Li in Taiwania, 1 (1948) 72.

1. **R. glutinosa** (Gaertn.) DC. Prodr. 9 (1845) 275; Maxim. in Bull. Soc. natur. Moscou, 54 (1879) 33; Forbes and Hemsley, Index Fl. Sin. 2 (1902) 193; Pai in Contribs Inst. Bot. Nat. Ac. Peiping, 2, 7 (1934) 203; Hao in Engler's Bot. Jahrb. 68 (1938) 636; Kitag. Lin. Fl. Mansh. (1939) 396; Walker in Contribs U.S. Nat. Herb. 28 (1941) 660; Li in Taiwania, 1 (1948) 75 [sub nom. *R. glutinosa* (Gaertn.) Fisch. et Mey.].   —*R. chinensis* Fisch. et Mey. Ind. sem. Horti Petrop. 1 (1835) 36.   —*R. chanetii* (Lévl.) Lévl. in Feddes repert. 9 (1911) 323.   —*Digitalis glutinosa* Gaertn. in Novi Comm. Ac. Petrop. 14, 1 (1770) 544.   —*Gerardia glutinosa* Bunge, Enum. pl. China bor. (1832) 49.   —**Ic.:** Bot. Reg. 22, tab. 1960 (sub nom. *R. chinensis*).

Described from China. Type in London.

Rocky slopes, rocks and pebble beds.

IA. Mongolia: *East. Mong.* (Muni-Ula, nor. slope, on slope and in foothills, on silty-rocky soil, rare, June 28, 1871; nor. slope of Muni-Ula, on rocks, May 3, 1872—Przew.; "Suiyuan: Wusuotu, No. 17, 1783, May 28, 1933, Hsia"—Pai, l.c.). *Alash. Gobi* (mouth of Hsi Jeh Kou [Ho Lan Shan—A La Shan], along exposed moist, rocky banks of irrigation ditches, common, No. 174, Ching"—Walker, l.c.).

General distribution: China (Dunbei, North, North-West, East, Cent.).

## 12. **Leptorhabdos** Schrenk
in Fisch. et Mey. Enum. pl. nov. 1 (1841) 23.

1. **L. parviflora** (Benth.) Benth. in DC. Prodr. 10 (1846) 510; Ivanina in Fl. SSSR, 22 (1955) 527; Fl. Kirgiz. 10 (1962) 190; Fl. Kazakhst. 8 (1965) 98. — *L. micrantha* Schrenk in Fisch. et Mey. Enum. pl. nov. 1 (1841) 23. —*L. brevidens* Schrenk in Fisch. et Mey. Ind. sem. hort. Petrop. 9, Suppl. (1843) 4, 13. —*Gerardia parviflora* Benth. ex Wall. Cat. (1829) No. 3888, nom. nud.; Benth. Scroph. Ind. (1835) 48. —**Ic.:** Fl. Uzbek. 5, Plate XV, fig. 3; Fl. Kazakhst. 8, Plate XI, fig. 3.

Described from Himalayas. Type in London (K).

Banks of rivers, brooks and irrigation ditches, on talus and sand, in steppes, common as weed in fields and on roadsides.

IB. **Kashgar:** *East.* ((north-west of Khami, July 6, 1956—Ching).

IIA. Junggar: *Tien Shan* (upper Ili near Kash river estuary, 750 m, Aug. 28, 1876—Przew.; between Urtaksary and Borotola, 1800–2100 m, Aug. 4, 1878—A. Reg.; environs of Fukan, Aug. 21, 1898—Klem.; around Urumchi town, Bogdo-Ula foothills, Sept. 14, 1929—Pop.; 10 km south-west of Guilyu [Karabura], along edge of canal, 1500–1800 m, No. 1044, Aug. 19, 1957—Kuan; left bank of Ili river valley, around Yamatu crossing, near oasis, Aug. 21; Ketmen' mountain range, 3–4 km above Sarbushin settlement, along Ili—Kzyl-Kure road, on south. slope, on talus, Aug. 23—1957, Yun. et al.).

General distribution: Fore Balkh., Jung.-Tarb., Nor. and Cent. Tien Shan, East. Pam.; Fore Asia, Caucasus, Mid. Asia, Himalayas (west., Kashmir).

## 13. **Castilleja** Mutis ex L. f.
Suppl. (1781) 47.

1. **C. pallida** (L.) Spreng. Syst. Veg. 2 (1825) 774; Kryl. Fl. Zap. Sib. 10 (1939) 2466; Kitag. Lin. Fl. Mansh. (1939) 391; Grubov, Konsp. fl. MNR (1955) 233; Gorschkova in Fl. SSSR, 22 (1955) 531, p.p.; Rebristaya in Nov. sist. vyssh. rast. (1964) 288; Fl. Kazakhst. 8 (1965) 98. —*Barisia pallida* L. Sp. pl. (1753) 602. —**Ic.:** Fl. Kazakhst. 8, Plate XI, fig. 4.

Described from Siberia. Type in London (Linn.).

Moist banks of rivers and brooks, meadows, larch forests, turf-covered rock screes and as weed in ploughed fields in forest and subalpine belt of mountains.

IA. **Mongolia:** *Khobd.* (Ulan-Nachin river valley and between Tszusylan and Sarynchi, July 15, 1879—Pot.). *Mong. Alt.* (V. Khobdo lake, forest meadows, June 27, 1906—Sap.). *Gobi-Alt.* (Baga Bogdo, moist terraces at 1950 m, No. 241—Chaney).

General distribution: Europe (South. Urals), West. Sib. (south. part and Altay), East. Sib., Far East, Nor. Mongolia, China (Dunbei).

## 14. Euphrasia L.

Sp. pl. (1753) 604, p.p.; Wettst. Monogr. Gatt. Euphr. (1896) 9.

1. Leaves and calyx densely pubescent with long-stalked (3–10-celled stalks), somewhat crispate glandular and simple hairs. ................... ........................................................................................... E. hirtella Jord.

+ Leaves and calyx without glandular pubescence or with short (with 1-2-celled stalks) glandular and simple hairs .............................. 2.

2. Leafy bracts and calyx with short-stalked glandular and simple hairs; stem pubescent with whitish simple hairs, sometimes mixed with short glandular hairs. Inflorescence compressed, slightly elongated in fruit .......................................................... 2. E. regelii Wettst.

+ Leafy bracts and calyx without glandular pubescence (glands seen sometimes at calyx base), more or less densely covered with simple hairs or bristles; stem subglabrous or with whitish hairs not mixed with glandular hairs. Inflorescence dense; greatly elongated in fruit and interrupted or few-flowered, short .......................................... 3.

3. Plant with slender green or whitish, 5–15 cm tall stem; caulous leaves few, interrupted, with 1–3 subobtuse teeth on each side. Inflorescence few-flowered, 0.5–1.5 cm long .. 3. E. syreitschikovii Govor.

+ Plant with strong stem turning red or brown, 8–50 cm tall; caulous leaves several, with 3–8 acute or subobtuse teeth on each side. Inflorescence many-flowered, greatly elongated and interrupted after anthesis, 1.5–15 cm long ................................................................ 4.

4. Caulous leaves ovate or broadly ovate, with 3–8 acute, not patent teeth; leafy bracts broadly ovate, with 5–8 acute, generally aristate teeth. Stem simple or branched at tip ..................................... .................................................................. 1. E. maximowiczii Wettst.

+ Caulous leaves cuneate or obovate, with 1–5 subobtuse or acute teeth (upper leaves with 4–7 acuminate teeth on each side, of which lower ones patent); leafy bracts ovate, with 7–9 acute chondroid, obliquely upturned teeth. Stem simple or branched in lower and middle parts ......................................................... 4. E. tatarica Fisch.

E. hirtella Jord. in Reuter in Compt. rend. d. 1. Soc. Haller. 4 (1854–1856) 120; Wettst. Monogr. Gatt. Euphr. (1896) 175; Kryl. Fl. Zap. Sib. 10 (1939) 2484; Juzepczuk in Fl. SSSR, 22 (1955) 635; Fl. Kazakhst. 8 (1965) 108. —Ic.: Fl. SSSR, 22, Plate XXXI, fig. 2; Fl. Kazakhst. 8, Plate XII, fig. 3.

Described from France. Type in Paris (?).

Alpine and subalpine meadows, forest grasslands, steppes, scrub and forests.

IA. Mongolia: *Mong. Alt.* (occurrence possible).

General distribution: Europe, Balk.-Asia Minor, Caucasus, Mid. Asia, West. and East. Sib., Nor. Mongolia (Hent., Hang.), China (Dunbei).

1. **E. maximowiczii** Wettst. Monogr. Gatt. Euphr. (1896) 87; Kitag. Lin. Fl. Mansh. (1939) 392; Juzepczuk in Fl. SSSR, 22 (1955) 568.  —Ic.: Fl. SSSR, 22, Plate XXVII, fig. 1.

Described from Japan. Type in Leningrad.

Meadows, among shrubs and in forest fringes.

IA. **Mongolia:** *Fore Hing.* (Khalkha-Gol river, meadow, under cover of purple osier willows, Aug. 7, 1899—Pot. and Sold.; Khuntu somon, 5 km west of Togë-Gol river, forb meadow, Aug. 7, 1949—Yun.).

General distribution: Far East, China (Dunbei), Japan.

2. **E. regelii** Wettst. Monogr. Gatt. Euphr. (1896) 81; Kryl. Fl. Zap. Sib. 10 (1939) 2484; Pampanini, Fl. Carac. (1930) 190; Juzepczuk in Fl. SSSR, 22 (1955) 589; Fl. Kirgiz. 10 (1962) 192; Fl. Kazakhst. 8 (1965) 105.  —? *E. subpetiolaris* Pugsley in J. Bot. (London) 74 (1936) 282.  —Ic.: Fl. SSSR, 22, Plate XXVIII, fig. 1.

Described from Junggar. Type in Vienna (?). Isotype in Leningrad.

Forest and alpine meadows and rubble talus.

IIA. **Junggar:** *Tien Shan* (Sairam-Nur lake, south. bank, July, 1877; Aryslyn, 2750 m, July 15, 1879—A. Reg., isotypus!; Manas river basin, Ulan-Usu river valley at confluence with Dzhartas, upper forest fringe, on rubble talus on left bank of valley, nor.-nor.-east. slope, July 18; same site, subalpine belt, on rubble talus, July 18; Sairam-Nur basin, 2 km north-east of Sairam crossing on road to Ili, meadow on slope below spruce groves, Aug. 19—Yun. et al.; "upper Koksu, Tian-Shan, 2600 m, No. 752, Aug. 8, 1930, Ludlow [BM]"—Pugsley, l.c.).

IIIC. **Pamir:** (Pas-Rabat settlement, on brim of irrigation ditch, rare, July 3, 1909—Divn.)

General distribution: Jung.-Tarb., Nor. and Cent. Tien Shan; Mid. Asia (Pamiro-Alay), Himalayas (west.).

3. **E. syreitschikovii** Govor. in Pavl. in Byull. Mosk. obschch. ispyt. prir. 38, 1–2 (1929) 126; Kryl. Fl. Zap. Sib. 10 (1939) 2477; Grubov, Konsp. fl. MNR (1955) 243; Juzepczuk in Fl. SSSR, 22 (1955) 574; Fl. Kazakhst. 8 (1965) 105.  —*E. pectinataeformis* Kryl. et Serg. in Tr. Biol. n.-i. inst. Tomsk. gos. univ. 1 (1935) 74.  —*E. officinalis* α. *pectinata* Kryl. Fl. Alt. (1907) 954.  —Ic.: Govor, l.c. 126.

Described from Nor. Mongolia (Hangay). Type in Moscow. Isotype in Leningrad.

Subalpine meadows, rock screes and pebble beds.

IA. **Mongolia:** *Khobd.* (Khashatu river, from Ulan-Daban to Kobdo, Aug. 9, 1899—Lad.). *Mong. Alt.* (nor. slope of Khara-Dzarga mountain range near Khairkhan-Dur, larch forest, Aug. 25; Khasagtu-Khairkhan mountains, Ulyasten-Gol river valley, coastal pebble bed, Sept. 20; same site, on turf, Sept. 21—1930, Pob.; Bulugun somon in upper Indertiin-Gol, marshy meadow in high-alpine belt, July 24; 25 km south of Tamchi-Daba crossing, Bidzhi-Gol river, left bank rocky slope of valley, birch groves near spring, Aug. 10—1947, Yun.). *Cent. Khalkha* (Dzhargalante river basin, Kharukhe river sources, Uste mountain, subalpine belt, nor. slope near peak, Aug. 12; Ubur-Dzhargalante river sources, marshy meadow, Aug. 12—1925, Krasch and Zam.).

General distribution: West. Sib. (Alt.) Nor. Mong. (Hang.).

Note. Specimens collected from Khasagtu-Khairkhan mountains differ from typical specimens (red stems, sometimes, branched from base, leaves suborbicular at base, 1-2-toothed on each side)—var. *pectinataeformsi* (Kryl. et Serg.). Ivanina comb. nova (*E. pectinataeformis* Kryl. et Serg. l.c.). Plants from Cent. Khalkha (Dzhargalante river basin, subalpine belt) are evidently hybrids between *E. syreitschikovii* and *E. tatarica* Fisch.

4. **E. tatarica** Fisch. in Spreng. Veg. 2 (1825) 777; Wettst. Monogr. Gatt. Euphr. (1896) 88; Pampanini, Fl. Carac. (1930) 190; Kitag. Lin. Fl. Mansh. (1939) 392; Kryl. Fl. Zap. Sib. 10 (1939) 2478; Juzepczuk in Fl. SSSR, 22 (1955) 570; Fl. Kirgiz. 10 (1962) 191; Fl. Kazakhst. 8 (1965) 103.   —*E. officinalis* auct. non L.: ? Sapozhn. Mong. Alt. (1911) 380; Pai in Contribs Inst. Bot. Nat. Ac. Peiping, 2, 7 (1934) 206; Walker in Contribs U.S. Nat. Herb. 28 (1941) 658; Grubov, Konsp. fl. MNR (1955) 243.   —Ic.: Fl. SSSR, Plate XXIII, fig. 2.

Described from Europe (around Saratova town). Type in Leningrad.

Meadow and steppe slopes, along banks of rivers and brooks, in groves, scrub, forests, marshy areas, marshy meadows and coastal pebble beds in forest, steppe and subalpine belts.

IA. **Mongolia:** *Khobd.* (Bukhu-Muren-Gol river floodplain, 4–5 km nor. of somon, in scrub bottom-land deciduous forest and meadow, July 31, 1945—Yun.). *Cent. Khalkha* (nor. part). *Bas. Lakes* (Ulyasutai—Kosh-A gach road, June 15-July 15, 1880—Pev.; Ulangom, Sept., 5, 1879—Pot.; 4 km south of Ulangom, Kharkhira river valley, solonchak-like meadow, July 27, 1945*—Yun.). *Gobi-Alt.* (Dzun-Saikhan, Yailo creek valley, on pebble bed of river, under shade of rocks, Aug. 20, 1931*—Ik.-Gal.). *East. Mong.* (Ourato, sous les arbres, en montagnes, No. 2758, July 1866*—David; "Wulashan, No. 3105, June 21, 1931, Hsia"—Pai, l.c.). *Alash. Gobi* "Ho Lan Shan, along shaded margins of streams, No. 1083, Ching"—Walker, l.c.).

IIA. **Junggar:** *Tien Shan* (lower Kunges, forest belt, 1050 m, in forest, in moist soil, June 17, 1877*—Przew.; "Bogdo Ula, Aug. 1930, Ting; Bogdo Ula, near Foshowze, No. 3274, Aug. 16, 1931, Liou"—Pai, l.c.).

IIIA. **Qinghai:** *Nanshan* (25 km south of Gulan, east. extremity of Nanshan, gentle slopes of lower mountains, mountain steppe with scrub, 2450 m, Aug. 12, 1958*—Petr.; "Kokonor, without precise locality, No. 1282, Sept. 14, 1960, Hao"—Pai, l.c.; "Hsi Mi Jai, along shaded streams, Ching"—Walker, l.c.).

General distribution: Jung.-Tarb., Nor. and Cent. Tien Shan; Europe, Mediterr., Balk.-Asia Minor, Fore Asia, Caucasus, Mid. Asia (mountains), West. and East. Sib., Nor. Mong., China (Dunbei, North, North-West, South-West), Himalayas, Japan.

Note. Most listed specimens (asterisked) resemble *E. frigida* Pugsl. in their dense and white hairy stems, late-falling leaves and generally 3–4 teeth on each side; they differ in very large corolla and other characteristics.

15. **Omphalothrix** Maxim.
Prim. Fl. amur. (1859) 208.

1. **O. longipes** Maxim. Prim. Fl. amur. (1859) 209; Komarov, Fl. Manchzh. (1907) 475; Pai in Contribs Inst. Bot. Nat. Ac. Peiping, 2, 7 (1934) 207; Kitag. Lin. Fl. Mansh. (1939) 324; Golubkova in Fl. SSSR, 22 (1955) 640.   —Ic.: Maxim. l.c. tab. X.

Described from Far East. Type in Leningrad.

Marshy meadows, mostly in river valleys, clogged streams and stagnant puddles.

IA. **Mongolia:** *Ordos* (Komarov, l.c.; Golubkova, l.c. specimen filed in Herbarium of the Komarov Botanical Institute, lost).

**General distribution:** Far East, China (Dunbei), Korean peninsula.

## 16. Odontites Ludw.

Inst. Reg. Veg. ed. 2 (1757) 120.

1. **O. serotina** (Lam.) Dum. Fl. Belg. (1827) 32; Pampanini, Fl. Carac. (1930) 191; Kryl. Fl. Zap. Sib. 10 (1939) 2488; Grubov, Konsp. fl. MNR (1955) 248; Golubkova in Fl. SSSR, 22 (1955) 650; Fl. Kirgiz. 10 (1962) 192; Fl. Kazakhst. 8 (1965) 111.  —*O. serotina* (Lam.) Reichb. Fl. Germ. exc. (1830–1832) 359; Kitag. Lin. Fl. Mansh. (1939) 394.  —*O. rubra* Pers. Syn. pl. 2 (1807) 150; Danguy in Bull. Mus. nat. hist. natur. 20 (1914) 81; Walker in Contribs U.S. Nat. Herb. 28 (1941) 658.  —*O. odontites* auct. non Wettst.: Pai in Contribs Inst. Bot. Nat. Ac. Peiping, 2, 7 (1938) 208.  —*Bartsia odontites* auct. non Benth. et Hook.: Kung et Wang in Contribs Inst. Bot. Nat. Ac. Peiping, 2, 8 (1934) 382; Chen and Chou Rast. pokrov r. Sulekhe (1957) 90.  —**Ic.:** Fl. SSSR, 22, Plate XXXIII, fig. 1.

Described from France. Type in Paris.

Solonchak-like and floodplain meadows, coastal willow groves as well as weed in crops and around canals.

IA. **Mongolia:** *Mong. Alt.* (Buyantu camp, crop, Aug. 28, 1930—Bar.; Bulugun river floodplain at confluence with Ulyaste-Gol, grassy meadow on wet alluvium, July 20; lower Uinchi-Gol 15–20 km below winter camp in somon, solonchak-like meadow on banks of streams, July 29—1947, Yun.). *Cis-Hing.* (Khalkhin-Gol river valley, 15 km south-east of Khamar-Daban, willow thickets along meander, Aug. 11, 1949—Yun. et al.,; Khuna district, near Trekhrech'e, mountain slope, 600–650 m, No. 2073, Aug. 24, 1951—Wang). *Cent. Khalkha, East. Mong.* (30 km away from Bain-Buridu somon, depression in steppes, alkali grass and forb meadow, Aug. 18, 1949—Yun.; Khailar, solonetz site, No. 3015, 1954—Wang; "Wulashan, No. 3104, Aug. 21, 1931, Hsia"—Pai, l.c.). *Bas. Lakes, Val. Lakes* (Bain-Gobi somon, Tsagan-Gol river, near somon camp, sedge grassland with barley and buttercup , on lower bank of river, July 27, 1948—Grub.). *Alash. Gobi* ("Ho Lan Shan, No. 1143, Ching"—Walker, l.c.). *Ordos* (25 km south-east of Otok town, solonchak meadow near Khaolaitunao lake, Aug. 1; 10 km south-west of Ushin town, willow and sea buckthorn scrub, on meadow, Aug. 4; 20 km west of Dzhasak town, meadow in valley, Aug. 17—1957, Petr.).

IIA. **Junggar:** *Jung. Alt.* (Tien Shan Laoba-Myaoergou, on wet site, No. 2397, Aug. 3, 1957—Kuan). *Tien Shan* ("Tulu-Fan to Ouroumtai riverside, No. 2656, Lion"—Pai, l.c.). *Jung. Gobi* ("montagnes entre l'Ouchte et l'Irtish, No. 1250, Aug. 19; steppes, No. 1249, Aug. 25, 1895, Chaff."—Danguy, l.c.). *Balkh.-Alak.* (Toli district, Uty, on intermontane plain, No. 4897, 4905, Aug. 3, 1957—Kuan).

**General distribution:** Aralo-Casp., Fore Balkh., Jung.-Tarb., Nor. and Cent. Tien Shan; Europe, Balk.-Asia Minor, Fore Asia, Caucasus, Mid. Asia, West. and East. Sib., Far East, Nor. Mong., China (Dunbei, North, North-West).

## 17. Rhinanthus L.

Sp. pl. (1753) 603. —*Alectrolophus* Zinn. Cat. pl. (1757) 288 (nom. superfl. illegit. pro *Rhinanthus*).

1. **Rh. songaricus** (Sterneck) B. Fedtsch. in Fedtsch. and Fler. Fl. Evrop. Ross. (1910) 880; Soó in Feddes repert. 26 (1929) 201; Kryl. Fl. Zap. Sib. 10 (1939) 2533; Vasil'chenko in Fl. SSSR, 22 (1955) 671; Fl. Kazakhst. 8 (1965) 114. —*Alectrolophus songaricus* Sterneck in Abh. zool.-bot. Gesellsch. Wien, 1, 2 (1901) 79. —*Rh. montanus* auct. non Sauter: Grubov, Konsp. fl. MNR (1955) 248. —Ic.: Fl. Kazakhst. 8, Plate XII, fig. 7.

Described from Kazakhstan ("Songoria"). Type in Vienna.

Meadows and tugais in river valleys.

IIA. Junggar: *Cis-Alt.* (Kran river, in floodplain, July 28, 1959 —A.R. Lee (1959)). —*Jung. Alt.* (Archaty-Gol river valley, poplar forest on floodplain, Aug. 17, 1957—Yun. et al.). *Tien Shan* (Sarbushin crossing, Ketmen' mountain range, on Ili—Kzyl-Kure road, steppe fied meadow on gentle nor. slope, Aug. 23; Ketmen' mountain range, Tekes river valley, 10 km south-east of Kalman-Kure settlement, along road to Aksu, hill meadow, Aug. 24—1957, Yun. et al.).

General distribution: Aralo-Casp., Jung.-Tarb., Nor. and Cent. Tien Shan; West. Sib., Nor. Mong. (Hang.).

Note. Hybrid *Rh. pseudo-songoricus* Vass. has been described for Kul'szha in Fl. SSSR, 22 (1955) 685 (*Rh. vernalis* x *Rh. songaricus*). Found in valleys of rivers and lakes.

## 18. Pedicularis L.

Sp. pl. (1753) 607.

1. Corolla tube comparatively short, not more than twice longer than calyx ......................................................................................... 2.

+ Corolla tube long, cylindrical, narrow, more than twice longer than calyx ......................................................................................... 34.

2. Leaves alternate (rarely upper ones opposite or whorled); plant invariably with distinct stem although sometimes very low ........... 3.

+ Leaves (and bracts) whorled (4, sometimes 3, leaves in whorl) or opposite; plants sometimes stem less (or stem weak, ascending)... ......................................................................................... 43.

3. All leaves undivided or incised-lobed or only upper caulous leaves and leaves of branches undivided (while rest pinnately lobed or pinnatipartite), linear, linear-oblong or lanceolate; plant more than 7–10 cm tall. ......................................................................... 4.

+ Leaves 1–2 or nearly 3 times pinnatisect or pinnatipartite; plant sometimes very low (1–5 cm tall) or moderate and tall ............. 10.

4. Flowers resupinate, subsessile, one each in axils of upper leaves; corolla purple, more or less pilose, with highly falcate hood and nearly as long beak. Radical leaves absent, caulous alternate or opposite, oblong-lanceolate or lanceolate, incised-serrate-crenate ..... ......................................................................... 60. P. resupinata L.

+ Flowers not resupinate, aggregated into racemose or terminal inflorescence; corolla yellow or purple ................................................ 5.

5. Biennial plant, 10–15 cm tall, stems generally branched from base; lower and middle leaves pinnatipartite, with linear-oblong acuminate lobes, upper caulous and branch leaves undivided, serrulate. Flowers in racemose inflorescence at tip of stems and branches; corolla yellow with reddish tinge along hood; beak short with 2 linear teeth ....................................................... **P. labradorica** Wirs.

+ Perennial plant 10–40 cm tall; stems not branched; all leaves undivided ......................................................................................................... 6.

6. Hood with long beak ...................................................................... 7.

+ Hood without or with short beak ................................................... 9.

7. Hood with long hairs on back, gradually transforming into falcate beak; calyx teeth oblong-ovate, obtuse, unevenly serrate-dentate. Root with highly thickened transversely rugose root fibers ............
................................................................ 66. **P. trichoglossa** Hook. f.

+ Hood puberulent, narrowed into beak, directed forward or forward and downward; calyx teeth lanceolate-deltoid, acute. Root stout, more or less thickened ....................................................................... 8.

8. Plant 30–80 cm tall, leaves acute, 5–8 cm long. Inflorescence elongated, racemose, 10–30 cm long, bracts longer than flowers; calyx glabrous; hood about 3 mm long, narrow, lower lip partite up to 2/3 into 3 ovate-lanceolate lobes ............ 63. **P. retingensis** Tsoong.

+ Plant 10–20 cm tall, leaves obtuse, 1.5–3 cm long. Inflorescence short, dense, sometimes interrupted at base, 2–5 cm long, bracts as long as calyx or shorter; calyx pilose; hood about 5–7 mm long, thickened, lower lip partite almost up to base into 3 orbicular lobes
................................................................ 60. **P. lasiophrys** Maxim.

9 (6). Hood obtuse or rounded at tip, diffusely pilose, villous-ciliate at front of mouth, lower lip 3-lobed; calyx teeth deltoid, subacute, entire or somewhat dentate. Leaves with incised-crenate and twice chondrondentate lobes. Rhizome slender, creeping, with filiform roots ......................................................................... 67. **P. tristis** L.

+ Hood subacute at tip, with long hairy cilia in middle along margin, gradually narrowing into short truncate beak, lower lip deeply tripartite; calyx teeth broadly deltoid, somewhat unevenly dentate. Leaves incised-dentate, with acuminate teeth. Root with very highly thickened transversely rugose roof fibres ....... 61. **P. ingens** Maxim.

10 (3). Annual or biennial; stem usually branched, with erect branches, 20–50 cm tall. Flowers on short pedicels, singly in axils of interrupted upper leaves; corolla pink, 14–16 mm long, with straight tube and somewhat falcate short beak ................ 59. **P. karoi** Freyn.

+ Perennial; stems unbranched. Flowers in more or less dense terminal inflorescences ................................................................................................ 11.
11. Hood with beak ....................................................................................................... 12.
+ Hood without beak ............................................................................................... 30.
12. Beak of hood without teeth ............................................................................... 13.
+ Beak of hood bidentate ...................................................................................... 16.
13. Corolla pink or purple, with narrow tube abute 2 mm broad, erect in calyx, hood with long circinate beak ........................................................ 14.
+ Corolla yellow, with 4–6 mm broad tube, curved in calyx, hood with slightly curved or suberect beak ........................................................ 15.
14. Corolla tube somewhat longer than calyx; latter 5–toothed with 10 nerves; corolla pink, lateral lobes of lower lip not folded, patent and not incumbent on hood and beak. Leaves linear-lanceolate, pinnatipartite, with winged rachis and dentate or pinnately lobed segments ....................................... 72. P. rhinanthoides Schrenk.
+ Corolla tube not longer than calyx; latter 3-toothed with 6 projecting nerves; corolla purple, lateral lobes of lower lip folded nearly double and incumbent on hood and beak. Leaves oblong, pinnatisect into elliptical pinnatipartite segments .......... 74. P. elwesii Hook. f.
15 (13). Hood not bent, villous along lower margin, gradually transforming into rather short beak, directed forward, corolla 16–17 mm long; calyx 5–6 mm long, inflorescence elongate, 10–20 cm long, bracts arachnoid-villous ........................................ 53. P. proboscidea Stev.
+ Hood at tip bent nearly at right angle and drawn into long straight beak, deflexed, corolla 17–20 mm long; calyx 9–12 mm long, inflorescence very compact, oblong-globose or capitate, 6–10 cm long, more or less coarsely villous or glabrous .....................................................
................................................................ 45. P. compacta Steph. ex Willd.
16 (12). Corolla yellow with purple nerves, 25–32 mm long, lower lip long-unguiform with short 3-lobed limb, subparallel to hood; inflorescence elongate, compact at anthesis, more or less lax in fruit
........................................................................................56. P. striata Pall.
+ Corolla purple or yellow ................................................................................... 17.
17. Hood uncinate at tip, with 2 teeth parallel to hood axis ............. 18.
+ Hood shortly falcate, teeth at tip of hood directed forward and downward, i.e., almost at right angle to hood axis ..................... 22.
18. Corolla short, about 14 mm long, its lower lip longer than hood, lateral lobes of lip compactly adhering to corolla tube. Leaves pinnatisect, segments narrowly lanceolate, pinnatisect, lobes acuminately chondrodentate ........ 44. P. breviflora Regel et Winkl.
+ Corolla more than 16 mm long, lateral lobes of lower lip more or less separate from corolla tube ......................................................... 19.

19. Corolla bright pink or white, 22–25 mm long, hood with short beak; calyx broadly campanulate, 5–6 mm long and 11–13 mm broad. Root shortened, with thick funiform root fibres; leaf oblong or ovate-oblong in general shape, pinnatipartite with deeply pinnatifid subobtuse segments ovate or lanceolate in general shape. Inflorescence dense, with white pubescence, surrounded by leaves at base ................................................................. 46. P. **dasystachys** Schrenk.

   \+ Corolla yellow ................................................................................ 20.

20. Radical leaves large, 18–25 cm long, 2.5–3 cm broad, pinnatisect into oblong pinnately lobed segments excurrent on rachis. Upper part of calyx somewhat narrow, inflated in fruit; corolla large, 28–32 mm long .................................................... 52. P. **physocalyx** Bunge.

   \+ Radical leaves 5–10 cm long, 1–3 cm broad, pinnatisect into lanceolate or linear pinnately lobed segments, hardly excurrent on rachis. Upper part of calyx not narrow, hardly inflated in fruit; corolla 20–26 (28) mm long. ........................................................................ 21.

21. Corolla puberulent on outer surface, pale yellow, with purple-teeth; inflorescence more or less crispate-hairy; calyx tubular-campanulate, almost herbaceous, glabrous or crispate-hairy with entire teeth .......................................................... 54. P. **pubiflora** Vved.

   \+ Corolla glabrous on outer surface, yellow; inflorescence arachnoid-villous; calyx broadly campanulate, nearly membranous, arachnoid-villous, with serrate teeth ......................... 50. P. **lasiostachys** Bunge.

22 (17). Calyx teeth broadly deltoid. Root fascicular or shortened, with more or less thickened root fibres; stems generally slender, tall .......... 23.

   \+ Calyx teeth narrowly deltoid. Root short, with somewhat thickened root fibres; stems more or less thick, short ..................................... 26.

23. Corolla purple-pink. Root fascicular, with thickened fusiform root fibres; stem strong, 30–50 cm tall; caulous leaves large, oblong or oblong-ovate in general shape, pinnatisect into linear, chondroid-acuminate, serrate segments. Calyx obliquely ovate, 5–7 mm long, with oblique deltoid acute teeth. ........................... 48. P. **elata** Willd.

   \+ Corolla yellow. Stem less strong, 10–40 cm tall; caulous leaves linear or oblong in general shape, pinnatipartite or pinnately lobed 24.

24. Lower lip almost as long or slightly than hood; calyx narrowly campanulated, 8–13 mm long, teeth uneven, 0.5–2 mm long (2 almost fully connate), broad, obtuse or subobtuse. Rachis more or less winged, caulous leaves rapidly reducing upward ............... 22.

   \+ Lower lip considerably shorter (almost by half) than hood, glabrous; calyx campanulate, 7–9 mm long, teeth more or less equal, 1–1.5 mm long, less broad, obtuse or acute. Rachis generally not winged, caulous leaves gradually reducing upward. Plant generally 10–25 cm tall. ................................................ 58. P. **venusta** (Bunge) Bunge.

25. Lip more or less ciliate along margin; calyx 8–10 mm long, shallowly split for 2–3 mm in front, more or less pubescent, sometimes with reddish-black spots. Lower caulous leaves narrowly winged, pinnatisect into oblong, oblong-lanceolate or lanceolate-obtuse segments. Capsule subsymmetrical ........... 43.P. **altaica** Steph. ex Stev.

  +  Lip glabrous; calyx 10–18 mm long, more or less deeply split for 4–5 mm in front, with white pubescence. Lower caulous leaves winged, pinnatipartite or pinnatisect into oblong-obtuse or rounded segments, their lobes acuminate-chondrodentate; upper caulous leaves undivided, pinnaticristate. Capsule oblong-lanceolate, oblique, 13–15 mm long ............................................................... 51. P. **mariae** Regel.

26 (22). Flowers pink or purple; calyx tubular-campanulate ................... 27.

  +  Flowers yellow; calyx campanulate or tubular-capanulate ........ 28.

27. Leaves twice or almost thrice pinnatisect into linear chondroid acuminate segments; caulous leaves 1–2; stem with long crispate hairs. Calyx 14–15 mm long, submembranous, teeth 2/5–1/2 length of tube; corolla pink ................................. 55. P. **rubens** Steph. ex Willd.

  +  Leaves once pinnatisect into lanceolate chondroid acuminate pinnately lobed segments; caulous leaves few; stem glabrous or villous under inflorescence. Calyx 10–14 mm long, almost herbaceous, teeth 1/3 length of tube; corolla purple ................................... ............................................................ 57. P. **uliginosa** Bunge.

28. Beak of hood narrow, i.e., longer than broad, truncate; calyx tubular-campanulate. Root with elongated-fusiform root fibres; leaves with winged rachis, pinnatipartite into oblong-lanceolate or lanceolate pinnately lobed segments; stem erect or curved, more or less villous, 10–80 cm tall ................... 47. P. **dolichorrhiza** Schrenk.

  +  Beak of hood broad and short; calyx campanulate. Stem erect or ascending, strong, with fine crispate hairs, 10–40 cm tall. .......... 29.

29. Root vertical, strong, branched; leaves pinnatisect twice into linear-lanceolate, interrupted, coarsely lobed segments; lobes of latter deltoid, acute, with acutely incised chondroid teeth, often recurved claw-like on drying. Corolla 28–32 cm long, tube nearly as long as hood; calyx subcoriaceous ....................................... 49. P. **flava** Pall.

  +  Root shortened, with thickened fusiform root fibres; leaves twice-thrice pinnatisect into linear chondroid acuminate lobes. Corolla 24–28 mm long, corolla tube 2/3 length of hood; calyx almost herbaceous ..................................... 42. P. **achilleifolia** Steph. ex Willd.

30 (11). Plant 30–80 cm tall. Hood lanate or ciliate along margin at front or most of corolla covered with glandular hairs; lower lip parallel to

hood and nearly as long ...................................................................31.

+ Plant short, 3–15 (20) cm tall. Corolla glabrous ...........................32.

31. Corolla very large, 30–40 mm long, yellow, but lobes of lower lip lilac at tip; lower lip appressed to hood and surrounding it laterally in developed flowers; calyx 12–14 mm long. Leaves alternate or whorled, uppermost ones small, entire at base, lower ones oblong, with ovate or oblong-ovate, obtuse or orbicular lobes, decurrent on rachis ............................ 65. P. sceptrum-carolinum L.

+ Corolla 20–25 mm long, wholly yellow, lower lip somewhat separate from hood; calyx 5–6.5 mm long. All leaves alternate, linear-lanceolate, with oblong double-dentate, chondroid acuminate lobes ............................................................................63. P. rudis Maxim.

32 (30). Plant very short, stem up to 1 cm tall, together with inflorescence barely 4 cm tall; leaves small, about 1 cm long, pinnatisect, segments oval, dentate. Inflorescence 2-3-flowered; calyx about 8 mm long; corolla about 23 mm long ......................... 69. P. muscoides Li.

+ Plant relatively tall, stem 3 cm or more tall; leaves 2–3 cm long, segments oblong-lanceolate or oblong-oval. Inflorescence many-flowered, elongated, 2–10 cm long. ................................................33.

33. Radical leaves linear in general shape, pinnatipartite, with oblong-oval or oval segments once or twice subobtusely deeply dentate, convoluted along margin. Calyx 5–6 mm long, tubular campanulate; corolla 18–23 mm long, yellowish, with purple-hood; capsule oblong-lanceolate, oblique, 15–20 mm long ........... 70. P. oederi Vahl.

+ Radical leaves lanceolate in general shape, pinnatisect into oblong-lanceolate segments slightly decurrent on winged rachis; lobes of segments acuminate, chondrodentate. Calyx 8–14 mm long, campanulate; corolla 15–17 mm long, pink-purple; capsule ovoid, oblique, about 12 mm long ................................68. P. albertii Regel.

34 (1). Corolla tube nearly twice as long as calyx; corolla red with white spots; calyx 4-toothed, canescent. Plant 10–15 cm tall, stem 1 or more, all leaves radical ................................................. 73. P. tibetica Franch.

+ Corolla tube 2–3 times as long as calyx; corolla pink, red or yellow; calyx 5- or 2-3-toothed, glabrous or pilose. Plant tall or, rarely short ........................................................................................................35.

35. Calyx 5-toothed: 4 teeth large and 1 rear one small or all 5 teeth somewhat unequal; corolla pink or red .........................................36.

+ Calyx 2-3-toothed: all teeth nearly similar or, of 3 teeth, 2 lateral ones foliaceous and 3rd at rear very severely reduced; corolla yellow or red ................................................................................................38.

36. Plant with strong branched root, multicaulis, branching, 15–30 cm tall; leaves pinnatisect. Flowers axillary. ... 75. P. muscicola Maxim.

+ Plant with short roots, almost stemless, not branching, 6–15 cm tall; leaves pinnatipartite. Flowers aggregated in small numbers at base of plant ......................................................................................37.

37. Hood straight, curved at end and transforming into suberect beak. Leaves oblong-lanceolate, pinnately lobed, lobes rounded, entire or crenate ......................................71. P. przewalskii Maxim.

+ Hood curved, rounded and transforming into long beak. Leaves linear-oblong, pinnatipartite, lobes, ovate, pinnate ........................ .................................................................. 821. P. siphonantha Don.

38 (35). Corolla pink; flowers relatively large, corolla tube 10–12 cm long, 2 mm broad; hood somewhat contorted, very densely pubescent in middle part, with prominently protruding crest at front end. Leaves pinnatipartite with obtuse-deltoid serrate lobes .............................. ..................................................................80. P. decorissima Diels.

+ Corolla yellow; hood not contorted. .......................................39.

39. Hood with slightly protruding crest at front end; corolla tube not more than twice longer than calyx, about 2 mm broad; calyx teeth 3, oblong or oblong-oval, pinnately lobed .....................................40.

+ Hood without crest at front end; corolla tube 4–10 times longer than calyx, about 1.5 mm broad; calyx teeth 3 or 2, ovate, sub-orbicular or broadly elliptical, dentate, sometimes 3-lobed ........................41.

40. Lower lip about 2.5 mm long; corolla tube about 40 mm long; lateral lobes of lower lip orbicular, entire. Leaves alternate, pinnatipartite, with narrow, oblong, acute- and acuminate-dentate lobes ...............................................78. P. cranolopha Maxim.

+ Lower lip considerably larger, about 4 cm long; corolla tube about 30 mm long; lateral lobes of lower lip more or less emarginate. Leaves alternate or sometimes opposite, pinnatisect into ovate or oblong, acute, pinnatipartite acuminate-dentate segments ............. .............................................................. 79. P. croizatiana Li.

41 (39). Calyx teeth broadly elliptical, spinescent; corolla tube almost 4–6 times longer than calyx, slender, filiform. Leaves pinnatipartite, with orbicular emarginate-dentate segments, their lobes with long chondroid cuspidate (spinescent) teeth .........76. P. armata Maxim.

+ Calyx not spinescent, beak curved nearly ring-like; corolla tube 6–10 times longer than calyx.................................................................42.

42. Lobes of lower lip emarginate, densely ciliate; beak bipartite at tip; calyx split for 1/2 or deeper, 3-toothed, teeth tripartite, lateral teeth almost foliaceous, with spatulate, acute, sharply toothed lobes, rear tooth severely reduced, deltoid. Leaves with lustrous petiole, pinnatipartite emarginately, with winged rachis and oblong, subobtuse, crenate-chondrodentate lobes; plant short, 3–5 (10) cm tall ...................................................... 81. P. longiflora Rudolph.

+ Lobes of lower lip orbicular, entire, glabrous; beak entire at tip; calyx split for 1/3-1/2, bidentate; teeth nearly equal, 3-lobed, lobes dentate. Leaves with dull (not lustrous) petiole, pinnatipartite or pinnately lobed, with orbicular unequally dentate lobes; plant up to 30 cm tall. ...................................................... 77. **P. chinensis** Maxim.

43 (2). Hood without or with vestigial beak ............................................. 44.

+ Hood with short or more or less long beak ................................. 66.

44. Annuals or biennials ........................................................................ 45.

+ Perennials .......................................................................................... 55.

45. Plant very small, 2–4 cm tall ......................................................... 46.

+ Plant generally taller than 5 cm. ................................................... 47.

46. Leaves entire, oblong-oval, coarsely crenate-dentate, opposite. Inflorescence racemose, few-flowered, bracts foliaceous; calyx tubular; corolla tube twice longer than calyx, hood curved at tip and slightly enlarged at end, with 3–4 teeth .............. 40. **P. lyrata** Prain.

+ Leaf linear-oblong in general shape, pinnatisect, with incumbent, orbicular or oval, sharply toothed segments; leaves whorls of 4. Inflorescence capitate, many-flowered, bracts tripartite; calyx orbicular-ovate; corolla tube barely longer than calyx, hood straight, nearly as long as tube rounded at end, without teeth ....................
............................................................................... 17. **P. pygmaea** Maxim.

47 (45). Corolla tube in calyx erect, corolla 15–28 mm long; calyx tubular, 8–11 mm long, not inflated at base .................................................. 48.

+ Corolla tube bent in calyx, corolla 8–16 mm long; calyx campanulate, 3–8 mm long, more or less inflated at base .................................... 52.

48. Leaves opposite, sometimes some in whorls of 3, deltoid-ovate, 0.6–1.2 cm long, not deeply pinnately lobed. Hood rounded-saccate at end, with 1 tooth below on both sides along margin ......................
................................................................ 39. **P. deltoidea** Franch. ex Maxim.

+ Leaves, in whorls of 3–4, oblong or oblong-ovate, 2–10 cm long, pinnatisect, pinnatipartite or deeply pinnately lobed. Hood not saccate at end and without teeth ...................................................... 49.

49. Leaves, in whorls of 3, oblong, deeply pinnately lobed or pinnatipartite with obtusely serrate elliptical lobes ............................
.................................................................................... 22. **P. sima** Maxim.

+ Leaves, in whorls of 4, oblong-ovate or ovate in general shape, pinnatisect with acuminate chondrodentate segments ............... 50.

50. Corolla red, tube with fold near throat; calyx teeth ovate or ovate-oblong, distinctly dentate ............................ 13. **P. moschata** Maxim.

+ Corolla yellow or light yellow, corolla tube without fold; calyx teeth deltoid, nearly entire ........................................................................ 51.

51. Hood with short (vestigial) beak, filaments of 2 stamens pilose ....
.................................................................................... 10. **P. ludwigii** Regel.

+ Hood almost without beak, filaments **glabrous**..............................
..................................................**1. P. abrotanifolia** M. B. ex Stev.
52 (47). Leaves oblong-ovate in general shape, pinnatisect.....................53.
+ Leaves linear-lanceolate or oblong in general shape, pinnatipartite or pinnately lobed ...........................................................................54.
53. Leaves glabrous, corolla about 18 mm long ..............................
.................................................................... 28. **P. kansuensis** Maxim.
+ Leaves with dense white pubescence, corolla about 13 mm long...
.............................................5. **P. densispica** Franch. ex Maxim.
54 (52). Hood short, 1/2 length of lower lip; calyx teeth very short, oblong-oval, chondrodentate; corolla 12–14 mm long. Caulous leaves linear-lanceolate in general shape, emarginately pinnatipartite or emarginately deeply pinnately lobed with obtuse semi-circular-oblong chondroserrate lobes .................................... 33. **P. spicata** Pall.
+ Hood nearly as long or slightly shorter than lower lip; calyx teeth short, oblong, enlarged at tip, generally somewhat dentate, Caulous leaves linear-lanceolate or oblong in general shape, pinnatipartite, their lobes ovate, entire or dentate..... 34. **P. szetschuanica** Maxim.
55 (44). Leaves, in whorls of 3. Flowers dark red, inflorescence with dense grey or white tomentose pubescence ......................................56.
+ Leaves in whorls of 4, lower ones sometimes opposite. Flowers purple, pink, white or yellow .............................................................57.
56. Plant up to 30 cm tall. Inflorescence sparse, canescent; calyx teeth oblong-deltoid...............................................23. **P. ternata** Maxim.
+ Plant short, less than 12 cm tall. Inflorescence dense, with white pubescence; calyx teeth short, deltoid ... 16. **P. pilostachya** Maxim.
57 (55). Corolla white or pink. Stems weak, ascending or plant stemless ..
..............................................................................................................58.
+ Corolla purple, violet or yellowish. Stem distinctly developed, although sometimes short .........................................................59.
58. Corolla white, 30–35 mm long, hood with very short beak, with 2 indistinct teeth under tip. Plant almost stemless but with long ascending branches; radical leaves pinnatisect into oblong pinnatipartite segments, with crenate lobes. .....................................
........................................................ 11. **P. maximoviczii** Krassn.
+ Corolla pink of pale pink with cherry-coloured spot in throat or spots on lower lip; hood obtuse without vestigial beak and teeth; Plant with weak, long, crispate hairy stem; radical leaves absent, lower caulous leaves reduced, middle ones opposite or whorled, pinnatipartite into oblong-ovate, obtuse, pinnately lobed segments, with dentate-chondroid lobes..................... 20. **P. semenovii** Regel.
59 (57). Corolla tube erect, extended into hood ..........................................
.............................................................. 12. **P. mollis** Wall. ex Benth.

+ Corolla tube bent at base (inside calyx) or at tip (outside calyx) ......................................................................................................60.

60. Corolla yellowish hood shorter than lower lip, with fold around throat, rounded; calyx teeth obtuse, unequal, 2 considerably smaller than rest. Leaves pinnatisect, with oblong pinnatifid segments, with acuminate dentate lobes ................................... 30. P. plicata Maxim.

+ Corolla purple or violate, without fold; calyx teeth almost equal or one smaller than the other, acute ......................................................61.

61. Calyx slightly inflated; teeth ovate or elliptical with broadly deltoid base; corolla purple. Root vertical, weak, branched. ..........62.

+ Calyx not inflated, teeth linear with narrow deltoid base; corolla pink-violet, rarely white. Root reduced, generally with thickened root fibres ........................................................................................63.

62. Calyx split in front for 1/3; lower lip of corolla slightly longer than hood. Leaf segments broadly oblong or almost semicircular, coarsely dentate .................................................................... 35. P. verticillata L.

+ Calyx not split in front; lower lip of corolla almost twice longer than hood. Leaf segments oblong or elliptical denticulate .............. ..........................................................................32. P. roylei Maxim.

63 (61). Lower lip 2/3 length of hood. Leaves pinnatipartite, with winged rachis and pinnately lobed obtuse chondroid segments .................. .......................................................36. P. violascens Schrenk.

+ Lower lip longer or as long as hood. Leaves pinnatisect, radical leaves long-petiolate with more or less acute segments. ............ 64.

64. Lip as long or slightly longer than hood. Leaf segments serrulate or almost entire ............................ 26. P. anthemifolia Fisch. ex. Colla.

+ Lip 1/2–2/3 length of hood. Leaf segments with sharply dentate or coarsely dentate lobes ................................................................65.

65. Stems 5–10 (15) cm tall, with 4 pubescent lines or ring of crispate pubescence below; whorls of caulous leaves 1–2, radical leaves linear-lanceolate in general shape with sharply dentate lobes segments .................................................. 24. P. amoena Adams ex Stev.

+ Stems 20–50 cm tall, glabrous or slightly pubescent, lustrous; whorls of caulous leaves 2–4; lower caulous leaves (radical absent) lanceolate in general shape with acuminate coarsely dentate segments .. ...................................................... 26. P. macrochila Vved.

66 (43). Beak longer than hood ........................................................................67.

+ Beak shorter than hood ........................................................................75.

67. Beak erect, horizontal, directed almost forward. Stem branched from base; leaves deeply pinnatipartite, in whorls of 3–4. ...................... ................................................................... 19. P. scolopax Maxim.

+ Beak strongly bent, directed laterally, upward or downward. Stem single or more, not branched; leaves pinnatisect or shallowly

pinnatipartite, opposite or in whorls ............................................... 68.
68. Beak cyclic or sigmoid ........................................................... 69.
 + Beak bent more or less angularly ...................................... 73.
69. Leaves undivided, linear-oblong, serrate-dentate ...........................
....................................................... 37. P. **integrifolia** Hook. f.
 + Leaves pinnatipartite or pinnatisect .................................... 70.
70. Corolla yellow, hood not pectinate, beak contorted ring-like, turned laterally. Annual plant. ...................... 21. P. **semitorta** Maxim.
 + Corolla pink or red, hood pectinate. Perennial plant .................. 71.
71. Calyx with dense white hairs, teeth linear, acuminate, unequal, rear one smaller than rest ........................ 6. P. **fetisowii** Regel.
 + Calyx glabrous or with hairs only along margin of teeth ............ 72.
72. Calyx teeth ovate or oblong-elliptical, obtuse; midlobe of lower lip nearly 1/2 length of lateral lobes. Inflorescence sparsely flowered, elongate ................................................ 15. P. **oliveriana** Prain.
 + Calyx teeth lanceolate, acute; midlobe of lower lip slightly smaller than lateral lobes. Inflorescence dense, capitate .....................................
....................................................... 9. P. **junatovii** Ivanina.
73 (68). Leaves opposite, 1–2 cm long, lanceolate, pinnatipartite. Inflorescence capitate, 1–2 cm long, with whitish pubescence. Plant 6–12 cm tall ................................................. 41. P. **pheulpinii** Bonati.
 + Leaves, in whorls of 3–4, rarely upper caulous leaves opposite, 2–5 cm long, oblong-ovate in general shape, bipinnatipartite. Inflorescence oblong, 3–6 cm long, somewhat pilose. Plant 15–40 cm tall 74.
74. Corolla purple; calyx tube not split, calyx teeth ovate, obtuse ........
.......................................................... 8. P. **gracilis** Wall. ex Benth.
 + Corolla yellow; calyx tube split almost up to 1/2 its length, calyx teeth lanceolate, acuminate .................... 18. P. **roborowskii** Maxim.
75 (66). Corolla tube erect in throat of calyx and above .......................... 76.
 + Corolla tube more or less bent in throat of calyx. ..................... 80.
76. Beak short and broad, shortly bidentate at tip, lower lip nearly 1/2 length of hood; calyx teeth lanceolate or linear-lanceolate. Plant 20–30 cm tall ....................................... 4. P. **chorgossica** Regel et Winkl.
 + Beak entire at tip; calyx teeth broadly ovate or oblong ovate ..... 77.
77. Perennial plant; inflorescence capitate ................................... 78.
 + Annual plant; inflorescence sparse, elongate ............................. 79.
78. Caulous leaves opposite. Calyx sparsely hairy, calyx teeth pinnately lobed; corolla about 28 mm long ............ 38. P. **chenocephala** Diels.
 + Caulous leaves in whorls of 4. Calyx densely hairy; calyx teeth entire; corolla about 14 mm long. .................... 7. P. **globifera** Hook. f.
79 (77). Calyx slightly split in front, with short-ciliate or glabrous teeth; corolla yellow, tip of hood more or less falcate, gradually

transforming into narrow beak and directed forward and downward. Stem branched throughout length; leaves and branches opposite or in whorls of 3–4; axis of branches and petiole broadly winged .......................................................... 2. P. alaschanica Maxim.

+ Calyx not split in front, with long crispate ciliate teeth; corolla yellow with reddish nerves, tip of hood slightly bent, transforming into short and broad beak, turned forward subhorizontally. Stem branched from base, rarely simple; leaves in whorls of 4, petiole and axis of branches not winged ................ 14. P. myriophylla Pall.

80 (75). Corolla red or speckled (bicoloured), corolla tube somewhat curved ................................................................................................ 81.

+ Corolla yellow, tube strongly curved (almost at right angle) ..... 82.

81. Hood falcate, beak truncate, lower lip slightly shorter than or 2/3 of hood, corolla bicoloured (hood purple or pink, lower lip white). Root branched fasciculate ................... 3. P. chelanthifolia Schrenk.

+ Hood suberect and shortly rounded only at end, beak narrow, lower lip 1/2 of hood, corolla red. Root vertical, short, with slender root fibres ........................................................ 25. P. anas Maxim.

82. Plant annual, tall, 30–50 cm, whorls of leaves and of lower flowers interrupted. Corolla about 15 mm long, calyx glabrous, teeth ovate, short, acute ................................................... 28. P. curvituba Maxim.

+ Plant perennial, low, up to 15, rarely 30 cm, radical leaves aggregated at base. Corolla about 17 mm long; calyx pilose along nerves, teeth oblong, pinnatipartite to dentate ............................................
............................................................... 31. P. pseudocurvituba Tsoong.

## Section Orthosiphonia Li

1. P. abrotanifolia M.B. ex Stev. in Mém. Soc. natur. Moscou, 6 (1823) 22; Maxim. in Bull. Ac. Sci. St.-Pétersb. 32 (1888) 591; Sapozhn. Mong. Alt. (1911) 380; Kryl. Fl. Zap. Sib. 10 (1939) 2498; Li in Proc. Ac. Natur. Sci. Philad. 100 (1948) 266; Grubov, Konsp. fl. MNR (1955) 244; Vved. in Fl. SSSR, 22 (1955) 780; Tsoong in Fl. R.P. Sin. 68 (1963) 204; Fl. Kazakhst. 8 (1965) 128. — Ic.: M.B. ex Stev. l.c. tab. 5, fig. 1; Ledeb. Ic. pl. fl. ross. 3 (1831) tab. 278; Maxim. l.c. tab. V, fig. 114.

Described from Altay. Type in Leningrad.

Rocky slopes in steppes, banks of rivers and brooks, pebble beds, talus and rocks in alpine belt and in upper forest zone.

IA. **Mongolia:** *Mong. Alt.* (Tsagan-Gol river, Prokhodnaya river confluence, meadow and arid moraine, June 29; upper Kalgutta river [Dzhirgalanty], steppe slopes, July 8, 1905, Sap. [TK]; upper Tsagan-Gol, forest belt near Prokhodnaya river, June 30, 1905—Sap.; 10 km south-east of Yusun-Bulak, midportion of trail of Khan-Taishiri mountain range, forb-wheat grass-feather grass steppe, July 14; south. slope of Tamchi-Daba pass, mountain steppe, July

16; nor. trail of Bus-Khairkhan mountain range, depression in upper portion of trail, July 17; same site, gently sloping face, about 3000 m, July 17—1947, Yun.). *Gobi-Alt.* (Ikhe-Bogdo mountain range, south-east. slope, Narin-Khurimt gorge, east. flank, on rocks, above 2900 m, July 28; same site, south-south-west. slope of mountain range, 2440–2500 m, July 30—1948, Grub.).

General distribution: West. Sib. (Altay), East. Sib. (Shira lake environs), Nor. Mongolia (Fore Hubs., Hang.).

2. P. alaschanica Maxim. in Bull. Ac. Sci. St.-Pétersb. 24 (1877) 59, 32 (1888) 578, p.p.; Prain in Ann. Bot. Gard. Calc. 3 (1890) 164; Hemsley in J. Linn. Soc. London (Bot.) 30 (1894) 118; Deasy, in Tibet and Chin. Turk. (1901) 403; Hemsley, Fl. Tibet (1902) 192; Diels in Futterer, Durch Asien, 3 (1903) 20; Limpr. Feddes repert. 20 (1924) 242; Rehder and Kobuski in J. Arn. Arb. 14 (1933) 32; Pain in Contribs Inst. Bot. Nat. Ac. Peiping, 2, 7 (1934) 808; Walker in Contribs U.S. Nat. Herb. 28 (1941) 659; Li in Proc. Ac. Natur. Sci. Philad. 100 (1948) 283; Tsoong in Acta Phytotax. Sin. 3, 3 (1954) 327; id. in Fl. R.P. Sin. 68 (1963) 211.   —Ic.: Maxim. l.c. (1888) tab. V. fig. 86; Prain, l.c. tab. 25A.

Described from Mongolia (Alashan mountain range). Type in Leningrad. Plate IX, fig. 5 Map 3.

Steppe and semi-desert rocky, stony and loessial mountain slopes, meadows in floodplains of rivers and on clayey and sandy banks of rivers and lakes, predominantly in midbelt of mountains, sometimes at high altitudes, up to 4000 m.

IA. Mongolia: *Alash. Gobi* west. part of Alashan mountains, in midbelt, on rocks, rare, July 2, 1873—Przew., lectotypus!; in Alashan mountains, July 12, 1873—Przew., Alashan mountain range, Khatu-Gol gorge, southeast. slope, midbelt, on humus soil, June 19, 1908—Czet.; "Ho Lan Shan, in Picea forest, 1923, Ching"—Walker, l.c.).

IB. Kashgar: *South.* (nor. slope of Russky mountain range, Achan village; loessial slopes of mountains, June 23, 1890—Rob.; "sandy plateau west of Polu, 3150 m, 1898, Deasy"— Deasy, l.c.).

IIIA. Qinghai: *Nanshan* (on North Tetungsk mountain range and on others running parallel, in desert among tall shrubs on clayey soil, very rare, July 3, 1872; Nanshan alps, in lower part, brook clays, July 11, 1879; South Kukunor mountain range and Kuku-Nor lake, 3200 m, June 7; Kuku-Nor lake. sandbanks, common, July 3—1880, Przew.; north-west. slope of Humboldt mountain range, Chansai- Lodtsy area, 3350–4000 m, June 22, 1895—Rob.; Kuku-Nor lake, marshy bank, 3650 m, Aug. 15; Uiyu area, Sept. 12—1908, Czet.; Altyntag mountain range, 15 km south of Aksai settlement, rocky slopes of gorge, 2800 m, Aug. 2, 1958; meadow on east. bank of Kuku-Nor lake, 3210 m, Aug. 5; 33 km west of Xining town, rocky slopes of knolls, mountain wormwood-forb steppe, 2950 m, Aug. 5, 1959—Petr.; "La Chang-Kou [Lien Ch'eng], sandy and gravelly beaches, 1923, Ching"—Walker, l.c.). *Amdo* (upper Huang He river, Churmyn river valley, May 18, 1880—Przew.).

IIIB. Tibet: *Chang Tang* (nor. Tibet, Assak-Gol river valley, July 11, 1884—Przew.), *South.* ("Gyantse Hills, Stewart; Gobshi, No. 793, Chapman"—Li, l.c.; "Sowgon, near Gyantze, No. 35, Oct. 3, 1934, Ludlow; Reting, 60 miles north of Lhasa, No. 8888, July 1924; Lhasa, 4300 m, No. 9752, July 11, 1943, Ludlow and Scherriff; Reting, Chakla, 30 miles north of Lhasa, No. 11138, Aug. 24, 1944, Ludlow, Scheriff and Elliot"—Tsoong, l.c. [1954]).

General distribution: China (North-West), Himalayas (east.).

3. **P. cheilanthifolia** Schrenk in Bull. phys.-math. Ac. Sci. St.-Pétersb. 1 (1842) 79; Maxim. in Bull. Ac. Sci. St.-Pétersb. 24 (1877) 58, 32 (1888) 584; Hemsley in J. Linn. Soc. London (Bot.) 30 (1894) 118; Alcock, Rep. nat. hist. results Pamir boundary miss. (1898) 25; Deasy, In Tibet and Chin. Turk. (1901) 398; Forbes and Hemsley, Index Fl. Sin. 2 (1902) 206; Hemsley, Fl. Tibet (1902) 192; Diels in Futterer, Durch Asien (1903) 20; Paulsen in Hedin, S. Tibet, 6, 3 (1922) 43; Limpr. in Feddes repert. 20 (1924) 211; Rehder and Kobuski in J. Arn. Arb. 14 (1933) 32; Pai in Contribs Inst. Bot. Nat. Ac. Peiping, 2, 7 (1934) 210; Li in Proc. Ac. Natur. Sci. Philad. 100 (1948) 271; Tsoong in Acta Phytotax. Sin. 3, 3 (1954) 332; id. in Fl. R.P. Sin. 68 (1963) 196; Vved. in Fl. SSSR, 22 (1955) 713; Ikonnikov in Dokl. AN Tadzh. SSR, 20 (1957) 56; Fl. Kirgiz. 10 (1962) 206; Ikonnikov, Opred. rast. Pamira (1963) 221; Fl. Kazakhst. 8 (1965) 122. —*P. abrotanifolia* auct. non M.B. ex Stev.: Henderson and Hume, Lahore to Jarkand (1873) 330. —*P. svenhedinii* Pauls. in Hedin, S. Tibet, 6, 3 (1922) 44. —*P. albida* Pennella in Monogr. Ac. Natur. Sci. Philad. 5 (1943) 123. —*P. purpurea* Pennell, l.c. 125. —**Ic.** Fl. Kirgiz. 10, Plate 24, fig. 2; Fl. Kazakhst. 8, Plate XIII, fig. 2

Described from Jung. Alatau (Ispuli hill). Type in Leningrad. Plate IX, fig. 1.

Turf-covered rocky and fine-rubble slopes, near steep fall of rivers, marshy meadows (mostly of cobresia), swamps and turf-covered moraines in alpine and subalpine belts, 2150–5500 m.

**IB. Kashgar:** *Nor.* (south. slope of Tien Shan, Kum-Aryk, conglomerates and rubble piles near steep fall of rivers, mid-June 1903—Merzbacher; Uch-Turfan, June 18, 1908—Divn.). *West.* Jarkand, 1870, Henderson; Bostan-Terek settlement, July 11, 1929—Pop.), *South.* (Nura river, 2745 m, July 23; same site, in alpine belt, Aug. 5—1885, Przew.).

**IIA. Junggar:** *Tien Shan* (Muzart river gorge, 3000–4600 m, Aug. 1877—A. Reg.; Bogdo-Ula mountain, camp on south. side, [May] 26, 1908—Merzbacher; Manas river basin, upper Ulan-Usu river, 8–10 km above confluence with Dzhartas river, alpine meadow, turf-covered moraine, July 19; same site, Danu-Daban pass, nival belt, fine-rubble slope, 3600 m, July 19; same site, upper Danu-Gol river, on Se-Daban pass, high-altitude belt, forb-cobresia meadow, July 21—1957, Yun, et al.; Kotyl' pass 40–45 km north of Balinte settlement, along Karashar-Yuldus road, cobresia meadow, No. 235, Aug. 1; crossing from M. Yuldus into Ulyasutai valley, cobresia meadow, 3200 m, No. 654, Aug. 15—1958, Lee et al.

**IIIA. Qinghai:** *Nanshan* (South Tetungsk mountain range, alpine meadow, common, July 13, 1872; Mudzhik river valley, June 30; marshes near Kuku-Nor lake, 3100 m, July 13—1880, Przew.; Kuku-Nor lake, marshy meadow, 3650 m, Aug. 15, 1908—Czet.; nor. bank of Kuku-Nor lake, Aug. 3; same site, Dege-chao, Aug. 4—1890, Gr.-Grzh.; Mon'yuan', Ganshiga river valley [left tributary of Peishikhe, Tetungkhe basin], 3350–3720 m, 1958; Kuku-Nor lake, meadow on east. bank of lake, 3210 m, Aug. 5; 86 km west of Xining, pass, forb-grassy steppe with alp. features, Aug. 5, 1959—Petr.; "am Küke-nur, in allen Randgebieten Tibets, Futterer"—Diels, l.c.). *Amdo* (on gravelly eastern slopes of Wago [Waro] Kangom pass [Grasslands between Labrang and Yellow River], alt. 4250 m, No. 14465, July 25; grassy slopes of Wanchennang valley, No. 14526, July 29—1926, Rock).

**IIIB. Tibet:** Chang Tang (Keriya mountain range, Kyuk-egil' river estuary, on slightly rocky and silty soil, 3800–4000 m, July 11, 1885—Przew.; "Camp 29, 33°54' N. lat., 82°26' E. long., 4950 m, Aug. 4, 1896, Deasy and Pike"—Hemsley, l.c. [1902], Deasy, l.c.; "northern

Tibet, Mandarlik, 3437 m, July 1900, Hedin"—Paulsen, l.c.). *Weitzan* (bank of [Dychu] river, 4150 m, June 26; Konchunchu river, alpine meadow, rare, 4000–4300 m, June 30; bank of Bychu river tributary, common, 4250 m, July 13; nor. slope of Burkhan-Budda mountain range, alpine marshy meadows, common, 4000–4750 m, Aug. 13—1884, Przew.; in isthmus separating Russkoe lake from Ekspeditsiya lake, in willow groves, on humus, 4150 m, June 28, 1900; nor. slope of Burkhan-Budda mountain range, Khatu gorge, 4600 m, July 12, 1901—Lad.; "Earthy water-logged soil in wide valleys at 5200 m, Thorold"—Hemsley, l.c. [1894]; "Balch pass, about 5200 m, Strachey and Winterbottom"—Hemsley, l.c. [1902]). *South.* ("S.W. Tibet, above source of Tsangpo, northern foot of Himalaya, 5015 m, July 13, 1907, Hedin"—Paulsen, l.c.; "Gyantse, No. 41; near Shigatze, No. 114, Cutting and Vernay"—Li, l.c.).

IIIC. Pamir (Chicheklik river valley, moist meadow, July 28, 1909—Divn.; Chigalyk village, on descent from Kok-Mainak pass, July 27, 1913—Knorring; Kanlyk river midcourse, 2500–3200 m, July 12; Kanlyk river upper course, mountain tundra, 4500–5000 m, July 14; moraine watershed between Atrakyr and Tyuzutek rivers, moss tundra, 4500–5000 m, July 20; Kara-Dzhilga river, 4000–4500 m, July 22; Taspestlyk area, 4000–5000 m, July 25; Shor-Luk river gorge, 4000–5500 m, July 28; Kulan-Aryk area, between Zaz settlement and Tash-Ui river, 3500–3800 m, July 29—1942, Serp.; "Common in damp ground all along banks of R. Aksu and its affluents, up to at least 4300 m, No. 17752"—Alcock, l.c.; "Eastern Pamir, Kamper-Kishlak, Mus-tagh-ata, about 4500 m, July 29, 1894, Hedin"—Paulsen, l.c.; "Kungur, southwest. slope, left Koksel' glacial moraine, Yaman-Yarsu river, 4450 m, Aug. 8, 1956, Pen Shu-li et al."—Ikonnikov [1957], l.c.).

General distribution: Jung.-Tarb., Nor. and Cent. Tien Shan, East. Pam.; Mid. Asia (Pamiro-Alay), China (North-West), Himalayas.

4. **P. chorgossica** Regel et Winkl. in Acta Horti Petrop. 6 (1879) 350 ("*chorgonica*"); Maxim. in Bull. Ac. Sci. St.-Pétersb. 32 (1888) 586. —Ic.: Maxim, l.c. tab. V, fig. 101.

Described from Tien Shan. Type in Leningrad.

River valleys and banks of brooks in midbelt of mountains, 1200–2450 m.

IIA. Junggar: *Tien Shan* (mid. Khorgos, 1500–1800 m, May 15, 1878—A. Reg.; lectotypus!; same site, May 16, 1878; middle section of Dzhin valley, tributary of Tsagan Usu, 1200–1800 m, June 7; Borgaty gorge, 2450–2750 m, June 7; Naryn-Gol brook near Tsagan-Usu, Iren-Khabirga mountain range, 1800–2450 m, June 10—1879, A. Reg.).

General distribution: endemic.

Note. On July 12, 1945, A.A. Yunatov collected in a larch forest in Mongolian Altay (Taishiri-Ula mountain range) 2 faded plant specimens probably belonging to *P. interrupta* Steph. ex Willd., closely related to *P. chorgossica*. The plants were badly damaged, however, and their precise identification not possible.

5. **P. densispica** Franch. ex Maxim. in Bull. Ac. Sci. St.-Pétersb. 32 (1888) 594; Forbes and Hemsley, Index Fl. Sin. 2 (1902) 208; Bonati in Notes Bot. Gard. Edinburgh, 13 (1921) 133, 15 (1926) 150; Limpr. in Feddes repert. 20 (1924) 209; Hand.-Mazz. Symb. Sin. 7 (1936) 850; Li in Proc. Ac. Natur. Sci. Philad. 100 (1948) 268; Tsoong in Acta Phytotax. Sin, 3, 3 (1954) 296, 326; id. in Bull. Brit. Mus. (Nat. Hist.) Bot. 2, 1 (1955) 21; id. in Fl. R.P. Sin. 68 (1963) 205. —*P. magninii* Bonati in Notes Bot. Gard. Edinburgh, 15 (1926) 149; Limpr. in Feddes repert. 22 (1927) 334. —Ic.: Maxim. l.c. tab. V, fig. 111; Fl. R.P. Sin. 68, tab. XLVII, figs. 4–5.

Described from South-West China (Yunnan province). Type in Paris.
Forests and wet meadows, 2750–4270 m.

IIIB. Tibet: *South.* ("Hills of Lhasa, 3650 m, No. 9749, July 11, 1943, Ludlow and Scherriff"—Tsoong, l.c. [1954]).
General distribution: China (South-West).

6. **P. fetisowii** Regel in Acta Horti Petrop. 6 (1879) 349; id. in Bull. Ac.
Sci. St.-Pétersb. 27 (1881) 512; Limpr. in Feddes repert. 20 (1924) 264; Li in
Proc. Ac. Natur. Sci. Philad. 100 (1948) 291; Tsoong in Fl. R.P. Sin. 68 (1963)
222. —Ic.: Fl. R.P. Sin. 68, tab. LI, figs. 8–9.
Described from Junggar (Tien Shan). Type in Leningrad. Plate IX, fig.
3. Map 3.
Slopes and foot of mountains, 2450–2750 m.

IIA. Junggar: *Tien Shan* (Yuldus river, Sept. 1878—Fet.; typus!; south of confluence of
Kara-Gol river, 2450–2750 m, June 16; Aryslyn river, 2450–2750 m, July 8; same site, 2750 m,
July 15; Iren-Khabirga mountain range, Mengute, Aug. 2—1879, A. Reg.).
General distribution: endemic.

7. **P. globifera** Hook. f. Fl. Brit. Ind. 4 (1884) 308; Maxim. in Bull. Ac.
Sci. St.-Pétersb. 32 (1888) 585; Prain in Ann. Bot. Gard. Calc. 3 (1890) 170;
Paulsen in Hedin, S. Tibet, 6, 3 (1922) 43; Limpr. in Feddes repert. 20 (1924)
210; Tsoong in Acta Phyto tax. Sin. 3, 3 (1954) 333; id. in Fl. R.P. Sin. 68
(1963) 198. —Ic.: Maxim. l.c. tab. V, fig. 95; Prain, l.c. tab. 32, D, figs. 7–9;
Fl. R.P. Sin. 68, tab. XLIV, figs. 6–10.
Described from East. Himalayas (Sikkim). Type in London (K).
Alpine belt, 3600–5150 m.

IIIB. Tibet: *Chang Tang* ("E. Tibet, Camp XLIV [33°32' N. lat., 88°52' E. long.], 5127 m,
Aug. 9, 1901, Hedin"—Paulsen, l.c.). *South.* ("Gyantse, 4000 m, No. 61, Ludlow; Lhasa, 4300
m, Nos. 8609, 8699, 9054, 9520, 9881, May 23–28, 1942; Nangtze, 20 miles N of Lhasa, No. 9823,
July 28, 1943; Reting, 60 miles N of Lhasa, Nos. 8808, 8838, 8919, July 20, 1942—Aug. 14, 1944,
Ludlow and Scherriff"—Tsoong, l.c. [1954]).
General distribution: Himalayas (east.).

8. **P. gracilis** Wall. ex Benth. Scroph. Ind. (1835) 52; Wall. Numer. List.
Ind. Mus. (1829) No. 413, nom.; Maxim. in Bull. Ac. Sci. St.-Pétersb. 32 (1888)
552; Prain in Ann. Bot. Gard. Calc. 3 (1889) 137; Forbes and Hemsley, Index
Fl. Sin. 2 (1902) 209; Bonati in Notes Bot. Gard. Edinburgh, 5 (1911) 84;
Limpr. in Feddes repert. 20 (1924) 259; Marquand in J. Linn. Soc. London
(Bot.) 48 (1929) 212; Li in Proc. Ac. Natur. Sci. Philad. 100 (1948) 279; Tsoong
in Acta Phytotax. Sin. 3, 3 (1954) 307, 330; id. in Fl. R.P. Sin. 68 (1963) 78. —
*P. stricta* Wall. Numer. List. Ind. Mus. (1829) No. 414, nom.; Maxim. in Bull.
Ac. Sci. St.-Pétersb. 27 (1881) 513. —Ic.: Li, l.c. tab. 16, fig. 15; Fl. R. P. Sin.
68, tab. XII, figs. 1–3.
Described from India. Type in London (K).
Alpine meadows and steep slopes, 2200–3800 m.

IIIB. Tibet: *South* ("Gyantse to Phari, 3650 m, Oct, 14, 1942, Ludlow and Scherriff"— Tsoong, l.c. [1954]).

General distribution: Fore Asia (Afghanistan), China (South-West), Himalayas.

### 9. P. junatovii Ivanina sp. nova.

Radix abbreviata, fibris vix fusiformiter incrassatis; caules pauci, simplices, recti, tenues, 23–35 cm alti; folia radicalia petiolata, petiolis appresse pilosis flexuosis laminae subaequilongis, ambitu oblongo-ovata in segmenta oblongo-ovata obtusa remota crenulato-dentata pinnatilobata dissecta vel nulla; caulina in verticillis 3–5 terna-quaterna (inferiora opposita) breviter petiolata vel sessilia, in segmenta lanceolata dissecta, suprema diminuta, lineari-lanceolata, in lobos orbiculares cartilagineo-dentatos pinnatipartita. Inflorescentia globosa 6–15-flora, glabra, bracteris lanceolatis, infimis foliaceis reliquis longioribus; calyx 5–5.5 mm longus, campanulatus, membranaceus, 10-costatus, vix fissus, dentibus lanceolatis cartilagineis margine tenuiter pilosis, ca 1 mm longis; corolla rosea 18–25 mm longa, tubo cylindrico ca 10 mm longo, galea reversa dorso cristata, 4–5 mm longa in rostrum sigmoideum 6–8 mm longum protracta, labio ambitu orbiculari ca 10 mm in diam., lobo medio obovato 2–3 mm longo; filamenta duo pilosa; capsula ignota.

Typus: Tjan-Schan orientalis, ad declive boreale, systema fl. Manas, vallis fl. Ulan-Ussu, 2–3 km infra osteum fl. Dzhartas, limes silvae superior, July 24 A.A. Junatov, Li-Schi-in, Yuanj-Y-fenj; in Herb. Inst. Bot. Acad. Sci. URSS (Leningrad) conservatur.

Affinitas. Species *P. fetisowii* Regel maxime affinis, a qua calyce subglabro (nec tomentoso), corolla majore 18–25 mm longa (nec 15–18 mm longa), labio majore ca. 10 mm longa (nec 6–7 mm longo), foliorum segmentis crenato-dentatis (nec acute dentatis) et notis aliis differt.

Plate IX, fig. 4.

IIA. Junggar: *Tien Shan* (Manas river basin, Ulan-Usu river valley at confluence with Dzhartas, subalpine meadow, on upper half of nor. rocky slope, No. 850, July 18; same site, Ulan-Usu river valley, 4–5 km above confluence of Dzhartas, high-altitude belt, on talus, July 23; same site, Ulan-Usu river valley, 2–3 km below confluence of Dzhartas, upper forest boundary, first upper spruce forest with aspen, July 24—1957, Yun. et al., typus!).

General distribution: endemic.

— Note. The species described above is closest to *P. fetisowii* Regel, but well distinguished by subglabrous (not tomentose) calyx; very large corolla, 18-25 (not 15–18) mm long; very large lower lip, about 10 (not 6–7) mm long; leaves with crenate-dentate segments.

### 10. P. ludwigii (Regel in Bull. Soc. natur. Moscou, 41 (1868) 107; Vved. in Fl. SSSR, 22 (1955) 729; Fl. Kazakhst. 8 (1965) 127. —*P. leptorhiza* Rupr. in Mém. Ac. Sci. St.-Pétersb. VII sér. 14 (1869) 62; Maxim. in Bull. Ac. Sci. St.-Pétersb. 32 (1888) 581; Limpr. in Feddes repert. 20 (1924) 211; Li in Proc. Ac. Natur. Sci. Philad. 100 (1948) 273; Ikonnikov, Opred. rast. Pamira (1963) 208. —*P. abrotanifolia* auct. non M.B. ex Stev.; Deasy, In Tibet and Chin.

Turk. (1901) 403; Paulsen in Hedin, S. Tibet, 6, 3 (1922) 43; Persson in Bot. notiser (1938) 302. —**Ic.:** Maxim. l.c. tab. V, fig. 92 (sub nom. *P. leptorhiza* Rupr.).

Described from Junggar Alatau (Keisy-Karachai pass). Type in Leningrad. Plate IX, fig. 2.

Rocky and rubble slopes of mountains and as weed along irrigation ditches and canals.

**IB. Kashgar:** *Nor.* (Uchturfan, July 17, 1908—Divn.). *West.* (upper Kizil-Su, above Kashgar, before Ulugchat, floodplain, July 2, 1929—Pop.).

**IIA. Junggar:** *Tien Shan* (Manas river basin, Ulan-Usu river valley near confluence with Dzhartas river, on south. rocky slope, subalp. belt, Aug. 18, 1957; 4–5 km below Balinte settlement, along road from Urumchi to Karashar, floodplain of Ulyasutai-Chagan river, wild rye meadow, July 1, 1958—Yun. et al.).

**IIIB. Tibet:** *Chang Tang* ("Aksu, 4800 m, 1898"—Deasy, l.c.).

**IIIC. Pamir** (west. vicinity of Tashkuragan town, along irrigation ditches, June 13, 1959—Yun. et al.; "Kara-jilga at Bassik-kul, 3727 m, July 24, 1894"—Paulsen, l.c.; "Tash-korghan, Jurgal, 3290 m, June 29, 1935; Jerzil, 3200 m, July 14, 1930"—Persson, l.c.).

**General distribution:** Jung.-Tarb., Nor. and Cent. Tien Shan, East. Pam.

11. **P. maximoviczii** Krassn. in Zap. Russk. geogr. obshch. 19 (1888) 339; id. in Script. Horti Univ. Petrop. 2 (1889) 18; Maxim. in Mél. biol. 12 (1888) 913; Vved. in Fl. SSSR, 22 (1955) 728; Fl. Kazakhst. 8 (1965) 127. —**Ic.:** Maxim. l.c. tab. VI, fig. 164.

Described from Tien Shan (upper Muzart river and Khan-Tengri). Type in Leningrad.

Rubble slopes of hills and talus in alpine belt.

**IIA. Junggar:** *Tien Shan* (near Muzart river source, 1886—Krasnov, typus!).
**General distribution:** Nor. Tien Shan.

12. **P. mollis** Wall. ex Benth. Scroph. Ind. (1835) 53; Wall. Numer. List. Ind. Mus. (1829) nom.; Maxim. in Bull. Ac. Sci. St.-Pétersb. 32 (1888) 602; Prain in Ann. Bot. Gard. Calc. 3 (1890) 176; Limpr. in Feddes repert. 20 (1924) 210; Tsoong in Acta Phytotax. Sin. 3, 3 (1954) 326; id. in Fl. R. P. Sin. 68 (1963) 202. —**Ic.:** Maxim. l.c. tab. IV, fig. 126; Prain, l.c. tab. 29, A, figs. 1–8; Fl. R. P. Sin. 68, tab. XLVI, figs. 1–3.

Described from Himalayas. Type in London.
Mountains, 3000–4500 m.

**IIIB. Tibet:** *South.* ("Nangtse, 20 miles W. of Lhasa, 3800 m, No. 9807, July 27, 1943, Ludlow and Scherriff"—Tsoong, l.c.).
**General distribution:** Himalayas (east.).

13. **P. moschata** Maxim. in Bull. Ac. Sci. St.-Pétersb. 27 (1881) 516, 32 (1888) 592; Limpr. Feddes repert. 20 (1924) 209; Li in Proc. Ac. Natur. Sci. Philad. 100 (1948) 266; Grubov, Konsp. fl. MNR (1955) 245.—**Ic.:** Maxim, l.c. (1888) tab. V, fig. 195.

Described from Mongolia (Mong. Altay). Type in Leningrad. Map 3.

Rubble and rocky slopes, talus and pebble beds.

IA. **Mongolia:** *Mong. Alt.* (Dzusylyn river estuary, on pebble bed between rocks, June 29, 1877—Pot., typus!; Tatal river source, on sand among *Artemisia fragrans*, July 8; Tsitsirin-Gol river, July 11, 1877—Pot.; Korum pass [from Kulagash to Saksai], July 7, 1906; east of Ak-Korum pass, Kutologoi, alpine tundra, July 13, 1908; Kak-Kul'lake, old moraine, July 17, 1909—Sap. [TK]; environs of Borogol-Daban, on rock talus, Aug. 18; Khara-Dzarga hills, Sakhir-Sala river valley, south. rubble slope of mountain, Aug. 23—1930, Pob.); South. slope of Adzhi-Bogdo mountain range, Indertiin-Gol, river valley, Aug. 6; nor. slope of Adzhi-Bogdo, Mainigtu-Ama creek valley, change of high-altitude steppes to talus coenosis, Aug. 7—1947, Yun.).

General distribution: endemic.

14. **P. myriophylla** Pall. Reise, 3 (1776) 737; Maxim. in Bull. Ac. Sci. St.-Pétersb. 24 (1878) 61, 32 (1888) tab. V; Forbes and Hemsley, Index Fl. Sin. 2 (1902) 213; Limpr. in Feddes repert. Beih. 12 (1922) 485, 20 (1924) 242; Kryl. Fl. Zap. Sib. 10 (1939) 2500; Li in Proc. Ac. Natur. Sci. Philad. 100 (1948) 282; Grubov, Konsp. fl. MNR (1955) 245; Vved. in Fl. SSSR, 22 (1955) 729; Tsoong in Fl. R. P. Sin. 68 (1963) 210. —Ic.: Pall. l.c. tab. S, fig. IA; Maxim. l.c. (1888) tab. V, fig. 91.

Described from Siberia. Isotype in Leningrad.

Rubble and rocky slopes, rocks under shade and creek valleys, in subalpine belt.

IA. **Mongolia:** *Cent. Khalkha* (Dzhargalante river basin, Kharukhe and Ara-Dzhargalante river sources, Uste mountain, subalp. zone, nor slope near top, Aug. 12; same site, Agit mountain, rocky steppes, south. slope, Aug. 26; same site, between sources and Agit hill, rocks under shade, Aug. 31—1925, Krasch. and Zam.). *Gobi Alt.* (Dzun-Saikhan hills, near spring in a creek valley of Yailo ravine, Aug. 21, 1931—Ik.-Gal.).

General distribution: West. Sib. (Altay), East. Sib. (south.), Nor. Mong. (Fore Hubs., Hent., Hang.), China (North).

15. **P. oliveriana** Prain in J. As. Soc. Bengal, 58 (1889) 257; id. in Ann. Bot. Gard. Calc. 3 (1890) 133; Limpr. in Feddes repert. 20 (1924) 261; Tsoong in Acta Phytotax. Sin. 3, 3 (1954) 300, 327 (sub nom. *P. olivoiana*); id. in Bull. Brit. Mus. (Bot.) 2, 1 (1955) 22; id. in Fl. R. P. Sin. 68 (1963) 223. —Ic.: Prain, l.c. (1889) tab. 22, A, figs. 1–7; Fl. R. P. Sin. 68, tab. LI, figs 6–7.

Described from East. Himalayas. Type in Calcutta.

Humid forests, willow thickets, river banks, 3400–4000 m.

IIIB. **Tibet:** *South.* ("Gyantse, 1924, Ludlow; mountains north of Lhasa, 4000 m, July 10, 1942, Ludlow and Scherriff"—Tsoong, l.c. [1954]).

General distribution: Himalayas (east.).

16. **P. pilostachya** Maxim. in Bull. Ac. Sci. St. Pétersb.-24 (1877) 64, 32 (1888) 593; Forbes and Hemsley, Index Fl. Sin. 2 (1902) 213; Limpr. in Feddes repert. 20 (1924) 209; Rehder and Kobuski in J. Arn. Arb. 14 (1933) 34; Li in Proc. Ac. Natur. Sci. Philad. 100 (1948) 266; Tsoong in Fl. R. P. Sin. 68 (1963) 297. —Ic.: Maxim. l.c. (1888) tab. V, fig. 109; Fl. R. P. Sin. 68, tab. LXV, figs. 4–5.

Described from Qinghai. Type in Leningrad.
Rocky slopes, pebble beds on river banks, alpine meadows, 2050–5070 m.

IIIA. Qinghai: *Nanshan* (nor. slope of South Tetungs mountain range, at peak of Sodi-Soroksum mountains, 4150 m, common, Aug. 12, 1872—Przew., typus!; Nanshan alps, on rocky soil, 3350 m, July 11, 1879—Przew.; nor. slope of Humboldt mountain range, Kuku-Usu river, on pebble bed along river, alpine meadow, 2750–4000 m, June 9, 1894—Rob.; "Jupar Range [No. 14331] Rock"—Rehder and Kobuski, l.c.).
General distribution: China (North-West).

17. P. pygmaea Maxim. in Bull. Ac. Sci. St.-Pétersb. 32 (1888) 595; Prain in Ann. Bot. Gard. Calc. 3 (1890) 95; Limpr. in Feddes repert. 20 (1924) 208; Li in Proc. Ac. Natur. Sci. Philad. 101 (1949) 273; Tsoong in Fl. R. P. Sin. 68 (1963) 165. —Ic.: Maxim. l.c. tab. V, fig. 114.
Described from Tibet. Type in Leningrad.
Marshy meadows on river banks, about 4000 m alt.

IIIB. Tibet: *Weitzan* (midcourse of Dzhagyn-Gol river, marshy meadows, river banks, common, 4000 m, July 16, 1884—Przew., typus!; marshy grasslands along Dzhagyn-Gol river banks, 4000 m, July 1, 1900—Lad.).
General distribution: endemic.

18. P. roborowskii Maxim. in Bull. Ac. Sci. St.-Pétersb. 27 (1881) 512, 32 (1888) 546; Forbes and Hemsley, Index Fl. Sin. 2 (1902) 215; Limpr. in Feddes repert. 20 (1924) 262; Li in Proc. Ac. Natur. Sci. Philad. 100 (1948) 291; Tsoong in Fl. R. P. Sin. 68 (1963) 218. —Ic.: Maxim. l.c. tab. I, fig. 29.
Described from Qinghai. Type in Leningrad.
Alpine belt.

IIIA. Qinghai: *Nanshan* (in east up to Gantsaga river, Tetung river basin, July 18; between Raka-Gol and Yusun-Khatyma rivers, July 23, 1880—Przew., syntypus!).
General distribution: China (South-West—nor. Sichuan).

19. P. scolopax Maxim. in Bull. Ac. Sci. St.-Pétersb. 27 (1881) 513, 32 (1888) 547; Forbes and Hemsley, Index Fl. Sin. 2 (1902) 216; Limpr. in Feddes repert. 20 (1924) 262; Rehder and Kobuski in J. Arn. Arb. 14 (1933) 34; Li in Proc. Ac. Natur. Sci. Philad. 100 (1948) 287; Tsoong in Fl. R. P. Sin. 68 (1963) 216. —Ic.: Maxim. l.c. (1888) tab. I, fig. 30.
Described from Qinghai. Type in Leningrad.
Alpine belt, 3500–4000 m.

IIIA. Qinghai: *Amdo* (in upper Huang He, in alps, May 21, 1880—Przew., typus!; "Radja and Yellow River Gorges, Deyang valley, east of Radja, No. 14132, Rock"—Li, l.c.).
IIIB. Tibet: *Weitzan* (Yangtze river basin, rare, June 25, 1884 —Przew.).
General distribution: endemic.

20. P. semenovii Regel in Bull. Soc. natur. Moscou, 41 (1868) 108; Maxim. in Bull. Ac. Sci. St.-Pétersb. 32 (1888) 602; Vved. in Fl. SSSR, 22 (1955) 725; Fl. Kirgiz. 10 (1962) 212; Ikonnikov, Opred. rast Pamira (1963) 222; Tsoong

in Fl. R. P. Sin. 68 (1963) 201; Fl. Kazakhst. 8 (1965) 126.   —*P. pycnantha* var. *semenovii* (Regel) Prain in Ann. Bot. Gard. Calc. 3 (1890) 180, p.p.; Limpr. in Feddes repert. 20 (1924) 213, p.p.; Tsoong in Acta Phytotax. Sin. 3, 3 (1954) 326.   —**Ic.:** Maxim. l.c. tab. IV, fig. 129; Fl. Kirgiz. 10, Plate 26, fig. 1; Fl. Kazakhst. Fl. Kazakhst. 8, Plate XIII, fig. 5.

Described from Jung Alatau (Bayan-Dzhuruk town). Type in Leningrad.

Steppe rocky and rubble mountains slopes, 2700–4500 m.

**IB. Kashgar:** *West.* (King-Tau mountain range, 3–4 km south-east of Kosh-Kulak settlement, in depression, steppe belt, 2900 m, June 10, 1959—Yun. et al.).

**IIA. Junggar:** *Tien Shan* (Khorgos, 1500–1800 m, May 15, 1878—A. Reg.).

**IIIB. Tibet** ("Tibet"—Tsoong, l.c.).

General distribution: Jung.-Tarb., Nor. and Cent. Tien Shan, East. Pamir. Fore Asia (Afghanistan), Mid. Asia (Pamiro-Alay), Himalayas.

21. **P. semitorta** Maxim. in Bull. Ac. Sci. St.-Pétersb. 32 (1888) 546; Forbes and Hemsley, Index Fl. Sin. 2 (1902) 216; Limpr. in feddes repert. 20 (1924) 262; Rehder and Kobuski in J. Arn. Arb. 14 (1933) 34; Pai in Contribs Inst. Bot. Nat. Ac. Peiping, 2, 7 (1934) 219; Li in Proc. Ac. Natur. Sci. Philad. 100 (1948) 290; Tsoong in Fl. R. P. Sin. 68 (1963) 221.   —**Ic.:** Maxim. l.c. tab. II, fig. 28; Fl. R. P. Sin. 68, tab. LI, figs. 1–3.

Described from North-West China (Gansu province). Type in Leningrad.

Alpine meadows, 2500–3900 m.

**IIIA. Qinghai:** *Amdo* ("Radja and Yellow River Gorges, Deyang valley, east of Radja, No. 14127; Radja, No. 14215, between Hetso and Chiu-ssu, between Labrang and Yellow River, No. 14548, Rock"—Li, l.c.).

General distribution: China (North-West and South-West—nor. Sichuan).

22. **P. sima** Maxim. in Bull. Ac. Sci. St.-Pétersb. 27 (1881) 514, 32 (1888) 592; Forbes and Hemsley, Index Fl. Sin. 2 (1902) 216; Limpr. in Feddes repert. 20 (1924) 223; Li in Proc. Ac. Natur. Sci. Philad. 100 (1948) 285; Tsoong in Fl. R. P. Sin. 68 (1963) 208.   **Ic.:** Maxim. l.c. (1888) tab. V, fig. 107.

Described from Qinghai. Type in Leningrad.

Alpine meadows.

**IIIA. Qinghai:** *Nanshan* (South Tetung mountain range, Yusun-Khatyma river region, on midsection of slope, alpine moist meadows, July 23, 1880—Przew., typus!).

General distribution: China (North-West and South-West—nor. Sichuan).

22. **P. ternata** Maxim. in Bull. Ac. Sci. St.-Pétersb. 24 (1877) 64, 32 (1888) 592; Forbes and Hemsley, Index Fl. Sin. 2 (1902) 218; Limpr. in Feddes repert. 20 (1924) 209; Li in Proc. Ac. Natur. Sci. Philad. 100 (1948) 266; Tsoong in Fl. R. P. Sin. 68 (1963) 298.   —**Ic.:** Maxim. l.c. (1888) tab. V, fig. 108.

Described from Mongolia (Alashan mountain range). Type in Leningrad.

In thickets and upper boundary of forests, along river valleys and gorges, in wet or marshy soil, 3200–4550 m.

**IA. Mongolia**: *Alash. Gobi* (Alashan mountain range, in midsection on west. slope, wet forest on upper boundary, common, July 10, 1873—Przew., typus!).

**IIIA. Qinghai**: *Nanshan* (between Nanshan and Donkyr mountain ranges, along Rako-Gol river, on wet soil between scrub, 3000 m, July 23, 1880—Przew.).

**IIIB. Tibet**: *Weitzan* (nor. slope of Burkhan-Budda mountain range, in midsection, on grassy slope between rocks, marshy sites, common, Aug. 14, 1884—Przew.; Yangtse basin, Donra area, along Ichu river, Khichu, in willow thickets, on humus, 4000 m, July 16; Burkhan-Budda mountain range, Ikhe-Gol gorge, in willow thickets, 3300–4000 m, July 23—1900, Lad.).

**General distribution**: endemic.

## Section Sigmantha Li

24. **P. amoena** Adam ex Stev. in Mém. Soc. natur. Moscou, 6 (1823) 25; Maxim. in Bull. Ac. Sci. St.-Pétersb. 24 (1877) 63, p.p., 32 (1888) 596; Sapozhn. Mong. Alt. (1911) 380; Limpr. in Feddes repert. 20 (1924) 208, p.p.; Kryl. Fl. Zap. Sib. 10 (1939) 2497, p. max. p.; Vved. in Fl. SSSR, 22 (1955) 709; Fl. Kazakhst. 8 (1965) 120. —*P. arctica* M.B. ex Stev. l.c. —*P. pulchella* Turcz. ex Bess. in Beibl. 1, Flora, 16 (1834) 21. —*P. hulteniana* Li in Proc. Ac. Natur. Sci. Philad. 100 (1948) 310, p.p.; Grubov, Konsp. fl. MNR (1955) 245, p.p. — Ic.: Stev. l.c. tab. 7; Maxim. l.c. (1888) tab. V, fig. 115; Fl. SSSR, 22, Plate XXXV, fig. 4.

Described from Siberia (Lena estuary). Type in Leningrad.

On montane-steppe rubble and rocky slopes, discontinuous alpine mats among rocks and granite screes, alpine meadows, 2100–3700 m.

**IA. Mongolia**: *Mong. Alt.* (Urmogaity lake [June] 1903—Gr.-Grzh.; upper Tsagan-Gol river, rock crests facing glaciers, placers, July 1, 1905, peak between Turgyun' and Sumdairyk rivers, alpine tundra, July 3, 1906; Aksu river [Belaya Kobdo], steep screes, July 22, 1909—Sap. [TK]; Kak-Kul' lake granite ravine, alpine tundra, June 22, 1906—Sap.; Bulugun somon, Khargatiin-Daba pass, alpine meadow July 23; west of summer camp in Bulugun somon, in upper Indertiin-Gol river, alpine meadow, July 24; Bulugun river basin, along upper ketsu-Saivin-Gol river, on moraine and slopes towards glacier, alpine meadow, July 26; Adzhi-Bogdo mountain range, Burgasin-Daba pass, between Indertiin-Gol and Dzuslangin-Gol, rubble placers in alpine belt, Aug. 6—1947, Yun.). *Gobi-Alt.* (Baga Bogdo, on slopes, 2100–2450 m, No. 263, 1925—Chaney; Ikhe-Bogdo mountain range, south. slope, upper part of Narin-Khurimt creek valley, montane sheep's fescue-sedge steppe, June 28; same site, upper belt, rock scree, June 29—1945, Yun.; same site, entrance of Narin-Khurimt gorge, south. exposure, site among rocks, about 3500 m, July 29; same site, plateau-shaped crest of mountain range, discontinuous alpine mat among granite screes, about 3700 m, July 29—1948, Grub.).

**General distribution**: Jung.-Tarb., Nor. Tien Shan; Arct. Europe (Urals), West. Sib. (Altay), East. Sib., Far East, Nor. Mongolia (Fore Hubs., Hent., Hang.).

25. **P. anas** Maxim. in Bull. Ac. Sci. St.-Pétersb. 32 (1888) 578; Forbes and Hemsley, Index Fl. Sin. 2 (1902) 205; Limpr. in Beih. Feddes repert. 12 (1922) 485; id. in Feddes repert. 20 (1924) 243; Walker in Contribs U.S. Nat.

Herb. 28 (1941) 659; Li in Proc. Ac. Natur. Sci. Philad. 100 (1948) 333; Tsoong
in Fl. R. P. Sin. 68 (1963) 199. —Ic.: Maxim. l.c. tab. V, fig. 87; Li, l.c. tab. 19,
fig. 56.

Described from South-West China (Sichuan province). Type in Leningrad.

Alpine meadows, 3000–4300 m.

IIIA Qinghai:*Nanshan* ("Ta P'an Shan, on exposed, moist grassy slopes, No. 645, Ching,
1923"—Walker, l.c.).

General distribution: China (North-West, South-West).

26. P. anthemifolia Fisch. ex Colla, Herb. Pedan. 4 (1835) 370; Tsoong
in Fl. R. P. Sin. 68 (1963) 165, p.p., pro subsp. *anthemifolia*. —*P. verticillata*
auct. non L.: Bunge in Ledeb. Fl. alt. 2 (1830) 427..—*P. amoena* auct. non
Adams ex Stev.: Maxim. in Bull. Ac. Sci. St.-Pétersb. 24 (1877) 63, p.p.; 32
(1888) 596; Limpr. in Feddes repert. 20 (1924) 208, p.p.; Kryl. Fl. Zap. Sib. 10
(1939) 2498, p. min. p. —*P. hulteniana* Li in Proc. Ac. Natur. Sci. Philad.
100 (1948) 310, p.p.; Grubov, Fl. MNR (1955) 245, p. min. p. —*P.
arguteserrata* Vved. in Fl. SSSR, 22 (1955) 706, 809. —*P. amoena* var.
*arguteserrata* (Vved.). Serg. in Kryl. Fl. Zap. Sib. 12, 2 (1964) 3451.

Described without reference to location (probably from Altay). Type in
Turin (TO).

Grasslands in upper part of forest belt and meadows in subalpine belt.

IA. Mongolia: *Khobd.* (Ulan-Daban pass, on nor.-east. slope, in forest gorge, June 22,
1879—Pot.). *Mong. Alt.* (Kran river, June 24, 1903—Gr.-Grzh.).

General distribution: Europe (Urals), West. Sib. (south-west. part and Altay), East. Sib.
(Ang.-Sayan).

27. P. curvituba Maxim. in Bull. Ac. Sci. St.-Pétersb. 24 (1877) 60, 32
(1888) 578, p.p., quoad pl. kansuen.; Forbes and Hemsley, Index Fl. Sin. 2
(1902) 207, p.p.; Limpr. in Feddes repert. 20 (1924) 242, p.p.; Li in Proc. Ac.
Natur. Sci. Philad. 100 (1948) 332, p.p.; Tsoong in Fl. R. P. Sin. 68 (1963)
214. —Ic.: Maxim. l.c. (1888) tab. V, fig. 85; Fl. R. P. Sin. 68, tab. XLIX, figs.
1–4.

Described from North-West China (Gansu province). Type in Leningrad.
Hilly grassy-forb and scrub steppes, 2400–3400 m.

IIIA. Qinghai: *Nanshan* (68 km west of Xining town, pass, grassy-forb steppe with alpine features, June 5; 108 km west of Xining town, 6 km west of Daudankhe settlement, hilly
scrub steppe, 3400 m, Aug. 5; 25 km south of Kulan [Gulan], east. extremity of Nanshan,
gentle slopes of low mountains, montne scrub steppe, 2450 m, Aug. 12—1958, Petr.).

General distribution: China (North-West).

28. P. kansuensis Maxim. in Bull. Ac. Sci. St.-Pétersb. 27 (1881) 516, 32
(1888) 596; Forbes and Hemsley, Index Fl. Sin. 2 (1902) 210; Diels in Futterer,
Durch Asien, 3 (1903) 21; id. in Filchner, Wissensch. Ergebn. 10, 2 (1908)
263; Limpr. in Feddes repert. 20 (1924) 208; Rehder and Kobuski in J. Arn.

Arb. 14 (1933) 33; Walker in Contribs U.S. Nat. Herb. 28 (1941) 659; Li in Proc. Ac. Natur. Sci. Philad. 100 (1948) 318, p.p.; Tsoong in Acta Phyto-tax. Sin. 3, 3 (1954) 311, 332; id. in Bull. Brit. Mus. (Bot.), 2, 1 (1955) 30; id. in Fl. R. P. Sin. 68 (1963) 167. —*P. verticillata* var. *chinensis* Maxim. in Bull. Ac. Sci. St.-Pétersb. 24 (1877) 63; Kanitz in Szechenyi, Wissensch. Ergebn. 2 (1898) 723. —*P. violascens* auct. non Schrenk: Maxim. in Bull. Ac. Sci. St.-Pétersb. 32 (1888) 594, p.p. —*P. goniantha* Bur. et Franch. in J. Bot. (Paris) 5 (1891) 128; Bonati in Bull. Herb. Boiss. sér. 2, 7 (1907) 545; Limpr. in Feddes repert. 20 (1924) 210. —*P. futtereri* Diels in Futterer, Durch Asien, 3 (1903) 20; Limpr. in Feddes repert. 20 (1924) 211. —*P. yargongensis* Bonati in Bull. Soc. Bot. France, 55 (1908) 312; Limpr. in Feddes repert. 20 (1924) 207; Li in Proc. Ac. Natur. Sci. Philad. 100 (1948) 312. —*P. szetschuanica* auct. non Maxim.: Rehder and Kobuski in J. Arn. Arb. 14 (1933) 34, p.p. quoad pl. Rock No. 14192. —Ic.: Maxim. l.c. (1888), tab. IV, fig. 116; Fl. R. P. Sin. 68, tab. XXXIII, figs. 5–8.

Described from Qinghai. Type in Leningrad.

Alpine, coastal and forest meadows, meadow and montane steppe slopes, moist grasslands, banks of rivers and springs, pebble beds, 2650–4250 m.

IIIA. Qinghai: *Nanshan* (South Tetungsk mountain range, hill slopes, common and dispersed in groups, July 8; South Tetung mountain range, in forest, common July 12—1872, Przew., lectotypus!; along Yusun-Khatyma river, alpine meadow, 2750–3050 m, July 29, 1880—Przew.; Xiningkhe river, May 16; Xining hills, Myndan'sha river, June 14—1890, Gr.-Grzh.; Dulan-khit temple, 3350 m, Aug. 8, 1901-Lad.; Mon'yuan, Ganshig river valley, left tributary of Peishikhe, 3350–3720 m, [Aug.] 1958; Kuku-Nor lake, meadow on east. bank of lake, 3210 m, Aug. 5; 33 km west of Xining, rocky slopes of knolls, hilly wormwood-grass-forb steppe, 2950 m, Aug. 5; 108 km west of Xining, 6 km west of Daudankhe settlement, hilly scrub steppe, 3450 m, Aug. 5; 27 km south of Xining, knoll slopes, grassy-forb steppe, 2650 m, Aug. 6—1959, Petr.; "Yao Chich, along moist road-sides, No. 280, Ching"—Walker, l.c.). *Amdo* (Huang He river [upper] on silty soil and around springs, 2150 m, common, June 9, 1880—Przew.; "Radja and Yellow River gorges, east of Radja, No. 14118; Radja and Yellow River gorges, south-east of Radja, No. 14192; Radja and Yellow River gorges, below Lungmen Valley along Yellow River, south-east of Radja, No. 14208; Ba Valley, banks of Ba streams, No. 14270, Rock"—Li, l.c.).

IIIB. Tibet: *Weitzan* (upstream of Konchunchu river, alpine belt, 4000–4250 m, June 30; bank of Razboinichya river, 4150 m, July 26—1884, Przew.; on nor.-west. bank of Russkoe lake, on clay and pebble bed, 4150 m, June 27, 1900; right bank of Yellow river, at its very source from Russkoe lake and on nor. bank of lake, on slopes, on clay and humus, June 3; nor. slope of Burkhan-Budda mountain range, Khatu gorge, moist grasslands, in clayey soil and sometimes in pebble bed, rarely on humus, 3200 m, June 15—1901, Lad.). *South.* ("Reting, 4000 m, edge of water channels, No. 8696, July 24, 1942; Reting, 4600 m, No. 9967, July 11, 1944, Ludlow and Scherriff"—Tsoong, l.c. [1954]).

General distribution: China (North-West, South-West).

Note. K.I. Maximovicz's analysed and sketched specimen is selected here as the lectotype.

29. P. macrochila Vved. in Byull. Sredneaz. gos. univ. 11 (1925) 24; Vved. in Fl. SSSR, 22 (1955) 705; Fl. Kirgiz. 10 (1962) 202; Fl. Kazakhst. 8 (1965)

119. —*P. amoena* auct. non Adams ex Stev.: Maxim. in Bull. Ac. Sci. St.-Pétersb. 24 (1877) 63, p.p. quoad pl. dzungar.; Limpr. in Feddes repert. 20 (1924) 208, p.p. —*P. amoena* var. *elatior* Regel in Acta Horti Petrop. 6 (1880) 348. —*P. hulteniana* Li in Proc. Ac. Natur. Sci. Philad. 100 (1948) 310, p.p. quoad pl. dzungar.—*P. anthemifolia* auct. non Fisch. ex Colla: Tsoong in Fl. R. P. Sin. 68 (1963) 166, p.p. pro subsp. *elatior* (Regel) Tsoong.

Described from Tien Shan (Kirgiz Alatau). Type in Tashkent (TAK).
Grassy slopes in midbelt of mountains.

IB. Kashgar: *West.* ("in Kashgar region"—Tsoong, l.c.).
IIA. Junggar: *Jung. Alt.* (Yugantash mountain range, 1800–2100 m, May 25, 1878—A. Reg.; on Kegen river, 2100 m, July 23, 1878—Fet.). *Tien Shan* (Khanakhai mountain, 1500–2100 m, June 16; Chapchal gorge, 1800–2100 m, June 28—1878, A. Reg.; Kutukshi (west of Kul'dzha], June 6, 1878—Fet.).
General distribution: Jung.-Tarb., Nor. and Cent. Tien Shan; Mid. Asia (Pamiro-Alay).

30. P. plicata Maxim. in Bull. Ac. Sci. St.-Pétersb. 32 (1888) 598; Forbes and Hemsley, Index Fl. Sin. 2 (1902) 213; Limpr. in Feddes repert. 20 (1924) 206, p.p.; Li in Proc. Ac. Natur. Sci. Philad. 100 (1948) 330; Tsoong in Acta Phytotax. Sin. 3, 3 (1954) 331; id. in Bull. Brit. Mus. (Bot.) 2 (1955) 28; id. in Fl. R. P. Sin. 68 (1963) 149. —*P. giraldiana* Diels ex Bonati in Bull. Soc. Bot. France, 57 (1911) Sess. Extraord. 60; Limpr. in Beih. Feddes repert. 12 (1922) 285; Tsoong, l.c. [1963] 150. —*P. cheilanthifolia* auct. non Schrenk: Marquand in J. Linn. Soc. London (Bot.) 48 (1929) 211. —*P. floribunda* auct. non Franch.: Pai in Contribs Inst. Bot. Nat. Ac. Peiping, 2, 7 (1934) 212. —**Ic.:** Maxim. l.c. tab. IV, fig. 120; Pl. R. P. Sin. 68, tab. XXX, figs. 1–3.

Described from South-West China. Type in Leningrad.
Rock screes and wet slopes in alpine belt, 2900–4150 m.

IIIA. Qinghai: *Nanshan* ("Ta Hwa, near Pinfan, No. 514, Ching"—Li, l.c.).
General distribution: China (North-West, South-West).

31. P. pseudocurvituba Tsoong in Fl. R. P. Sin. 68 (1963) 212, 412. —*P. curvituba* auct. non Maxim. (1877): Maxim. in Bull. Ac. Sci. St.-Pétersb. 32 (1888) 578, p.p.; Limpr. in Feddes repert. 20 (1924) 242, p.p.; Li in Proc. Ac. Natur. Sci. Philad. 100 (1948) 332, p.p.—**Ic.:** Fl. R. P. Sin. 68, tab. XLIX, figs. 5–8.

Described from Tibet. Type in Leningrad.
Wet alpine meadows, clay and sand.

IC. Qaidam: *montane* (Ritter's mountain range, Olun-Nor area, on sand, 4000–4300 m, June 25, 1894—Rob.).
IIIA. Qinghai: *Nanshan* (nor. slope of Humboldt mountain range, alpine meadow, 2750–3650 m, June 30, 1894—Rob.).
IIIB. Tibet: *Weitzan* (Dzhagyn-Gol, on sand, common, 4500 m, June 23; Assak-Gol river valley, Aug. 11—1884, Przew.; Russkoe lake, nor.-east. bank, on loose dry clay, along lake banks and Yellow river, at its source from this lake, 4400 m, June 20–23, 1900; nor. slope of

Burkhan-Budda mountain range, Khatu gorge, on wet [clayey] grasslands, at gorge opening, 3200 m, June 15, 1901—Lad., typus!).
General distribution: endemic.

32. P. roylei Maxim. in Bull. Ac. Sci. St.-Pétersb. 27 (1881) 517, 32 (1888) 599; Prain in Ann. Bot. Gard. Calc. 3 (1890) 173; Daguy in Bull. Mus. nat. hist. natur. 5 (1911) 14; Limpr. in Feddes repert. 20 (1924) 207; Marquand in J. Linn. Soc. London (Bot.) 48 (1929) 213; Pampanini, Fl. Carac. (1930) 192; Li in Proc. Ac. Natur. Sci. Philad. 100 (1948) 321; Tsoong in Acta Phytotax. Sin. 3, 3 (1954) 332; id. in Bull. Brit. Mus. (Bot.) 2, 1 (1955) 29; id. in Fl. R. P. Sin. 68 (1963) 157. —P. amoena auct. non Adams: Maxim. l.c. (1877) 63, p.p. quoad pl. hymal. —P. shawii Tsoong in Acta Phytotax. Sin. 3, 3 (1954) 309, 331; id. in Bull. Brit. Mus. (Bot.) 2, 1 (1955) 29. —P. likiangensis auct. non Franch.: Li, l.c. 306, p.p. —Ic.: Maxim. l.c. (1888) tab. VII, fig. 122; Prain, l.c. tab. 33B-C, figs. 5–11; Fl. R. P. Sin. 68, tab. XXXI, figs. 1–3.
Described from Himalayas. Type in Leningrad.
Grassy slopes in alpine belt, 4100–4500 m.

IIIA. Qinghai: Nanshan (South Tetungsk mountain range, on meadows in uppermost alpine belt, common, July 13, 1872—Przew.; "Hong-Chouli-Kan-sou, No. 344, July 4, 1908, Vaillant"—Danguy, l.c.).
IIIB. Tibet: Weitzan (left bank of Yangtze river, 4000 m, June 20; crossing of Talachu and Bychu rivers, marshy meadow, 4500 m, common, July 5; same site, July 6; watershed between Dyaochu and Konchunchu rivers, July 13—1884, Przew.).
General distribution: China (South-West), Himalayas.

33. P. spicata Pall. Reise, 3 (1776) 738; Maxim. in Bull. Ac. Sci. St.-Pétersb. 24 (1877) 64, 32 (1888) 597; Palibin in Acta Horti Petrop. 14 (1895) 135; Forbes and Hemsley, Index Fl. Sin. 2 (1902) 395; Bonati in Bull. Sco. Bot. Genève, sér. 2, 1 (1912) 328; Limpr. in Beih. Feddes repert. 12 (1922) 485; id. in Feddes repert. 20 (1924) 205; Pai in Contribs Inst. Bot. Nat. Ac. Peiping, 2, 8 (1934) 384; Kitag. Lin. Fl. Mansh. (1939) 395; Li in Proc. Ac. Natur. Sci. Philad. 100 (1948) 304; Grubov, Konsp. fl. MNR (1955) 247; Vved. in Fl. SSSR, 22 (1955) 731; Tsoong in Fl. R. P. Sin. 68 (1963) 177. —Ic.: Pall. l.c. tab. S, fig. 2; Maxim. l.c. (1888) tab. IV, fig. 117; Fl. SSSR, 22, Plate 37, fig. 4.
Described from Dauria. Type lost.
Forb meadows and among shrubs.

IA. Mongolia: Cis-Hing. (Yaksha railway station, mountain slope, Aug. 19, 1902—Litw.; Khuntu somon, 5 km west of Toge-Gol river, forb meadow, Aug. 7, 1949—Yun.).
General distribution: East. Sib., Far East, China (Dunbei, North, North-West), Korean peninsula.

34. P. szetschuanica Maxim. in Bull. Ac. Sci. St.-Pétersb. 32 (1888) 601; Forbes and Hemsley, Index Fl. Sin. 2 (1902) 217; Bonati in Bull. Herb. Boiss. sér. 2, 7 (1907) 545; Limpr. in Beih. Feddes repert. 12 (1922) 285; id. in Feddes repert. 20 (1924) 205, p.p.; Rehder and Kobuski in J. Arn. Arb. 14 (1933) 34, p.p.; Pai in Contribs Inst. Bot. Nat. Ac. Peiping, 2, 7 (1934) 221, p.p.; Li in

Proc. Ac. Natur. Sci. Philad. 100 (1948) 316, p.p.; Tsoong in Acta Phytotax. Sin. 3, 3 (1954) 332; id. in Fl. R. P. Sin. 68 (1963) 179.   —Ic.: Fl. R. P. Sin. 68, tab. XXXIX, fig. 1–3.

Described from North-West China (Gansu province). Type in Leningrad.

In alpine belt, 3380–4450 m.

IIIA. Qinghai: *Amdo* ("Radja and Yellow River gorges, No. 14191; Ba Valley, No. 14416; Amnyi Machen range, No. 14415, Rock"—Rehder and Kobuski, l.c.).

General distribution: China (North-West—south. Gansu, South-West—nor. Sichuan).

35. P. verticillata L. Sp. pl. (1753) 608; Maxim. in Bull. Ac. Sci. St.-Pétersb. 24 (1877) 62, 32 (1888) 600, p.p.; Diels in Bot. Jahrb. 29 (1900) 572; Forbes and Hemsley, Index Fl. Sin. 2 (1902) 219; Danguy in Bull. Mus. nat. hist. natur. 17, 5 (1911) 554, 17, 7 (1914) 82; Limpr. in Beih. Feddes repert. 12 (1922) 485; id. in Feddes repert. 20 (1924) 203; Hulten in Svensk. Vet. Ak. Handl. 3, 8 (1930) 125; Pai in Contribs Inst. Bot. Nat. Ac. Peiping, 2, 7 (1934) 223, p.p.; Kryl. Fl. Zap. Sib. 10 (1939) 2495; Kitag. Lin. Fl. Mansh. (1939) 396; Li in Proc. Ac. Natur. Sci. Philad. 100 (1948) 322; Grubov, Konsp. fl. MNR (1955) 247; Vved. in Fl. SSSR, 22 (1955) 714; Tsoong in Fl. R. P. Sin. 68 (1963) 162; Fl. Kazakhst. 8 (1965) 122.   —*P. stevenii* Bunge in Ledeb. Fl. alt. 2 (1829) 427.   —*P. tangutica* Bonati in Bull. Soc. Bot. Genève, 2, 1 (1912) 328; Limpr. in Feddes repert. 20 (1924) 204.   —*P. sikiangensis* Li in Proc. Ac. Natur. Sci. Philad. 50 (1948) 323.   —*P. calosantha* Li, l.c. 324.   —*P. bonatiana* Li, l.c. 325.   —*P. szetschuanica* auct. non Maxim.: Limpr. in Beih. Feddes repert. 12 (1922) 485, p.p.; Rehder and Kobuski in J. Arn. Arb. 14 (1933) 34, p.p.   —*P. kansuensis* auct. non Maxim.: Li in Proc. Ac. Natur. Sci. Philad. 100 (1948) 318, p.p.   —Ic.: Maxim. l.c. (1888) tab. IV, fig. 123; Fl. R. P. Sin. 68, tab. XXXIV, figs. 4–8.

Described from Siberia, Switzerland and Austria. Type in London (Linn.).

Moist and marshy meadows, pebble beds on river banks, marshy placers, in coniferous forests and their fringes, 2000–3350 m.

IIIA. Qinghai: *Nanshan* (Kuku-Nor lake, July 8; between Nanshan and Donkyru mountain ranges on Rako-Gol river, marshy sites, July 22—1880, Przew.; Xining mountains, Myndan'sha river, May 27, 1890—Gr.-Grzh.; Dulan-khit temple, moss-covered spruce forest, 3350 m, July 8, 1901—Lad.; "Gol de Ta-Pan-Chan, alt. 4000 m, July 10, 1908, Vaillant"—Danguy, l.c.). *Amdo* (alpine region between Radja and Jupar range, meadows of Wajola, alt. 4300 m, No. 14161, June 1926—Rock).

IIIB. Tibet: *Weitzan* (Huang He river basin, Talachu and Bychu rivers, marshy meadows, July 6, 1884—Przew.).

General distribution: Arct., Europe, West. Sib. (Altay), East. Sib., Far East, Nor. Mongolia, China (Dunbei, North, North-West, South-West), Japan, Nor. America (Alaska).

36. P. violascens Schrenk in Bull. phys.-math. Ac. Sci. St.-Pétersb. 1 (1842) 79; id. in Fisch. et Mey. Enum. pl. nov. 2 (1842) 22; Maxim. in Bull.

Ac. Sci. St.-Pétersb. 32 (1888) 594, p.p. excl. pl. Przewalsk. e Tibet bor.; Limpr. in Feddes repert. 20 (1924) 208, p.p.; Kryl. Fl. Zap. Sib. 10 (1939) 2496; Li in Proc. Ac. Natur. Sci. Philad. 100 (1948) 309; Grubov, Konsp. fl. MNR (1955) 247; Vved. in Fl. SSSR, 22 (1955) 709; Fl. Kirgiz. 10 (1962) 206; Tsoong in Fl. R. P. Sin. 68 (1963) 161; Fl. Kazakhst. 8 (1965) 120.  —*P. amoena* var. *violascens* Regel in Bull. Soc. natur. Moscou, 41 (1868) 108.  —*P. tenuicalyx* Tsoong in Kew Bull. (1954) 448.  —Ic.: Maxim. l.c. tab. V, fig. 112; Fl. R.P. Sin. 68, tab, XXXIII, figs. 1–4.

Described from Jung. Alatau (Dzhabyk hill). Type in Leningrad.
Alpine and subalpine meadows and rocky slopes, 2150–3050 m.

IA. **Mongolia:** *Mong. Alt.* (in Taishiri-Ula forests, July 19, 1877—Pot.; at Tsagan-Sair pass, July 29, 1895; slopes at Shadzagain-Suburga pass, July 22, 1898—Klem.).

IIA. **Junggar:** *Tarb.* (Saur mountain range, south. slope, Karagaitu river valley, Bain-Tsagan right creek valley, subalpine meadow belt, June 23, 1957—Yun. et al.). *Tien Shan* (Nilki river, 2150 m, June 8; same site, 2750–3050 m, July 12; Naryn-Gol river near Tsagan-Usu, around Dzhin village, June 10; Aryslyn river, 2750–3050 m, July 16; same site, 2450–2750 m, July 19; Borgaty river 2450–2750 m, July 19—1879, A. Reg.; Ardyn-Daban second pass, from upper Kunges to Yuldus, north-west. slope of pass, upper subalpine belt, Aug. 3; Dagit-Daban pass, Narat mountain range, between Yuldus and upper Ili [Tsanma] valley, 2900 m, cobresia meadow, Aug. 6—1958, Yun.; "Kok-su Valley, 83° E. long. and 43° N. lat. Littledale"—Tsoong, l.c. [1954]).

**General distribution:** Jung.-Tarb., Nor. and Cent. Tien Shan, Mid. Asia (Pamiro-Alay), West. Sib. (Altay).

## Section H o l o p h y l l u m  Li

37. **P. integrifolia** Hook. f. Fl. Brit. Ind. 4 (1884) 308; Prain in Ann. Bot. Gard. Calc. 3 (1890) 128; Forbes and Hemsley, Index Fl. Sin. 2 (1902) 210; Bonati in Bull. Herb. Boiss. sér. 2, 7 (1907)  544; id. in Notes Bot. Gard. Edinburgh, 15 (1926) 168; Limpr. in Feddes repert. 20 (1924) 264, 23 (1927) 339; Marquand in J. Linn. Soc. London (Bot.) 48 (1929) 212; Pai in Contribs Inst. Bot. Nat. Ac. Peiping, 2, 7 (1934) 213; Tsoong in Acta Phytotax. Sin. 3, 3 (1954) 324; id. in Fl. R. P. Sin. 68 (1963) 324.  —*P. integerrima* Pennell et Li ex Li in Proc. Ac. Natur. Sci. Philad. 100 (1948) 351.  —Ic.: Maxim. in Bull. Ac. Sci. St.-Pétersb. 32 (1888), tab. II, fig. 23; Prain, l.c. tab. 5D, figs. 22–27; Fl. R. P. Sin. 68, tab. LXII, figs. 1–2.

Described from East. Himalayas (Sikkim). Type in London (K).
Alpine meadows and rocky steppe slopes.

IIIB. **Tibet:** *South.* ("Gangtze, No. 173, 1925, Ludlow; mountains of Lhasa, 4300 m, No. 9782, July 11, 1943; Reting, No. 11016, July 17, 1944, Ludlow and Scherriff"—Tsoong, l.c. [1954]).
**General distribution:** China (South-West), Himalayas (west., east.).

## Section Brachyphyllum Li

38. P. **chenocephala** Diels in Notizbl. Bot. Gart. Berlin, 10 (1930) 892; Rehder and Kobuski in J. Arn. Arb. 14 (1933) 33; Li in Proc. Ac. Natur. Sci. Philad. 100 (1948) 375; Tsoong in Fl. R. P. Sin. 68 (1963) 307. —Ic.: Li, l.c. tab. V, fig. 87; Fl. R. P. Sin. 68, tab. LXVIII, figs. 3–4.

Described from North-West China (south-west. Gansu). Type in London (K).

Alpine meadow slopes.

IIIB. Tibet: *Weitzan* ("Amnyi Machen range [west of Yellow River], grassy slopes of Mt. Druggu, 3950 m, No. 14435, July 1926, Rock"—Diels, l.c.).

General distribution: China (North-West).

39. P. **deltoidea** Franch. ex Maxim. in Bull. Ac. Sci. St.-Pétersb. 32 (1888) 604; Forbes and Hemsley, Index Fl. Sin. 2 (1902) 208; Bonati in Notes Bot. Gard. Edinburgh, 5 (1911) 89, 7 (1912) 241; Limpr. in Feddes repert. 20 (1924) 226; Li in Proc. Ac. Natur. Sci. Philad. 100 (1948) 357; Tsoong in Acta Phytotax. Sin. 3, 3 (1954) 323; id. in Fl. R. P. Sin. 68 (1963) 274. —Ic.: Maxim. l.c. tab. IV, fig. 133; Fl. R. P. Sin. tab. LXI, figs. 6–7.

Described from South-West China (Yunnan province). Type in Paris.

Exposed wet rocky wastelands, 2600–4300 m.

IIIB. Tibet: *South.* ("Reting, 4300 m, No. 11127, Aug. 14, 1944, Ludlow and Scherriff"— Tsoong, l.c. [1954]).

General distribution: China (South-West).

40. P. **lyrata** Prain ex Maxim. in Bull. Ac. Sci. St.-Pétersb. 32 (1888) 606; Prain in J. As. Soc. Bengal. 58, 2 (1889) 265; id. in Ann. Bot. Gard. Calc. 3 (1890) 165; Tsoong in Fl. R. P. Sin. 68 (1963) 279. —Ic.: Maxim. l.c. tab. IV, fig. 135; Prain, l.c. tab. 31B, figs. 7–13.

Described from East. Himalayas. Type in Calcutta.

Alpine meadows, 3650–4750 m.

IIIB. Tibet: *Weitzan* (nor. slope of Chamudug-la pass, Yangtze river basin, 4730 m, July 26, 1900—Lad.).

General distribution: China (South-West), Himalayas (east.).

41. P. **pheulpinii** Bonati in Bull. Soc. Bot. France, 55 (1908) 247; Limpr. in Feddes repert. 20 (1924) 260; Li in Proc. Ac. Natur. Sci. Philad. 100 (1948) 371, p.p.; Tsoong in Fl. R. P. Sin. 68 (1963) 291, 416.

Described from South-West China (Sikang province). Type in Paris.

Moist rocky slopes in spruce forests and on alpine meadows up to 4000 m alt.

IIIB. Tibet: *Weitzan* (Yantszytszyan river basin, Donra area, on Dychu river [Khichu river], 4000 m, July 17, 1900—Lad.; "Chieh-lien Hsia, Tza-ma-szu in wet stony places in *Picea* forest, No. 8639, July 30, 1958, Tsoong"—Tsoong, l.c.).

General distribution: China (South-West).

## Section A p o c l a d u s Li

42. **P. achilleifolia** Steph. ex Willd. Sp. pl. 3 (1800) 219; Maxim. in Bull. Ac. Sci. St.-Pétersb. 24 (1877) 79, 32 (1888) 611; Sapozhn. Mong. Alt. (1911) 381; Danguy in Bull. Mus. nat. hist. natur. 20 (1914) 81; Limpr. in Feddes repert. 20 (1924) 222; Kryl. Fl. Zap. Sib. 10 (1939) 2520; Li in Proc. Ac. Natur. Sci. Philad. 101 (1949) 21; Grubov, Konsp. fl. MNR (1955) 244; Vved. in Fl. SSSR, 22 (1955) 761; Tsoong in Fl. R. P. Sin. 68 (1963) 232; Fl. Kazakhst. 8 (1965) 139. —Ic.: Maxim. l.c. (1888) tab. VI, fig. 157; Fl. Kazakhst. 8, Plate XIV, fig. 8; Fl. R. P. Sin. 68, tab. LIV, figs. 3–4.

Described from Siberia. Type in Berlin. Isotype in Leningrad.

Sheep's fescue-forb, forb-feather grass and scrub steppes, rock steppe slopes and talus.

IA. **Mongolia:** *Khobd.* (Bairimen-Daban pass, rocky soil, June 20; Ulan-Daban pass, June 22—1879, Pot.). *Mong. Alt., Cent. Khalkha* (60 km south-west of Ulan-Bator, nor. slope of knoll, sheep's fescue-forb steppe, No. 250, July 7, 1941—Yun).

IIA. **Junggar:** *Tarb.* (Saur mountain range, May 20–June 28, 1876—Pev.; Saur mountain range, south. slope, Karagaity valley at its exit from hills onto trail, south. rocky slope, scrub steppe, June 23, 1957—Yun. et al.). *Tien Shan* (vicinity of Sairam-Nur lake, Talki brook, July 19, 1877—A. Reg.).

General distribution: Fore Balkh., Jung.-Tarb.; West. and East. Siberia, Nor. Mongolia (Fore Hubs., Hang.).

43. **P. altaica** Steph. ex Stev. in Mém. Soc. natur. Moscou, 6 (1823) 48, in obs. tab. 14A; Steph. ex Spreng. Syst. Veg. ed. 16, 2 (1825) 779; Maxim. in Bull. Ac. Sci. St.-Pétersb. 24 (1877) 78, 32 (1888) 608; Limpr. in Feddes repert. 20 (1924) 218; Kryl. Fl. Zap. Sib. 10 (1939) 2516; Li in Proc. Ac. Natur. Sci. Philad. 101 (1949) 18; Grubov, Konsp. fl. MNR (1955) 244; Tsoong in Fl. R. P. Sin. 68 (1963) 234; Fl. Kazakhst. 8 (1965) 141. —Ic.: Ledeb. Ic. pl. fl. ross. tab. 442; Maxim. l.c. (1888) tab. VI, fig. 147; Fl. Kazakhst. 8, Plate XVI, fig. 2.

Described from Altay. Type in Helsinki. Isotype in Leningrad.

Solonetz meadows on banks and around springs, sandy-pebble banks of rivers and shoals in steppe and forest belts of mountains.

IA. **Mongolia:** *Mong. Alt.* (Tsitsiriin-Gol river, on sandy-pebble soil, July 22, 1877—Pot.; Bodonchi river floodplain, 2–3 km south of Bodonchiin-Khure, environs of silted brook changing above into solonchak, July 19, 1947—Yun.).

General distribution: Fore Balkh.; West. Sib. (Altay).

44. **P. breviflora** Regel et Winkl. in Acta Horti Petrop. 6 (1879) 352; Maxim. in Bull. Ac. Sci. St.-Pétersb. 32 (1888) 612; Bonati in Bull. Soc. natur. France, 61 (1916) 232; Limpr. in Feddes repert. 20 (1924) 221; Li in Proc. Ac. Natur. Sci. Philad. 101 (1949) 18: Tsoong in Fl. R. P. Sin. 68 (1963) 228. —*P. dolichorrhiza* auct. non Schrenk: Vved. in Fl. SSSR, 22 (1955) 757, p.p. — Ic.: Maxim. l.c. tab. VI, fig. 158.

Described from Junggar. Type in Leningrad.

IIA. Junggar: *Tien Shan* (Sary-Bulak, north-west of Kul'dzha, 1200–1800 m—A. Reg., typus!).

General distribution: Mid. Asia (Pamiro-Alay).

Note. Flowers of this species resemble buds of *Pedicularis comosa* s. l. and *P. dolichorrhiza* Schrenk and hence the existence of *P. breviflora* Regel et Winkl. was questioned (Maximowicz, l.c.; Li, l.c.; Vvedensky, l.c.). A study of the buds of *P. breviflora* showed that they differ from those of *P. dolichorrhiza* in beak structure (beak of *P. breviflora* much shorter) and from those of *P. comosa* in structure of beak teeth (teeth of *P. breviflora* much broader and shorter).

45. **P. compacta** Steph. ex Willd. Sp. pl. 3 (1800) 219; Maxim. in Bull. Ac. Sci. St.-Pétersb. 24 (1878) 166, 32 (1888) 576; Sapozhn. Mong. Alt. (1911) 380; Limpr. in Feddes repert. 20 (1924) 231; Kryl. Fl. Zap. Sib. 10 (1939) 2504; Li in Proc. Ac. Natur. Sci. Philad. 101 (1949) 22; Grubov, Konsp. fl. MNR (1955) 244; Vved in Fl. SSSR, 22 (1955) 748; Fl. Kazakhst. 8 (1965) 131. —**Ic.:** Maxim. l.c. (1888) tab. IV, fig. 81; Fl. Kazakhst. 8, Plate XIV, fig. 3.

Described from Siberia. Isotype in Leningrad.

Meadow and rocky slopes in alpine tundra and in placers.

IA. Mongolia: *Khobd.* (near Kharkhira river sources, on pebble bed, July 23, 1879— Pot.). *Mong. Alt.* (Kongeita river valley, Sept. 18, 1876—Pot.; Shadzagain-Suburga, July 22, 1898—Klem.; Tsagan-Gol river, flat peak near Kharsala river estuary, alpine tundra and placer, July 2, 1905; source of Bzau-Kul', alpine tundra and placer, July 11, 1906; Ui-chilika river valley, July 2; east of Ak-Korum pass, alpine tundra, July 13; Karatyr, forest near lake, Aug. 1; Karatyr river source, alpine tundra, Aug. 3—1908; Dain-Gol lake, west. bank, July 27–29, 1909— Sap. [TK]; "upper Aksu, Onkattu, B. Khobdos lake, Sumdairyk, Kutologoi, M. Kairy"—Sap. l.c.).

IIA. Junggar: *Jung. Alt.* (Barlyk mountains, Kertau mountain range, Ku-Karagai pass, 2660 m, July 11, 1905—Obruchev [TK]).

General distribution: Jung.-Tarb. (Tarbagatai); Arct. (Asian), West. and East. Sib., Nor. Mongolia (Fore Hubs., Hang).

46. **P. dasystachys** Schrenk in Bull. phys.-math. Ac. Sci. St.-Pétersb. 2 (1844) 195; Kryl. Fl. Zap. Sib. 10 (1939) 2513; Grubov, Konsp. fl. MNR (1955) 244; Vved. in Fl. SSSR, 22 (1955) 749; Tsoong in Fl. R. P. Sin. 68 (1963) 228; Fl. Kazakhst. 8 (1965) 131. —*P. laeta* Stev. ex Claus. in Goebel, Reise, 2 (1838) 296, nom. nud.; Bunge in Ledeb. Fl. Ross. 3 (1849) 289; Maxim. in Bull. Ac. Sci. St.-Pétersb. 32 (1888) 610; Limpr. in Feddes repert. 20 (1924) 220; Li in Proc. Ac. Natur. Sci. Philad. 101 (1949) 13. —*P. tanacetifolia* auct. non Adams: Bunge in Bull. phys.-math. Ac. Sci. St.-Pétersb. 1 (1843) 337. — **Ic.:** Maxim. l.c. tab. VI, fig. 150 (sub nom. *P. laeta*); Fl. Kazakhst. 8, Plate XIV, fig. 4.

Described from West. Siberia (Ishim river). Type in Leningrad.

Solonetz and floodplain meadows.

IA. Mongolia: *Mong. Alt.* (environs of lake near Urmogaity pass, June 24, 1903—Gr.-Grzh.).

General distribution: Jung.-Tarb.; Europe, Mid. Asia, West. Siberia.

47. P. dolichorrhiza Schrenk in Bull. phys.-math. Ac. Sci. St.-Pétersb. 1 (1842) 80; id. in Fisch. et Mey. Enum. pl. nov. 2 (1842) 23; Maxim. in Bull. Ac. Sci. St.-Pétersb. 32 (1888) 609; Hemsley in J. Linn. Soc. London (Bot.) 30 (1894) 124; Limpr. in Feddes repert. 20 (1924) 223; Pampanini, Fl. Carac. (1930) 192; Persson in Bot. notiser (1938) 302; Kryl. Fl. Zap. Sib. 10 (1939) 2515; Li in Proc. Ac. Natur. Sci. Philad. 101 (1949) 18; Vved. in Fl. SSSR, 22 (1955) 757, p.p.; Fl. Kirgiz. 10 (1962) 219; Tsoong in Fl. R. P. Sin. 68 (1963) 235; Fl. Kazakhst. 8 (1965) 135.    —P. jugentassi Semiotr. in Bot. mat. gerb. Inst. bot. AN KazSSR, 2 (1964) 44; Fl. Kazakhst. 8 (1965) 136.    —Ic.: Maxim. l.c. tab. VI, fig. 146; Fl. R. P. Sin. 68, tab. LIV, figs. 1–2; Fl. Kazakhst. 8, Plate XIV, fig. 7, Plate XV, fig. 5 (sub nom. P. jugentassi Semiotr.).

Described from Junggar Alatau (Dzhabyk hill). Type in Leningrad.

Meadow and steppe slopes, 900–4000 (5200) m.

IB. Kashgar: West. (Jarkand, 1870—Henderson; Sarykol mountain range, Bostan-Terek, July 1, 1929—Pop.). East. (Algoi river, near Turfan, Sept. 12, 1879—A. Reg.).

IIA. Junggar: Tarb. (Saur mountain range, south. slope, Karagaitu river valley, Bain-Tsagan right creek valley, subalpine belt, meadow, June 23; same site, steppe on south. slope, June 23—1957, Yun. et al.). Tien Shan.

IIIB. Tibet: Chang Tang ("Tibet, Kuen-lun, Plains at about 5200 m, 1892, Pico"—Hemsley, l.c.).

IIIC. Pamir: Charlysh river basin, Ulugtuz gorge, on gentle slope, near brook, June 21, 1909—Divn.; Issyk-Su river, 3000–3100 m, June 21; Kashka-Su river, under moraine, 3500 m, July 5— 1942, Serp.).

General distribution: Jung.-Tarb., Nor. and Cent. Tien Shan; Fore Asia (Afghanistan), Mid. Asia (West. Tien Shan and Pamiro-Alay), Himalayas (Kashmir).

48. P. elata Willd. Sp. pl. 3 (1800) 210; Maxim. in Bull. Ac. Sci. St.-Pétersb. 24 (1877) 76; Sapozhn. Mong. Alt. (1911) 380; Limpr. in Feddes repert. 20 (1924) 217; Kryl. Fl. Zap. Sib. 10 (1939) 2511; Li in Proc. Ac. Natur. Sci. Philad. 101 (1949) 41; Grubov, Konsp. fl. MNR (1955) 244; Vved. in Fl. SSSR, 22 (1955) 744; Tsoong in Fl. R.P. Sin. 68 (1963) 225; Fl. Kazakhst. 8 (1965) 130.    —Ic.: Maxim. in Bull. Ac. Sci. St.-Pétersb. 32 (1888) tab. IV, fig. 141; Fl. R. P. Sin. 68, tab. LII, figs. 1–3; Fl. Kazakhst. 8, Plate XIV, fig. 1.

Described from East. Siberia (environs of Krasnoyarsk town). Isotype in Leningrad.

Larch forests and their fringes, alpine meadows.

IA. Mongolia: Khobd. (Ulan-Daba pass, in forest, June 23, 1879—Pot.).

General distribution: Jung.-Tarb.; West. and East. Siberia, Nor. Mongolia (Hang.).

49. P. flava Pall. Reise, 3 (1776) 736; Maxim. in Bull. Ac. Sci. St.-Pétersb. 24 (1877) 81, 32 (1888) 611; Danguy in Bull. Mus. nat. hist. natur. 17, 7 (1911) 554; Limpr. in Feddes repert. 20 (1924) 221; Kitag. Lin. Fl. Mansh. (1939) 394; Li in Proc. Ac. Natur. Sci. Philad. 101 (1949) 20; Grubov, Konsp. fl. MNR (1955) 245; Vved. in Fl. SSSR, 22 (1955) 760; Tsoong in Fl. R. P. Sin. 68

(1963) 231.   —Ic.: Maxim, l.c. (1888) tab. VI, fig. 153; Fl. R. P. Sin. 68, tab. LIII, figs. 1–2.

Described from Dauria (interfluve of Onon and Borza). Isotype in Leningrad.

Rocky steppe slopes and solonetz meadows.

IA. **Mongolia:** *Khobd.* (mountains between Kobdo and Ukha rivers, on rock talus, July 4, 1894—Klem.). *Mong. Alt.* (Dolon-Nor, pebble bed, July 8; Tatal river gorge, pebble bed, July 8; Tsitsirin-Gol river valley, July 9—1877, Pot.; Urtu-Gol river valley, larch forest on east. slope of mountain, Aug. 19; Khara-Dzarga mountain range, Sakhir-Sala river valley, on east. rubble slope of Imertsik mountain, Aug. 22—1930, Pob.). *Cent. Khalkha* (Dzhirgalante river basin, between Bogota and Agit mountains, near Dol'che-Gegen monastery, steppe zone, trails of depression on plateau-like elevation, Aug. 29, 1925—Krasch. and Zam.; near Choiren settlement, rocky steppes, July 26, 1926—Kondr.; environs of Ikhe-Tukhum-Nor lake, Modkho mountain, July 26, 1926—Zam.; Bain-Gol river, Aug. 31, 1926—Pavl.; Nalaikha settlement, Aug. 2, 1927—Zam.; Choiren-Ula, 1940, Sanzha; along Ulan-Bator—Dalan-Dzadagad road, Aug. 11, 1950—Lavrenko and Kal.). *East. Mong.* (Manchuria railway station, steppe belt, June 5, 1902—Litw.; same site, 1915—Nechaeva; Khukh-Khoto, rocky slope, July 24, 1926—Lis.; Nanshan mountain, 700 m, No. 833, April 24, 1951—Wang et al.). *Bas. Lakes* (between Guuta and Dzabkhyn rivers, on granite rocks, July 18; Khudzhirte area, Shuryk river, among shrubs on rocky soil, July 20—1877, Pot.). *Val. Lakes* (on Tui river, around Boro-Khoto town dumps [Sept. 7] 1886—Pot.; between Ongin-Gol and Tsagantek-Tal rivers, Aug. 21, 1926—Glag.; Khairkhan-Dulan somon, 40 km south-west on Bain-Khongor road, feather grass and feather-grass-like steppe on rubble chestnut soils, June 27; Narin-Del' somon, 30 km west of Tatsiin-Gol on South Hangay road, wheat grass-feather grass dry steppe, June 28—1941, Tsatsenkin; 7–8 km west of Bain-Khongor, wormwood-feather grass steppe on heavy loamy sand, Aug. 28, 1943; along Baishintu—Tszag-Baidarik somon road, feather grass steppe, Aug. 29; 40–45 km west of Bain-Khongor, Aug. 29—1945; 40 km east of Bain-Khongor, feather grass steppe on loamy sand, July 11, 1947—Yun.). *Gobi-Alt.* (Dundu-Saikhan mountains, south. slopes of upper and middle belt, July 5, 1909—Czet.; Ikhe-Bogdo mountain range, nor. slopes, June 18, 1926—Kozlova; Dundu-Saikhan mountains, rubble slope, Aug. 17, 1931—Ik.-Gal.; Dundu- and Dzun-Saikhan mountain ranges, slopes of mountains and gorges from trail to upper belt, July–Aug. 1933—M. Simukova; Dundu-Saikhan range, steppe on slopes of upper belt, June 19, 1945; pass between Dzun- and Dundu-Saikhan, July 22; east. fringe of Dundu-Saikhan mountain range, lower hill belt, rocky slope, talus with *Artemisia procera* shrubs, July 22; Ikhe-Bogdo range, south. slope, upper Gogeri-Gol, wormwood-feather grass dry steppe, Oct. 10—1943, Yun.).

General distribution: East. Sib. (Daur.), Nor. Mong. (Hent., Hang., Mong.-Daur), China (Dunbei).

50. **P. lasiostachys** Bunge in Ledeb. Fl. alt. 2 (1830) 434; Maxim. in Bull. Ac. Sci. St.-Pétersb. 32 (1888) 611; Limpr. in Feddes repert. 20 (1924) 221; Kryl. Fl. Zap. Sib. 10 (1939) 2521; Li in Proc. Ac. Natur, Sci. Philad. 101 (1949) 21; Grubov, Konsp. fl. MNR (1955) 245; Vved. in Fl. SSSR, 22 (1955) 759; Fl. Kazakhst. 8 (1965) 138.   —Ic.: Maxim, l.c. tab. VI, fig. 156.

Described from Altay. Isotype in Leningrad.

Alpine meadows and high-altitude tundra, on rocky slopes and placers in alpine belt.

IA. **Mongolia:** *Khobd.* (Tszusylan, above forest boundary, July 13, 1879—Pot.). *Mong. Alt.* (Tsagan-Gol river, flat peak near Kharkhala river estuary, alpine tundra and placers, July 2, 1905; Onkottu lake, Chingistei sentry post, alpine meadow, June 25; peak between Turgyun' and Sumdairyk rivers, alpine tundra, July 3—1906, Sap. [TK]).

General distribution: West. Sib. (Altay).

51. **P. mariae** Regel in Acta Horti Petrop. 6 (1879) 351; Vved. in Fl. SSSR, 22 (1955) 772; Tsoong in Fl. R. P. Sin. 68 (1963) 234; Fl. Kazakhst. 8 (1965) 141. —*P. altaica* auct. non Steph.: Maxim. in Mél. biol. 12 (1888) 908; Limpr. in Feddes repert. 20 (1924) 218, p.p.; Li in Proc. Ac. Natur. Sci. Philad. 101 (1949) 18, p.p. —Ic.: Fl. Kazakhst. 8, Plate XVI, fig. 3.

Described from Kazakhstan. Type in Leningrad.

Wet meadows and in tugais.

IB. **Kashgar:** *Nor.* (south. slope of Keinsk basin, upper Kyzyl river, 3–4 km south of Kein settlement, solonchak meadow at edge of spring, Sept. 2, 1958—Yun. et al.; "Kucha"—Tsoong, l.c.).

IIA. **Junggar:** *Tien Shan* (south. slope, upper South. Muzart valley, Sazlik area, reed grass meadow on right-bank terrace above meadow, Sept. 9, 1958—Yun. et al.). *Dzhark.* (Ili river bank near Kul'dzha, May 1877; Suidun, July 1878—A. Reg.).

General distribution: Fore Balkh., Nor. Tien Shan.

52. **P. physocalyx** Bunge in Bull. Ac. Sci. St.-Pétersb. 8 (1841) 252; Limpr. in Feddes repert. 20 (1924) 222; Kryl. Fl. Zap. Sib. 10 (1939) 2520, p.p.; Vved. in Fl. SSSR, 22 (1955) 750; Tsoong in Fl. R. P. Sin. 68 (1963) 229; Fl. Kazakhst. 8 (1965) 132. —*P. flava* auct. non Pall.: Bunge in Ledeb. Fl. alt. 2 (1830) 433. —*P. flava* var. *altaica* et var. *conica* Bunge in Mém. Ac. Sci. St.-Pétersb. Sav. Etrang. 2 (1835) 570. —*P. fedtschenkoi* Bonati in Bull. Soc. Bot. France, 59 (1914) 233. —Ic.: Ledeb. Ic. pl. fl. ross. tab. 439 (sub nom. *P. flava*); Maxim. in Bull. Ac. Sci. St.-Pétersb. tab. VI, fig. 155; Bonati, l.c. tab. 4 (sub nom. *P. fedtschenkoi*); Fl. Kazakhst. 8, Plate XIV, fig. 5.

Described from Altay. Isotype in Leningrad.

Steppes, meadow and steppe mountain slopes.

IIA. **Junggar:** *Tien Shan* ("Kul'dzha region"—Tsoong, l.c.).

General distribution: Aralo-Casp., Fore Balkh., Jung.-Tarb., Nor. Tien Shan; Europe (South-East.), West. Sib.

Note. We did not have specimens of this species from Central Asia but have plants from USSR regions bordering Chinese Junggar (Borokhudzhir, May 1878—Fet.; Bel-Bulyk gorge, 1800 m, May 20, 1878—A. Reg. et al.).

53. **P. proboscidea** Stev. in Mém. Soc. natur. Moscou, 6 (1823) 33, excl. syn.; Maxim. in Bull. Ac. Sci. St.-Pétersb. 24 (1877) 66, 32 (1888) 563; Limpr. in Feddes repert. 20 (1924) 230; Kryl. Fl. Zap. Sib. 10 (1939) 2502; Grubov, Konsp. fl. MNR (1955) 246; Vved. in Fl. SSSR, 22 (1955) 745; Tsoong in Fl. R. P. Sin. 68 (1963) 66; Fl. Kazakhst. 8 (1965) 130; Fl. Kirgiz. Dop. 1 (1967) 107. —Ic.: Maxim, l.c. tab. III, fig. 56; Fl. SSSR, 22, Plate XXXVI, fig. 2; Fl. Kazakhst. 8, Plate XIV, fig. 2.

Described from Altay (environs of Zmeinogorsk). Type locality not known.

Meadows and meadow slopes in alpine and subalpine belts.

IA. **Mongolia:** *Mong. Alt.* (south. Altay, 1876—Pot.).
IIA. **Junggar:** ("Sinkiang"—Tsoong, l.c.).
**General distribution:** Fore Balkh. (Zaisan), Jung.-Tarb., Cent. Tien Shan; West. Sib. (Altay).

54. **P. pubiflora** Vved. in Fl. SSSR, 22 (1955) 754, 812; Fl. Kirgiz. 10 (1962) 218; Fl. Kazakhst. 8 (1965) 134. —*P. songarica* auct. non Schrenk: Limpr in Feddes repert. 20 (1924) 217, p.p. quoad pl. tianschan.; Tsoong in fl. R. P. Sin. 68 (1963) 226, p.p. quoad pl. tianschan.—Ic.: Fl. SSSR, 22, Plate XXXVII, fig. 1; Fl. Kirgiz. 10, Plate 26, fig. 2; fl. R. P. Sin. 68, tab. LII, figs. 4–5 (sub nom. *P. songarica*).

Described from Kazakhstan (Sonkul'-tau). Type in Tashkent.

Alpine and subalpine meadows, 2450–3050 m.

IIA. **Junggar:** *Tien Shan* (Kumbel', 2750–3050 m, May 30; Borgaty brook, 2450–2750 m, June 7; Naryn-Gol near Tsagan-Usu, June 10; Borborogusun, along river, 2750 m, June 15; Aryslyn river [Kash], 2450–2750 m, July 8—1879, A. Reg.).
**General distribution:** Jung.-Tarb. (Jung. Alatau), Nor. and Cent. Tien Shan; Mid. Asia (Pamiro-Alay, West. Tien Shan).

55. **P. rubens** Steph. ex Willd. Sp. pl. 3 (1800) 219; Maxim. in Bull. Ac. Sci. St.-Pétersb. 24 (1877) 79, 32 (1888) 610; Limpr. in Feddes repert. 20 (1924) 220, p.p.; Kitag. Lin. Fl. Mansh. (1939) 395; Li in Proc. Ac. Natur. Sci. Philad. 101 (1949) 20, p.p.; Grubov, Konsp. fl. MNR (1955) 246; Vved. in Fl. SSSR, 22 (1955) 760; Tsoong in Fl. R. P. Sin. 68 (1963) 231. —*P. venusta* auct. non Bunge: Limpr. l.c. 218, p.p. quoad pl. chin.; Li, l.c. 18, p.p. quoad pl. chin. — Ic.: Maxim. l.c. (1888) tab. VI, fig. 152; Fl. SSSR, 22, Plate XXXIX, fig. 3; Fl. R. P. Sin. 68, Plate LIII, figs. 3–4.

Described from Siberia. Isotype in Leningrad.

Forest and flooded meadows, swamps, wet cliffs and talus.

IA. **Mongolia:** *Cis-Hing.* (near Yaksha railway station, mountain slope, June 13, 1902—Litw.; Arshan mountains, montane slopes, meadow, No. 417, June 13, 1950—Chang and Noda).
**General distribution:** East. Sib., Nor. Mongolia, China (Dunbei, North).

56. **P. striata** Pall. Reise, 3 (1776) 737; Maxim. in Bull. Ac. Sci. St.-Pétersb. 24 (1877) 82, 32 (1888) 613; Franch. Pl. David. 1 (1884) 226; Forbes and Hemsley, Index Fl. Sin. 2 (1902) 216; Danguy in Bull. Mus. nat. hist. natur. 20 (1914) 82; Limpr. in Feddes repert. 20 (1924) 215; Pai in Contribs Inst. Bot. Nat. Ac. Peiping, 2, 7 (1934) 220; Kitag. Lin. Fl. Mansh. (1939) 395; Walker in Contribs U.S. Nat. Herb. 28 (1941) 659; Li in Proc. Ac. Natur. Sci. Philad. 101 (1949) 12; Grubov, Konsp. fl. MNR (1955) 247; Vved. in Fl. SSSR, 22 (1955) 742; Tsoong in Fl. R. P. Sin. 68 (1963) 64. —Ic.: Pall. l.c. tab. R, fig. 2 c: Maxim, l.c. (1888) tab. VI, fig. 159.

Described from East. Siberia. Isotype in Leningrad.

Feather grass-forb and meadow steppes and thickets.

**IA. Mongolia:** *Cis-Hing.* (Abderiin-Gol river, in sandy steppe, June 25, 1899—Pot. and Sold.; near Yaksha railway station, June 13, 1902—Litw.; Khuntu somon, 17–20 km east-southeast of Bain-Tsagan, feather grass-tansy steppe on chestnut loamy sand, Aug. 6; Khalkha-Gol somon, 30 km from Bain-Buridu, depression among steppes, alkali grass and forb meadow, Aug. 18—1949, Yun.). *Cent. Khalkha* (Tsinkir-Mandal somon, Tsinkir-Gol river valley, opposite Tsinkir-Dugang, montane steppe, on gentle trails of creek valley, July 23; Muren somon, 20–25 km nor. of underkhan, wormwood snakeweed-feather grass steppe, July 25; 20–25 km north-west of Bain-Ul somon, along Ul'khun-Maikhan road, south-east. slope of Eren-Daba mountain range, thickets on talus, July 28—1949, Yun.). *East. Mong.* (Buir-Norsk plain, Khoren-Bulyn river valley, July 11, 1899—Pot. and Sold.; "Kailar, monticules de sables, altitude 800 m, June 25, 1896, Chaff."—Danguy, l.c.; Muni-Ula, nor. slope on humid soil, rare, July 24, 1871—Przew.; "Sui-yuan, Wutachao, No. 2969, Aug. 5, 1931, Hsia; Tatsingshan, Halochingkow, No. 2814, July 23, 1931, Hsia"—Pai, l.c.). *Alash. Gobi* (south. part of Alashan hills, on wet soil, common, July 10, 1873—Przew.; Alashan mountain range, Yamata gorge, south-east. slope of midbelt, in thickets of height up to 45 cm, humus soil, May 2, 1908—Czet.; "Ning-Hsia: Ya-sze-kow mountains, alt. 1900 m, No. 128, Aug. 28, 1933"—Pai, l.c.; "Ho Lan Shan, on steppes, Ching"—Walker, l.c.).

General distribution: East. Sib., Far East, Nor. Mongolia (Hent., Hang., Mong.-Daur.), China (Dunbei, North, North-West).

57. *P. uliginosa* Bunge, Ind. Sem. Hort. Dorpat. (1839) 8; id. in Bull. Ac. Sci. St.-Pétersb. 8 (1841) 251; Maxim. in Bull. Ac. Sci. St.-Pétersb. 24 (1877) 78, 32 (1888) 610; Alcock, Rep. nat. hist. results Pamir boundary com. (1898) 25; Sapozhn. Mong. Alt. (1911) 381; Paulsen in Hedin, S. Tibet, 6, 3 (1922) 44; Limpr. in Feddes repert. 20 (1924) 220; Kryl. Fl. Zap. Sib. 10 (1939) 2515; Li in Proc. Ac. Natur. Sci. Philad. 101 (1949) 19; Grubov, Konsp. fl. MNR (1955) 247; Vved. in Fl. SSSR, 22 (1955) 742; Fl. Kirgiz. 10 (1962) 217; Ikonnikov, Opred. rast. pamira (1963) 222; Tsoong in Fl. R. P. Sin. 68 (1963) 230; Fl. Kazakhst. 8 (1965) 129. —*P. rubens* auct. non Steph. ex Willd.: Ledeb. Fl. alt. 2 (1830) 435. —*P. rubens* var. *altaica* Bunge in Mém. Ac. Sci. St.-Pétersb. Sav. Etrang. 2 (1835) 571 —*P. rubens* var. *alatavica* Kar. et Kir. in Bull. Soc. Natur. Moscou, 15 (1842) 419; Sapozhn. Mong. Alt. (1911) 380. —**Ic.:** Ledeb. Ic. pl. fl. ross. tab. 441 (sub nom. *P. rubens*); Fl. R. P. Sin. 68, tab. LVIII, figs. 3–4.

Described from Altay (Charysh river). Type in Leningrad.

Sasa grasslands, marshy meadows, banks of brooks and rubble slopes.

**IA. Mongolia:** *Khobd.* (Tszusylan, alpine meadows, July 11; same site, in forest, July 13; near Kharkhira river sources, on grassy nor. slope, July 24—1879, Pot.). *Mong. Alt.* (Taishiri-Ula mountain range, in gorge, alpine belt, July 8; same site, July 18; same site, in forest, July 19—1877, Pot.; same site, on nor. slope of montane, July 16, 1894—Klem.; Kak-Kul' lake, between Tsagan-Gol and Kobdo, dry rubble ravines, June 22, 1906—Sap.; Elangash plateau, north of Dain-Gol, alpine steppe, July 4; Bzau-Kul' river source, alpine tundra, placers, July 11; Ulan-Daba pass toward Bulugun source, rocky alpine tundra, July 22—1906; Urmogaity pass, rubble slopes, July 11; Oigur river valley, July 15; Tsagan-Gol river valley, July 16; Chigirtei river valley, washed moraines, July 21; Dain-Gol lake, south-west. bank, July 29, 1909, Sap. [TK]; Taishiri-Ula mountain range, Tszasaktu-Khan area, larch forest on nor. slope of montane,

Aug. 9, 1930—Pob.; Taishiri-Ula mountain range, nor. slope, larch forest, July 12, 1945; upper Indertiin-Gol, swampy meadow in high-alpine belt, July 24, 1947—Yun.; nor. slope of Taishiri-Ula mountain range, site in larch forest 15 km south-east of Yusun-Bulak, Sept. 1, 1948—Grub.). *Gobi-Alt.* (Ikhe-Bogdo mountain range, watershed between Narim-Khurimt-Ama and Ketsu-Ama, cobresia-sedge meadow on rock scree, June 28, 1945—Yun.).

IIIC. Pamir ("marshy ground by R. Aksu, 4000–4300 m, No. 17751"—Alcock, l.c. "Eastern Pamir, Kamper-kishlak, Mus-tagh-ata, ab. 4500 m, July 29, 1894, Hedin"—Paulsen, l.c.).

General distribution: Jung.-Tarb., Nor. and Cent. Tien Shan, East. Pam.; Mid. Asia (Pamiro-Alay), West. Sib. (Altay), East. Sib., Nor. Mongolia.

58. *P.* venusta (Bunge) Bunge in Bull. Ac. Sci. St. Pétersb. 8 (1841) 252, nom. nud.; Bunge in Bull. phys.-math. Ac. Sci. St.-Pétersb. 1 (1842) 380; Maxim. in Bull. Ac. Sci. St.-Pétersb. 24 (1877) 80, 32 (1888) 610, excl. var. (Sachal. et Japon.); Palibin in Acta Horti Petrop. 14 (1895) 135; Danguy in Bull. Mus. nat. hist. natur. 17, 7 (1911) 554; Sapozhn. Mong. Alt. (1911) 381, p.p.; Limpr. in Feddes repert. 20 (1924) 218, p.p.; Kryl. Fl. Zap. Sib. 10 (1939) 2519; Kitag. Lin. Fl. Mansh. (1939) 396; Li in Proc. Ac. Natur. Sci. Philad. 101 (1949) 18, p.p.; Grubov, Konsp. fl. MNR (1955) 247; Vved. in Fl. SSSR, 22 (1955) 769; Tsoong in Fl. R. P. Sin. 68 (1963) 233; Fl. Kazakhst. 8 (1965) 140. —*P. comosa* var. *venusta* Bunge in Mem. Ac. Sci. St.-Pétersb. Sav. Etrang. 2 (1835) 570. —Ic.: Maxim. l.c. (1888) tab. VI, fig. 148; Fl. Kazakhst. 8, Plate XVI, fig. 1.

Described from Siberia. Isotype in Leningrad.

Moist floodplain and solonchak meadows along river banks.

IA. Mongolia: *Khobd.* (narrow valley of Kobdo river, rubble steppe, June 19, 1906—Sap. [TK]; Bukhu-Muren-Gol river, 4–5 km north of Bukhu-Muren somon, scrub bottomland deciduous forest and meadow, July 31; same site, 5–6 km nor. of somon, solonchak-like meadow July 31—1945, Yun.). *Mong. Alt.* (upper Tsagan-Gol river, between Kharsalai and Prokhodnaya rivers, standing moraines, June 30, 1905; Kak-Kul' lake, between Tsagan-Gol and Kobdo, dry rubble ravine, June 22; Ulan-Daban pass in Bulugun river sources, rocky alpine tundra, July 22, 1906; upper Saksai river, bank, July 14; Sumdairyk river bank, between moraines, July 30—1908; Tsagan-Gol river, midcourse, standing moraines, July 17; Karaganty river, Taldy-Bulak river estuary, rubble steppe, July 30—1909, Sap. [TK]). *Cent. Khalkha* (Dzhargalante river basin [47° N. lat., 104–105° E. long.], midcourse of Ubur-Dzhargalante river, near Dol'che-Gegen monastery, meadow, Aug. 29; same site, Ubur-Dzhargalante river, between sources and Agit montane, meadow, Aug. 30—1925, Krasch. and Zam.; Kholt area, in Hangay foothills, May 23; environs of Kholt, Aug. 1–3, 1926, Gus.; environs of Ikhe-Tukhum-Nor lake [46°30' N. lat., 104–105° E. long.], Kairkhan valley, June; same site, Khalzangin-Nor lake bank, June—1926, Zam.; between Ongiin-Gol river and Tsagantek-Tala valley, Aug. 21, 1926—Glag.). *Val. Lakes* (Tatsain-Gol river, 12 versts [1 verst = 1.067 km] beyond Tatsa urton, moist bank, July 19; Ongiin-Gol river, dry solonchak, July 27—1893, Klem.; 15 km nor.-west of Saikhan-Obo somon, upward along Ongiin-Gol river valley, solonchak meadow, July 9, 1941—Tsatsenkin).

General distribution: West. and East. Siberia, Far East, Nor. Mongolia, China (Altay, Dauria).

Note. Hybrids of this species with *P. achilleifolia* Steph. ex Willd. are common in Mongolian Altay.

## Section Cladomania Li

59. **P. karoi** Freyn in Oesterr. bot. Z. 46 (1896) 26; Limpr. in Feddes repert. 20 (1924) 205; Vved. in Fl. SSSR, 22 (1955) 776; Fl. Kazakhst. 8 (1965) 142. —*P. palustris* auct. non L.: Maxim. in Bull. Ac. Sci. St.-Pétersb. 24 (1877) 75, 32 (1888) 607; Danguy in Bull. Mus. nat. hist. natur. 20 (1914) 81; Kryl. Fl. Zap. Sib. 10 (1939) 2508; Kitag. Lin. Fl. Mansh. (1939) 395; Li in Proc. Ac. Natur. Sci. Philad. 101 (1949) 23; Grubov, Konsp. fl. MNR (1955) 246; Tsoong in Fl. R. P. Sin. 68 (1963) 116. —*P. palustris* subsp. *karoi* (Freyn) Tsoong, l.c. 117. —*P. pseudo-karoi* Bonati in Bull. Ac. Geogr. Bot. 15 (1905) 11; Limpr. in Feddes repert. 20 (1924) 215. —**Ic.:** Li, l.c. tab. 2, fig. 102 (sub nom. *P. palustris*).

Described from East. Siberia (environs of Nerchinsk town). Type in Leningrad.

Marshy meadows and wet river banks.

IA. Mongolia: *Mong. Alt.* (Buyantu somon, in lower courses of Buyantu-Gol river, 1941—Kondratenko). *Cis-Hing.* (near Yaksha railway station, marsh, Aug. 19, 1902 Litw.). *Ordos* (in Huang He river valley, on meadows, common, Aug. 18, 1871—Przew.).

IIA. Junggar: *Tien Shan* (south. slope of East. Tien Shan, Urte area, Sept. 9, 1895—Rob.). *Jung. Gobi* ("Altai, steppes entre l'Ouentre et l'Irtish, 22 août, 1895"—Danguy, l.c.).

General distribution: Jung.-Tarb., Europe (South-East., Urals), Mid. Asia, West. and East. Siberia, Nor. Mongolia, China (Dunbei).

**P. labroadorica** Wirs. Ecol. Bot. 2 (1778) sub tab. 10; Li in Proc. Ac. Natur. Sci. Philad. 101 (1949) 24; Vved. in Fl. SSSR, 22 (1955) 738; Tsoong in Fl. R. P. Sin. 68 (1963) 118. —*P. euphrasioides* Steph. ex Willd. Sp. pl. 3 (1800) 204; Maxim. in Bull. Ac. Sci. St.-Pétersb. 24 (1877) 74, 32 (1888) 606; Limpr. in Feddes repert. 20 (1924) 214; Kryl. Fl. Zap. Sib. 10 (1939) 2510; Grubov, Konsp. fl. MNR (1955) 244. —**Ic.:** Maxim. l.c. (1888) tab. IV, fig. 136 (sub nom. *P. euphrasioides*); Fl. SSSR, 22, Plate XXXVIII, fig. 1; Fl. R. P. Sin. 68, tab. XXIII, figs. 8–11.

Described from North America (Labrador). Type locality not known.

Larch and larch-birch forests and their fringes, and sub-alpine meadows.

IA. Mongolia: *Cent. Khalkha* (occurrence possible).

General distribution: Arct., West. and East. Sib., Far East, Nor. Mongolia (Fore Hubs., Hent., Mong.-Daur.), China (Dunbei), North America.

Note. Collections from adjoining regions of Cent. Khalkha (environs of Ulan-Bator) are known.

## Section Haplophyllum Li

60. **P. resupinata** L. Sp. pl. (1753) 608; Maxim. in Bull. Ac. Sci. St.-Pétersb. 24 (1877) 70, 32 (1888) 558; Franch. Pl. David. 1 (1884) 226; Forbes and

Hemsley, Index Fl. Sin. 2 (1902) 214; Limpr. in Feddes repert. 20 (1924) 237; Pai in Contribs Inst. Bot. Nat. Ac. Peiping, 2, 7 (1934) 217; Kung et Wang, ibid. 2, 8 (1934) 383; Kryl. Fl. Zap. Sib. 10 (1939) 2506; Kitag. Lin. Fl. Mansh. (1939) 395; Li in Proc. Ac. Natur. Sci. Philad. 101 (1949) 49; Grubov, Konsp. fl. MNR (1955) 246; Vved. in Fl. SSSR, 22 (1955) 737; Tsoong in Fl. R. P. Sin. 68 (1963) 120; Fl. Kazakhst. 8 (1965) 129. —*P. crassicaulis* Veniot ex Bonati in Bull. Ac. Geogr. Bot. 13 (1904) 241; Limpr. in Feddes repert. 20 (1924) 239. —*P. galeobdolon* Diels in Bot. Jahrb. 36, Beibl. 82 (1905) 96. —Ic.: Maxim. l.c. (1888) tab. IV, fig. 121; Fl. Kazakhst. 8, Plate XIII, fig. 8.

Described from Siberia. Type in London (Linn.).

Moist and marshy meadows, in willow and birch groves, along banks of brooks and swamps in forest and subalpine belts, 300–1600 m.

IA. **Mongolia:** *Cis-Hing.* (left bank tributary of Numurygin-Gol river, Toge-Gol river valley, marshy meadow, Aug. 9, 1949—Yun.). *Cent. Khalkha* (Dzhargalante river basin, Kharukhe river sources, Uste mountain, subalpine zone, willow scrubs on northern slopes and at top, Aug. 12, 1925—Krasch. and Zam.). *East. Mong.* (Muni-Ula hills, 1871—Przew.; Ul'gen-Gol river, moist meadow, July 24; Duchin-Gol river, Aug. 7—1899, Pot. and Sold.). *Bas. Lakes* (environs of Ubsa lake, Ulangom area, forest meadows along river banks, July 3, 1879—Pot.; south. slope of Tannuol, Kherul'ma river, Aug. 1, 1903—Gr.-Grzh.; Kharkhira river valley, birch grove, Sept. 1, 1931—Bar. and Shukh.). *Gobi-Alt.* (Dzun-Saikhan hills, top of Yailo creek valley, among willow and birch groves on slope, Aug. 29, 1931—Ik.-Gal.).

**General distribution:** Jung.-Tarb. (Tarb.); Europe (Urals), West. Siberia, East. Siberia, Far East, Nor. Mongolia, China (Dunbei, North, East, South-West, Cent., South), Korean peninsula, Japan.

## Section Lasioglossa Li

61. **P. ingens** Maxim. in Bull. Ac. Sci. St.-Pétersb. 32 (1888) 565; Forbes and Hemsley, Index Fl. Sin. 2 (1902) 210; Limpr. in Feddes repert. 20 (1924) 230; Marquand in J. Linn. Soc. London (Bot.) 48 (1929) 212; Rehder and Kobuski in J. Arn. Arb. 14 (1933) 33; Pai in Contribs Inst. Bot. Nat. Ac. Peiping, 2, 7 (1934) 212; Hao in Engler's Bot. Jahrb. 68 (1938) 637; Li in Proc. Ac. Natur. Sci. Philad. 101 (1949) 68; Tsoong in Fl. R. P. Sin. 68 (1963) 50. —Ic.: Maxim. l.c. tab. III, fig. 61.

Described from South-West China (Sichuan province). Type in Leningrad.

Alpine meadows, rocky slopes and rocks and montane steppes, 3400–4000 m.

IIIA. **Qinghai:** *Nanshan* (108 km west of Xining town and 6 km west of Daudankhe settlement, montane scrub steppe, 3400 m, Aug. 5, 1959—Petr.). *Amdo* ("Radja and Yellow River gorges"—Rehder and Kobuski, l.c.; "grasslands between Labrang and Yellow River; on rocky slopes near Htsechu, place called Sengle Kanchak, alt. 3550 m, No. 14481, July 1926, Rock"—Li, l.c.; "Kokonor, Dahoba, 4000 m, No. 1177, Sept. 7; Shalakutu [Schalakutu], 3400 m, No. 864, Aug. 18, 1930, Hao"—Pai, l.c.; Hao, l.c.).

**IIIB. Tibet:** *Weitzan* (Yantszytszyan river basin, environs of Kabchzha-Kamba village, 3700 m, July 21; same site, slopes of Ichu river valley, height 3800 m, July 28, 1900—Lad.).
General distribution: China (South-West).

62. **P. lasiophrys** Maxim. in Bull. Ac. Sci. St.-Pétersb. 24 (1877) 68, 32 (1888) 564; Kanitz in Szechenyi, Wissensch. Ergebn. 2 (1898) 723; Forbes and Hemsley, Index Fl. Sin. 2 (1902) 211; Limpr. in Feddes repert. 20 (1924) 230; Rehder and Kobuski in J. Arn. Arb. 14 (1933) 33; Pai in Contribs Inst. Bot. Nat. Ac. Peiping, 2, 7 (1934) 214; Li in Proc. Ac. Natur. Sci. Philad. 101 (1949) 79; Tsoong in Fl. R. P. Sin. 68 (1963) 57. —Ic.: Maxim. l.c. (1888) tab. III, fig. 57; Fl. R. P. Sin. 68, tab. V, figs. 3–5.
Described from Qinghai. Type in Leningrad.
Alpine meadows, 2750–5000 m.

**IIIA. Qinghai:** *Nanshan* (South Tetungsk mountain range, alpine meadows, common, July 25, 1872—Przew.; typus!; between Nanshan and Donkyr ranges; in alpine belt, common, 3050 m, July 22; along Yusun-Khatyma river, common 2750 m, July 23—1880, Przew.).
**IIIB. Tibet:** *Weitzan* (nor. slope of Burkhan-Budda mountain range, alpine belt, marshy site, 4750–5200 m, Aug. 13, 1884—Przew.; Yantszytszyan river basin, on Khichu river, alpine meadows, humus soil, 4300 m, July 11; nor. slope of Burkhan-Budda mountain range, Khatu gorge, alpine belt in dense meadows on humus and clayey soil, 4000–4300 m, July 17—1900, Lad.; Amnyi Machen range, moist alpine meadows of Mt. Druggu, alt. 4000 m, No. 14439, July 1926—Rock).
General distribution: China (North-West, South-West).

63. **P. retingensis** Tsoong in Acta Phytotax. Sin. 3, 3 (1954) 305, 329; id. in Bull. Brit. Mus. (Bot.) 2, 1 (1955) 26; id. in Fl. R. P. Sin. 68 (1963) 59.
Described from Tibet. Type in London (K).
Dry rocky slopes, about 4300 m alt.

**IIIB. Tibet:** *South.* ("Reting, 60 miles north of Lhasa, 4300 m, on dry stony hillsides, No. 11060, July 24, 1944, Ludlow and Scherriff, typus"—Tsoong, l.c. [1954]).
General distribution: endemic.

64. **P. rudis** Maxim. in Bull. Ac. Sci. St.-Pétersb. 24 (1877) 67, 32 (1888) 568; Diels in Bot. Jahrb. 29 (1900) 572; Forbes and Hemsley, Index Fl. Sin. 2 (1902) 215; Limpr. in Feddes repert. 20 (1924) 228; Rehder and Kobuski in J. Arn. Arb. 4 (1933) 34; Pai in Contribs Inst. Bot. Nat. Ac. Peiping, 2, 7 (1934) 219; Hand.-Mazz. Symb. Sin. 7 (1936) 857; Walker in Contribs U.S. Nat. Herb. 28 (1941) 659; Li in Proc. A. Natur. Sci. Philad 101 (1949) 64; Tsoong in Fl. R. P. Sin. 68 (1963) 43. —Ic.: Maxim. l.c. (1888) tab. III, fig. 63; Fl. R. P. Sin. 68, tab. IV, figs. 4–6.
Described from Inner Mongolia. Type in Leningrad.
Grassy slopes, spruce forests, scrub, 2350–3350 m.

**IA. Mongolia:** *Alash. Gobi* (west. extremity of midportion of Alashan mountains, in forest gorge, common, July 12, 1873—Przew., lectotypus!).
**IIIA. Qinghai:** *Nanshan* (South Tetung mountain range, upward along Tetung river, in valley, common, July 11; around Cheibsen temple, in belt of low shrubs, common, July 19, 1872—Przew.; same site, 2600 m, July 29; on Yusun-Khatyma river 2750 m, among shrubs,

alpine to forest [deciduous forest] belt, July 23—1880, Przew.; "Ch'ing Kang Yai [Ping Fan Hsien], in woods, Ching"—Walker, l.c.).

General distribution: China (North-West, South-West).

Note. Specimen marked "n. sp." by K.I. Maximovicz was selected as lectotype.

65. **P. sceptrum-carolinum** L. Sp. pl. (1753) 608; Maxim. in Bull. Ac. Sci. St.-Pétersb. 24 (1877) 84, 32 (1888) 614; Limpr. in Feddes repert. 20 (1924) 199; Kryl. Fl. Zap. Sib. 10 (1939) 2526; Kitag. Lin. Fl. Mansh. (1939) 395; Li in Proc. Ac. Natur. Sci. Philad. 101 (1949) 56; Grubov, Konsp. fl. MNR (1955) 246; Vved. in Fl. SSSR, 22 (1955) 793; Tsoong in R. P. Sin. 68 (1963) 36; Fl. Kazakhst. 8 (1965) 145. —*P. pubescens* Pai in Contribs Inst. Bot. Nat. Ac. Peiping, 2, 5 (1934) 125, 2, 7 (1934) 216. —**Ic.:** Reichb. Ic. fl. Germ. tab. 1763; Fl. Kazakhst. 8, Plate XVI, fig. 6.

Described from Europe. Type in London (Linn.).

Wet and marshy meadows, marshy banks of rivers and swamps.

IA. Mongolia: *East. Mong.* (Argun' district, near Tszilalin' village, wet meadow in river valley, 600 m, No. 1578, June 27, 1951—Wang).

General distribution: Arct., Europe, West. and East. Siberia, Far East, Nor. Mongolia (Hent.), China (Dunbei, North), Korean peninsula, Japan.

66. **P. trichoglossa** Hook. f. Fl. Brit. Ind. 4 (1884) 310; Maxim. in Bull. Ac. Sci. St.-Pétersb. 32 (1888) 566; Limpr. in Feddes repert. 20 (1924) 222, 23 (1927) 336; Marquand in J. Linn. Soc. London (Bot.) 48 (1924) 214; Bonati in Notes Bot. Gard. Edinburgh, 15 (1926) 157, 17 (1929) 87; Pai in Contribs Inst. Bot. Nat. Ac. Peiping, 2, 7 (1934) 222; Hand.-Mazz. Symb. Sin. 7 (1936) 857; Li in Proc. Ac. Natur. Sci. Philad. 101 (1949) 74; Tsoong in Fl. R. P. Sin. 68 (1963) 55. —**Ic.:** Maxim. l.c. tab. III, fig. 63.

Described from Himalayas. Type in London (K).

Rocky forest slopes, 3550–5000 m.

IIIB. Tibet: *Weitzan* (Goluboi [Yangtze] river basin, Bounchin rocks, Darindo area, near Chzherku temple, among willow bushes, 3550 m, Aug. 8, 1900—Lad.).

General distribution: China (South-West), Himalayas.

67. **P. tristis** L. Sp. pl. (1753) 608; Maxim, in Bull. Ac. Sci. St.-Pétersb. 32 (1888) 567; Forbes and Hemsley, Index Fl. Sin. 2 (1902) 218; Sapozhn. Mong. Alt. (1911) 381; Limpr. in Feddes repert. 20 (1924) 227; Rehder and Kobuski in J. Am. Arb. 14 (1933) 34; Pai in Contribs Inst. Bot. Nat. Ac. Peiping, 2, 7 (1934) 223; Kryl. Fl. Zap. Sib. 10 (1939) 2525; Li in Proc. Ac. Natur. Sci. Philad. 101 (1949) 61; Grubov, Konsp. fl. MNR (1955) 247; Vved. in Fl. SSSR, 22 (1955) 736; Tsoong in Fl. R. P. Sin. 68 (1963) 40; Fl. Kazakhst. 8 (1965) 128. —**Ic.:** Maxim. l.c. tab. III, fig. 65; Fl. Kazakhst. 8, Plate XIII, fig. 7.

Described from Siberia. Type in London (Linn.).

Marshy meadows, swamped larch forests, willow groves, along banks of rivers and near springs, in forb-feather grass and scrub steppes, 2000–3050 m.

**IA. Mongolia:**_Khobd._ (Dzusylan and Saryngi, July 15; Kharkhira river valley, near mountain top, July 21; near Kharkhira river sources, pebble bed, July 22—1879, Pot.). _Mong. Alt._ (Tsagan-Gol, upper camp near Kholodnyi spring, rocks, July 30, 1905—Sap.; B. Khobdosskoe lake, forest meadow, June 27–30, 1906; Oiguriin-Gol river valley [Oigur], July 15, 1909—Sap. [TK]; Taishiri-Ula mountain range, 10 km south-east of Yusun-Bulak, between nor. trails, forb-wheat grass-feather grass steppe, July 14, 1947—Yun.). _Cent. Khalkha_ (Dzhargalante river basin, Kharukhe river sources [Ara-Dzhargalante river], subalpine zone, willow groves, Aug. 12, 1925—Krasch. and Zam.).

**IIIA. Qinghai:** _Amdo_ (Mudzhik mountain, alpine belt, 3500 m, June 22; same site, June 27—1880, Przew.).

**IIIB. Tibet:** _Weitzan_ ("Amnyi Machen range, Rock"—Rehder and Kobuski, l.c.).

**General distribution:** Jung.-Tarb.; Arct. (Asian); West. and East. Siberia, Far East, Nor. Mongolia (Fore Hubs, Hent., Hang.), China (North, North-West, South-West).

## Section Botryantha Li

68. **P. albertii** Regel in Acta Horti Petrop. 6 (1879) 353; Maxim. in Bull. Ac. Sci. St.-Pétersb. 32 (1888) 617; Limpr. in Feddes repert. 20 (1924) 203; Vved. in Fl. SSSR, 22 (1955) 786; Fl. Kirgiz. 10 (1962) 222; Fl. Kazakhst. 8 (1965) 144.   —Ic.: Maxim, l.c. tab. VII, fig. 175; Fl. SSSR, 22, Plate XXXIX, fig. 2; Fl. Kazakhst. 8, Plate XVI, fig. 5.

Described from Nor. Tien Shan (environs of Alma-Ata town). Type in Leningrad.

Spruce and deciduous forests, 900–3050 m.

**IIA. Junggar:** _Tien Shan_ (Dzhirgalan river [tributary of lower Tekes], Nov. 1876; near sources of Dzhirgalan and Pilyuchi rivers, 1800 m, April 24; Pilyuchi river gorge, 2100–2400 m, April 26; Taldy river, 2400–2750 m, May 17; same site, 2750–3050 m, May 20—1879, A. Reg.; in Bogdo-Ula mountains, near lake, grassy spruce grove, in grassland on granites, April 26, 1959—Yun. et al.).

**General distribution:** Nor. and Cent. Tien Shan.

69. **P. muscoides** Li in Proc. Ac. Natur. Sci. Philad. 101 (1949) 91; Tsoong in Acta Phytotax. Sin. 3, 3 (1954) 321; id. in Fl. R. P. Sin. 68 (1963) 309.  — Ic.: Li, l.c. tab. 8, fig. 151; Fl. R. P. Sin. 68, tab. LXIX, figs. 1–4.

Described from South-West China (Sikang province). Type in London (K). Isotype in Philadelphia.

Alpine belt, 3950–5350 m.

**IIIB. Tibet:** _South._ ("Nyenchen-tang La, 4300 m, No. 9680, June 13; hills of Lhasa, 4300 m, No. 9549, 9578, May–June 14, 1943, Ludlow and Scherriff"—Tsoong, l.c. [1954]).

**General distribution:** China (South-West).

70. **P. oederi** Vahl in Hornem. Dansk. Pl. ed. 2 (1806) 380; Prain in Ann. Bot. Gard. Calc. 3 (1890) 181; Alcock, Rep. nat. hist. results Pamir boundary commiss. (1898) 25; Hemsley, Fl. Tibet (1902) 193; Paulsen in Hedin, S. Tibet (1922) 44; Limpr. in Feddes repert. 20 (1924) 202; Marquand in J. Linn. Soc. London (Bot.) 48 (1929) 213; Kryl. Fl. Zap. Sib. 10 (1939) 2524; Hurus.

in J. Jap. Bot. 22 (1948) 73; Li in Proc. Ac. natur. Sci. Philad. 101 (1949) 86; Tsoong in Acta Phytotax. Sin. 3, 3 (1954) 316; id. in Fl. R. P. Sin. 68 (1963) 331; Grubov, Konsp. fl. MNR (1955) 246; Vved. in Fl. SSSR, 22 (1955) 785; Fl. Kirgiz. 10 (1962) 221; Ikonnikov, Opred. rast. Pamira (1963) 220; Fl. Kazakhst. 8 (1965) 142. —*P. versicolor* Wahlenb. Veg. Helvet. (1813) 118; Maxim. in Bull. Ac. Sci. St.-Pétersb. 24 (1877) 88, 32 (1888) 618; Hemsley in J. Linn. Soc. London (Bot.) 30 (1894) 138; Forbes and Hemsley, Index Fl. Sin. 2 (1902) 219; Sapozhn. Mong. Alt. (1911) 381; Limpr. in Beih. Feddes repert. 12 (1922) 485; Rehder and Kobuski in J. Arn. Arb. 14 (1933) 34; Pai in Contribs Inst. Bot. Nat. Ac. Peiping, 2, 7 (1934) 223; Kung et Wang, ibid. 2, 8 (1934) 384; Persson in Bot. notiser (1938) 302. —**Ic.:** Maxim. l.c. (1888) tab. VI, fig. 177a, b; Fl. R. P. Sin. 68, tab. LXXV, figs. 1–5.

Described from Norway. Type in Copenhagen.

Alpine marshy meadows, rubble and melkozem slopes of mountains and shaded moist sites in forest, 2600–4000 m.

IA. **Mongolia:** *Khobd.* (Tszusylan, in forest, July 13, 1879—Pot.). *Mong. Alt.* (Tsagan-Gol river, rocky crests around glaciers, on talus, July 1, 1905—Sap. [TK]).

IB. **Kashgar:** *Nor.* (Uchturfan, June 18, 1909—Divn.). *West.* (Sarykol' mountain range, west of Kashgar, Bostan-Terek area, July 11, 1929—Pop.; nor. slope of King-Tau mountain range, 4 km south-east of Kosh-Kulak settlement, upper forest belt, juniper groves, June 10, 1959—Yun.).

IIA. **Junggar:** *Tien Shan.*

IIIA. **Qinghai:** *Nanshan* (North Tetungsk mountain range, alpine meadows, rare, on wet soil, June 4, 1872; along Dankhe river, 3350 m, July 18; Nanshan alps—1879, Przew.; along lower Tashitu river, 2900 m, June 27, 1866—Pot.; Humboldt mountain range [Argalin-Ula], alpine meadow, 3050 m, June 3, 1894—Rob.; Nanshan range, common below pass of mt. Kuang kl., alt. 3350 m, No. 12413, June 1925—Rock). *Amdo* (Syan'sibei hills, grassy slope, May 28; mountains along Mudzhik river valley, 3200 m, June 24—1880, Przew.; Radja and Yellow River gorges, grassy alpine meadows south of river valley, road to Ngolok country, alt. 3350 m, May 25; alpine region between Radja and Jupar range; meadows above Woti la, alt. 4400 m, July—1926, Rock).

IIIB. **Tibet:** *Chang Tang* (Keriya [Russky] mountain range, Kyuk-Egil' river gorge, July 11, 1885—Przew.; N. Tibet, between Camp 17 [37°01' N. lat., 90°01' E. long.] and 18, 4175 m, July 31, 1900"—Hedin, l.c.; "Tibet, Camp 94, 35°39' N. lat., 82°E. long., 4900 m, July 25, 1898"—Deasy, l.c.). *Weitzan* (mountain peak between Huang He and Yangtze rivers, on rocks, common, June 12; Yangtze river bank, in alpine belt, June 20—1884, Przew.; in isthmus separating Russkoe lake from Ekspeditsii lake, on humus, in willow groves, 4150 m, June 28, 1900; Yantszytszyan river basin, Makhmukhchu river valley and its right tributary, wet meadows, in humus, 4150–4300 m, May 21; Huang He river basin, Russkoe lake and Sergchu river, sandy-rocky banks of rivers and lake, 4150 m, May 27—1901, Lad.; "Valley of Murus, at 4750 m, 33°44' N. lat., 91°18' E. long., June 23, 1892, Rockhill"—Hemsley, l.c. [1902]). *South.* ("hills north of Lhasa, 4600 m, No. 8740, June 24, 1941; hills south of Lhasa, 4300 m, June 19, 1944, Ludlow and Scherriff"—Tsoong, l.c. [1954]; "S.W. Tibet, Camp 211, Tokchen, east of lake Manasarovar [30°44' N. lat., 81°42' E. long.], alt. 4634 m, Aug. 24, 1907"—Hedin, l.c.).

IIIC. **Pamir** (Billuli river, at confluence with Chumbus river, June 12, 1909—Divn.; Kokat pass, along sources of Yazag and Bolung rivers, 3200–4000 m, June 16; Pil'nen gorge 4500–5000 m, July 1; moraine watershed between Atrakyr and Tyuzutek rivers, mossy tundra, 4500–

5000 m, July 20—1942, Serp.; "marshy land along Aksu river, 4000–4300 m, No. 17750"—Alcock, l.c.).

General distribution: Jung.-Tarb., Nor. and Cent. Tien Shan; Arct., Europe, West. and East. Siberia, Far East, Nor. Mongolia (Fore Hubs., Hent., Hang.), China (North, North-West, South-West), Himalayas (west., east.), North America.

71. **P. przewalskii** Maxim. in Bull. Ac. Sci. St.-Pétersb. 24 (1877) 55, 32 (1888) 528; Hemsley in J. Linn. Soc. London (Bot.) 30 (1894) 138; id. in Kew Bull. 119 (1896) 213; Forbes and Hemsley, Index Fl. Sin. 2 (1902) 214; Hemsley, Fl. Tibet (1902) 194; Limpr. in Feddes repert. 20 (1924) 254, p.p.; Rehder and Kobuski in J. Arn. Arb. 14 (1933) 34; Li in Proc. Ac. Natur. Sci. Philad. 101 (1949) 112; Tsoong in Acta Phytotax. Sin. 3, 3 (1954) 317; id. in Bull. Brit. Mus. (Bot.) 2, 1 (1955) 6; id. in Fl. R. P. Sin. 68 (1963) 351. —*P. microphyton* Bur. et Franch. in J. Bot. (Paris) 5 (1891) 107; Bonati in Bull. Soc. Bot. France, 54 (1907) 184; 55 (1908) 244; Limpr. in Beih. Feddes repert. 12 (1922) 484; id. in Feddes repert. 20 (1924) 255; Hand.-Mazz. Symb. Sin. 7 (1936) 865. —*P. microphyton* var. *purpurea* Bonati in Bull. Soc. Bot. France, 55 (1908) 244. — *P. coppeyi* Bonati in Bull. Soc. Bot. France, 57 (1911) Sess. extr. 58; Limpr. in Feddes repert. 20 (1924) 255. —**Ic.:** Maxim. l.c. (1888), tab. I, fig. 2; Fl. R. P. Sin. 68, tab. LXXXI, figs. 3–4.

Described from Qinghai. Type in Leningrad.

Alpine wet meadows, 3000–4400 m.

IIIA. **Qinghai:** *Nanshan* (South Tetung mountain range, alpine meadows, high-altitude belt, common, July 13, 1872—Przew., typus!; South Tetungsk mountain range, alpine belt, 3050–3650 m, July 31, 1880—Przew.).

IIIB. **Tibet:** *Weitzan* (Dyaochu river, marshy bank, rare, July 11, 1884—Przew.; left bank of Dzhagyn-Gol river, in hummocky marshes, 4150 m, July 3; Yantszytszyan river basin, Khichu river, wet and moist alpine meadows, 4300 m, July 12—1900, Lad.; "basin of Suchu, valley north side, Draya'lamo pass, at 4300 m, 31°52' N. lat., 93°17' E. long., Aug. 2, 1892, Rockhill"—Hemsley, l.c. [1894]; "Amnyi Machen range, Rock"—Rehder and Kobuski, l.c.). —*South.* ("Gooring valley, 30°12' N. lat., 92°25' E. long., 5050 m, Littledale"—Hemsley, l.c. [1902]; "Reting, 60 miles north of Lhasa, 4350 m, July 24, 1942; hills of Lhasa, 4000–4300 m, No. 9707, June 35; [sic]; hills of Lhasa, Sha La, 4300 m, July 11—1943, Ludlow and Scherriff"—Tsoong, l.c. [1954]).

General distribution: China (North-West, South-West).

72. **P. rhinanthoides** Schrenk in Fisch. et Mey. Enum. pl. nov. 1 (1841) 22; Maxim. in Bull. Ac. Sci. St.-Pétersb. 32 (1888) 527; Prain in J. As. Soc. Bengal, 58, 2 (1889) 272; id. in Ann. Bot. Gard. Calc. 3 (1890) 110; Hemsley in Kew Bull. 119 (1896) 213; Alcock, Rep. nat. hist. results Pamir boundary commiss. (1898) 25; Hemsley, Fl. Tibet (1902) 195; Danguy in Bull. mus. nat. hist. natur. 14 (1908) 132; Sapozhn. Mong. Alt. (1911) 380; Bonati in Bull. Soc. Bot. Genève, 2, 5 (1913) 113; Pampanini, Fl. Carac. (1930) 191; Pai in Contribs Inst. Bot. Nat. Ac. Peiping, 2, 7 (1934) 218; Hao in Bot. Jahrb. 68 (1938) 637; Kryl. Fl. Zap. Sib. 10 (1939) 2500; Walker in Contribs U.S. Nat. Herb. 18 (1941) 659; Li in Proc. Ac. Natur. Sci. Philad. 101 (1949) 126; Tsoong in Acta Phytotax. Sin. 3, 3 (1954) 325; id. in Fl. R. P. Sin. 68 (1963) 262; Grubov,

Konsp. fl. MNR (1955) 246; Vved. in Fl. SSSR, 22 (1955) 700; Fl. Kirgiz. 10 (1962) 200; Ikonnikov, Opred. rast. Pamira (1963) 221; Fl. Kazakhst. 8 (1965) 118. —*P. labellata* Jacq. Voy. Inde, Bot. (1844) 118; Maxim. in Bull. Ac. Sci. St.-Pétersb. 24 (1877) 54, 32 (1888) 532; Forbes and Hemsley, Index Fl. Sin. 2 (1902) 210; Bonati in Notes Bot. Gard. Edinburgh, 5 (1911) 79, 7 (1912) 81; Limpr. in Feddes repert. 20 (1924) 248; Pai in Contribs Inst. Bot. Nat. Ac. Peiping, 2, 7 (1934) 213. —**Ic.**: Prain, l.c. (1890) tab. 1, fig. A-B; Fl. R. P. Sin. 68, tab. LIX, figs. 1–4; Fl. Kazakhst, 8, Plate XIII, fig. 1.

Described from Jung. Alatau (Baskan river). Type in Leningrad.

Moist and marshy meadows, swamps and marshy placers at foot of slopes, floodplains of rivers, banks of lakes and meanders, near springs, 3000–5000 m.

**IA. Mongolia:** *Mong. Alt.* (Kalgutty river valley [Dzhirgalanty], steppe slopes, July 8, 1905—Sap.; Belaya Kobdo [Aksu] river source, alpine tundra, June 29, 1906—Sap. [TK]; "upper Tsagan-gol"—Sapozhn. l.c.).

**IB. Kashgar:** *West.* (Taret pass, marsh of brook or spring overflow, several, June 9, 1909—Divn.).

**IIA. Junggar:** *Jung. Alt.* (Urtaksary, July 20, 1878—Fet.). *Tien Shan* (south. bank of Sairam-Nur lake, 2100–2750 m, July 20, 1877; Dzhagastai, 2450–2750 m, June 20, 1878; Aryslyn, 2750–3050 m, July 16; Mengute, 2750 m, Aug. 2—1879, A. Reg.; Urten-Muzart, Aug. 2, 1877—Fet.; Muzart [1886]—Krasnov; Muzart river upper valley, July 1907—Merzbacher; Manas river basin, Ulan-Usu river upper valley at ascent to Danu-Daban pass, marshy grassland near base of slope, high-altitude belt, July 19; same site, Danu-Gol river upper course, at ascent to Se-Daban pass, sedge marshy meadow at foot of slope, high-altitude belt, July 21—1957, Yun. et al.; Kutul' pass [3200 m], along road to Karashar from M. Yuldus basin, sedge marshy meadow, high-altitude belt, No. 645, Aug. 15, 1958—Lee et al.; "Tien Shan, Kensu Valley, Upper Koksu, Tekkes, 3350 m, No. 717, July 15, 1930, Ludlow"—Tsoong, l.c. [1954]).

**IIIA. Qinghai:** *Nanshan* (South-Kukunor mountain range, alpine meadow, July 7, 1872; Nanshan alps, Kuku-Usu river midcourse, 3350 m, July 15; same site, 600 m, July 26—1879, between Nanshan and Donkyr, on Raka-Gol river, alps, 3050–3350 m, July 21, 1880— Przew.; Yamatyn-Umru mountains, alpine meadow, 4000 m, June 17; Humboldt mountain range [Argalin-Ula], near springs on alpine meadows, 3650–4000 m, June 30—1894, Rob.; "Liu Fu Jai, P'ing Fan Hsien, on moist grasslands; Lang Tzu T'and Kou, in dense shaded woods, 1923, Ching"—Walker, l.c.). —*Amdo* ("alpine region between Radja and Jupar ranges, Rock"— Rehder and Kobuski, l.c.).

**IIIB. Tibet:** *Weitzan* (south. bank of Russkoe lake, on wet sandy soil, common, July 27; south. slope of Burkhan-Budda mountain range, alpine belt, 4400 m, Aug. 13—1844, Przew.; Yantszytszyan river basin, Chzhabu-Vrun area, on humus, 4300 m, July 11, 1900; nor. slope of Burkhan-Budda mountain range, Khatu-Gol gorge, on moist grasslands, 3200 m, July 1, 1901— Lad.). *South.* ("Gooring valley, 30°15' N. lat., 90°25' E. long., 5050 m, Littledale"—Hemsley, l.c. [1896]; "Reting, 60 miles north of Lhasa, No. 8912, July 28; same loc., No. 8938, July 18—1942; same loc. 4000 m, No. 11011, July 15, 1944, Ludlow and Scherriff"—Tsoong, l.c. [1954]).

**IIIC. Pamir** (Issyk-Su river, riverine meadows, 3100 m, July 3; Taspestlyk area, 4000–5000 m, July 25—1942, Serp.; "in marshy land along Aksu river, common at places, 4000–4300 m, No. 17749"—Alcock, l.c.; "terrains rocailleux, conterforts du Mouz-tag-Ata, alt. 4300 m, Sept. 30, 1906, Lecomte"—Danguy, l.c.).

**General distribution:** Jung.-Tarb., Nor. and Cent. Tien Shan, East. Pam.; Fore Asia (Afghanistan), Mid. Asia (West. Tien Shan and Pamiro-Alay), West. Sib. (south.), China (North-West, South-West), Himalayas.

## Section Macrostachys Li

73. **P. tibetica** Franch. in Bull. Soc. Bot. France, 47 (1900) 24; Limpr. in Beih. Feddes repert. 12 (1922) 483; id. in Feddes repert. 20 (1924) 246; Pai in Contribs Inst. Bot. Nat. Ac. Peiping 2, 7 (1934) 222; Li in Proc. Ac. Natur. Sci. Philad. 101 (1949) 137; Tsoong in Fl. R. P. Sin. 68 (1963) 253. —*P. dielsiana* Limpr. in Beih. Feddes repert. 12 (1922) 483. —*P. limprichtiana* Fedde in Feddes repert. 18 (1922) 122. —*P. ludovicii* Limpr. in Feddes repert. 20 (1924) 247. —**Ic.:** Li, l.c. tab. II, fig. 185.

Described from South-West China (Sichuan province). Type in Paris.
Alpine belt, 2500–4600 m.

**IIA. Junggar:** *Tien Shan* ("Shingkiang: Ouroumtai, Pai-yang-kow, along river, alt. 2500 m, No. 2919, July 30, 1931; Bogdo Ula, No. 3546, Liou" Pai, l.c.).
**General distribution:** China (South-West).

## Section Phanerantha Li

74. **P. elwesii** Hook. f. Fl. Brit. Ind. 4 (1884) 312; Maxim. in Bull. Ac. Sci. St.-Pétersb. 32 (1888) 532; Forbes and Hemsley, Index Fl. Sin. 2 (1902) 208; Bonati in Notes Bot. Gard. Edinburgh, 8 (1913) 37, 15 (1926) 164, 17 (1929) 66; Limpr. in Feddes repert. 20 (1924) 256; Marquand in J. Linn. Soc. London (Bot.) 48 (1929) 212; Hand.-Mazz. Symb. Sin. 7 (1936) 864; Li in Proc. Ac. Natur. Sci. Philad. 101 (1949) 144; Tsoong in Acta Phytotax. Sin. 3, 3 (1954) 325; id. in Fl. R. P. Sin. 68 (1963) 323. —**Ic.:** Maxim. l.c. tab. I, fig. 8; Fl. R. P. Sin. 68, tab. LXXIII, figs. 1–3.

Described from East. Himalayas. Type in London (K).
Alpine meadows, 3200–4300 m.

**IIIB. Tibet:** *South.* ("Nangtze, 20 miles west of Lhasa, 4400 m, No. 9833, July 30, 1943, Ludlow and Scherriff"—Tsoong, l.c. [1959]).
**General distribution:** China (North-West), Himalayas.

## Section Dolichomischus Li

75. **P. muscicola** Maxim. in Bull. Ac. Sci. St.-Pétersb. 24 (1877) 54, 32 (1888) 535; Hance in J. Linn. Soc. London (Bot.) 20 (1882) 292; Kanitz in Szechenyi, Wissensch. Ergebn. 2 (1898) 723; Forbes and Hemsley, Index Fl. Sin. 2 (1902) 213; Limpr. in Feddes repert. 20 (1924) 253; Rehder and Kobuski in J. Arn. Arb. 14 (1933) 34; Walker in Contribs U.S. Nat. Herb. 28 (1941) 659; Li in Proc. Ac. Natur. Sci. Philad. 101 (1949) 175; Tsoong in Fl. R. P. Sin. 68 (1963) 104. —*P. macrosiphon* auct. non Franch.: Rehder and Kobuski in J. Arn. Arb. 14 (1933) 33. —**Ic.:** Maxim. l.c. (1888) tab. I, fig. 13; Fl. R. P. Sin. 68, tab. XVIII, figs. 3–4.

Described from Inner Mongolia. Type in Leningrad.

Wet and shaded forests, meadows along river valleys, 1750–2650 m.

IA. Mongolia: *Alash. Gobi* (midsection of Alashan mountains, west. slope, among moss in forest, rare, July 2, 1873—Przew., lectotypus!; Alashan mountain range, Khote-Gol gorge, north-east, and north-west. slopes, on humus soil, June 17; same site, June 19—1908, Czet.; "T'u Er P'ing, Ho Lan Shan, in dense wet woods and swampy grasslands, Ching"—Walker, l.c.).

IIIA. Qinghai: *Nanshan* (South Tetunga mountain range, in coniferous forest, on wet mossy soil, common, July 7, 1872; between Nanshan and Donkyr mountain ranges, along Rako-Gol river, under moss cover, in coniferous forest, common 3000 m, July 21, 1880—Przew.; Tsilin'shan', 70 km south-east of Chzhan'e town, Matisy temple, thickets with grasslands in mountain valley, 2600 m, July 12, 1958—Petr.; "Altin-gomba in ditione Si-ning-fu, 2700 m, July 7, 1879, Szecheny"—Kanitz, l.c.). *Amdo* (Ba Valley No. 14279, Rock"—Rehder and Kobuski, l.c.).

General distribution: China (North, North-West, South-West).

Note. K.I. Maximovicz's analysed and sketched specimen is selected here as lectotype.

## Section Schizocalyx Li

76. P. armata Maxim. in Bull. Ac. Sci. St.-Pétersb. 24 (1877) 56, 32 (1888) 533; Forbes and Hemsley, Index Fl. Sin. 2 (1902) 205; Limpr. in Feddes repert. 20 (1924) 249; Rehder and Kobuski in J. Arn. Arb. 14 (1933)) 32; p.p.; Li in Proc. Ac. Natur. Sci. Philad. 101 (1949) 188; Tsoong in Fl. R. P. Sin. 68 (1963) 366. —*P. cranolopha* auct. non Maxim.: Pai in Contribs Inst. Bot. Nat. Ac. Peiping, 2, 7 (1934) 210, p.p. —Ic.: Maxim, l.c. (1888) tab. I, fig. 9; Fl. R. P. Sin. 68, tab. LXXXIV, figs. 1–2.

Described from Qinghai. Type in Leningrad.

Alpine meadows.

IIIA. Qinghai: *Nanshan* (South Tetung mountain range, in meadows, common, July 26, 1872—Przew., typus!).

General distribution: China (North-West, South-West).

77. P. chinensis Maxim. in Bull. Ac. Sci. St.-Pétersb. 24 (1877) 57, 32 (1888) 534, p.p.; Diels in Futterer, Durch Asien, 3 (1901) 20; Forbes and Hemsley, Index Fl. Sin. 2 (1902) 206; Danguy in Bull. Mus. nat. hist. natur. 5 (1911) 14; Limpr. in Beih. Feddes repert. 12 (1922) 483; id. in Feddes repert;. 20 (1924) 250, p.p.; Li in Proc. Ac. Natur. Sci. Philad. 101 (1949) 191; Tsoong in Fl. R. P. Sin. 68 (1963) 362. —*P. armata* auct. non Maxim.: Rehder and Kobuski in J. Arn. Arb. 14 (1933) 32, p.p.; Pai in Contribs Inst. Bot. Nat. Ac. Peiping, 2, 7 (1934) 209, p.p.; Walker in Contribs U.S. Nat. Herb. 28 (1941) 659. —Ic.: Fl. R. P. Sin. 68, tab. LXXXIII, figs. 4–6.

Described from Qinghai. Type in Leningrad.

Alpine meadows in river valleys, wet meadows and grasslands, marshy and wet pebble beds, 1700–2900 m.

IIIA. Qinghai: *Nanshan* (South Tetung mountain range, in gorges on moist pebble beds and usually in iris associations, July 7, 1872—Przew., lectotypus!; marshes along Kuku-Nor lake, usually 3600 m, July 13; along Yusun-Khatyma river, 2750 m, July 23—1880, Przew.; nor. bank of Kuku-Nor, mountains surrounding Ara-Gol river valley on right, July 20, 1890—Gr.-Grzh.; high Nanshan foothills [nor. slope of Rikhthofen mountain range], 55 km south of Chzhan'e town, small meadows in mountain valley, 2300 m, July 19, 1958—Petr.; "bei Kloster Kum-bum östlich von Sining-fu; am Küke-Nur, Futterer"—Diels, l.c.; "gol de Ta-Pan-chan, alt. 4000 m, July 10, 1908, Vaillant"—Danguy, l.c.; "Hsia Mi Jai [P'ing Fan Hsien], in large dense patches by shady streams in gorges, 1923, Ching"—Walker, l.c.). *Amdo* (hills in Mudzhik river valley, June 29, 1880—Przew.).

General distribution: China (North, North-West).

Note. K.I. Maximovicz's analysed and sketched specimen was selected as lectotype.

78. **P. cranolopha** Maxim. in Bull. Ac. Sci. St.-Pétersb. 24 (1877) 55, 32 (1888) 533; Forbes and Hemsley, Index Fl. Sin. 2 (1902) 207; Limpr. in Feddes repert. 20 (1924) 249; Rehder and Kobuski in J.Am.Arb. 14 (1933) 33; Pai in Contribs Inst. Bot. Nat. Ac. Peiping, 2, 7 (1934) 210; Li in Proc. Ac. Natur. Sci. Philad. 101 (1949) 186; Tsoong in Fl. R. P. Sin. 68 (1963) 358.    —*P. birostris* Bur. et Franch. in J. Bot. (Paris) 5 (1891) 107; Bonati in Bull. Herb. Boiss. sér. 2, 7 (1907) 541; Limpr. in Beih. Feddes repert. 12 (1922) 483; Pai l.c. 209.    — *P. garnieri* Bonati in Bull. Soc. Bot. France, 55 (1908) 243.    —**Ic.**: Maxim. l.c. (1888) tab. I, fig. 10; Hook. Ic. pl. 23 (1894) tab. 2208 A; Fl. R. P. Sin. 68, tab. LXXXII, fig. 4.

Described from Qinghai. Type in Leningrad.

Alpine meadows, about 3000 m alt.

IIIA. Qinghai: *Nanshan* (South Tetung mountain range, in valley and around Cheibsen-khit temple, July 13, 1872—Przew., typus!). *Amdo* (mountains in Mudzhik river valley, alpine meadow, 2750–3050 m, June 18, 1880—Przew.; "Radja and Yellow River gorges, mountains south-west of Radja, No. 14186, June 1926, Rock"—Rehder and Kobuski, l.c.).

General distribution: China (North-West, South-West, Sichuan—north).

79. **P. croizatiana** Li in Proc.Ac. Natur. Sci. Philad. 101 (1949) 187; Tsoong in Fl. R. P. Sin. 68 (1963) 357.    —*P. garnieri* auct. non Bonati: Tsoong in Acta Phytotax. Sin. 3, 3 (1954) 278, 318; id. in Bull. Brit. Mus. (Bot.) 2, 1 (1955) 32.    —**Ic.**: Li, l.c. tab. 14, fig. 218; Fl. R. P. Sin. 68, tab. LXXXII, figs. 1–3.

Described from South-West China (Sichuan province, Muli). Type in Philadelphia.

Coniferous forests and alpine meadows, 3700–4200 m.

IIIB. Tibet: *South.* ("Hills of Lhasa, 4000 m, No. 9908, Aug. 31, 1943; Reting, 4000 m, No. 11079, July 26, 1944, Ludlow and Scherriff"—Tsoong, l.c. [1954]).

General distribution: China (South-West).

80. **P. decorissima** Diels in Notizbl. Bot. Gart. Berlin. 10 (1930) 891; Rehder and Kobuski in J. Am. Arb. 14 (1933) 33; Pai in Contribs. Inst. Bot. Nat. Ac. Peiping 2, 7 (1934) 211; Li in Proc. Ac. Natur. Sci. Philad. 101 (1949) 192; Tsoong in Fl. R. P. Sin. 68 (1963) 367.    —**Ic.**: Li, l.c. tab. 15, fig. 222.

Described from Qinghai. Type in Berlin. Isotype in Philadelphia.

Alpine meadows, 2900–3500 m.

IIIA. Qinghai: *Amdo* ("Grasslands between Labrang and Yellow River, meadows between Hetso and Chinssu, 2900–3000 m, No. 14546, Aug. 1926—Rock, typus!; grassy slopes below Jobsha-nira, 3500 m, No. 14357, July 30, 1926, Rock"—Diels, l.c.).
General distribution: China (North-West, South-West).

81. P. longiflora Rudolph in Mém. Ac. Sci. St.-Pétersb. 4 (1811) 345; Maxim. in Bull. Ac. Sci. St.-Pétersb. 24 (1877) 56, 32 (1888) 317; Franch. in J. Bot. (Paris) 4 (1890) 317; Deasy, In Tibet and Chin. Turk. (1901) 398; Hemsley, Fl. Tibet (1902) 193; Sapozhn. Mong. Alt. (1911) 380; Paulssen in Hedin, S. Tibet, 6, 3 (1922) 44; Limpr. in Feddes repert. 20 (1924) 250, p.p.; Pampanini, Fl. Carac. (1930) 191; Pai in Contribs Inst. Bot. Nat. Ac. Peiping, 2, 7 (1934) 214; Hao in Engler's Bot. Jahrb. 68 (1938) 637; Kryl. Fl. Zap. Sib. 10 (1939) 2499; Li in Proc. Ac. Natur. Sci. Philad. 101 (1949) 189; Tsoong in Acta Phytotax. Sin. 3, 3 (1954) 278, 318; Grubov, Konsp. fl. MNR (1955) 245; Vved. in Fl. SSSR, 22 (1955) 699; Tsoong in Fl. R. P. Sin. 68 (1963) 364.    —*P. tubiflora* Fisch. in Mém. Soc. natur. Moscou, 3 (1812) 58.    —*P. chinensis* auct. non Maxim. (1877): Maxim. in Bull. Ac. Sci. St.-Pétersb. 32 (1888) 534 p.p.; Bonati in Bull. Soc. Bot. France, 54 (1907) 183; Limpr. in Feddes repert. 20 (1924) 250; p.p.; Rehder and Kobuski in J. Arn. Arb. 14 (1933) 33.    —Ic.: Rudolph. l.c. tab. 3; Fl. R. P. Sin. 68, tab. LXXXIII, figs. 1–3.

Described from East. Siberia (Lake Baikal). Type in Leningrad.

Wet and marshy alpine and subalpine meadows, 2700–5300 m.

IA. Mongolia: *Mong. Alt.* (N. Khobdos lake, on moist meadow bank of island, Aug. 2, 1899—Lad.; East. Sumdairyk river bank, between moraines, July 30, 1908—Sap.; Korumdy-Bulak river, rocky steppe, June 17; Dain-Gol lake, west. slope, July 27–29—1909, Sap. [TK]; "V. Khobdos lake, wet meadows"—Sapozhn. l.c.).
IIIA. Qinghai: *Nanshan* (nor. slope of South Kukunor mountain range, in marsh, near Naion-Khutun-Gol river, Aug. 2, 1894—Rob.; nor. bank of Kuku-Nor lake, mountains surrounding Ara-Gol river valley from right, July 20, 1890—Gr.-Grzh.; South Kukunor mountain range, Usubin-Gol river, on marshy banks, 3200 m, Aug. 16, 1901—Lad.; Xining town vicinity in marsh, July 30, 1909—Czet.; high Nanshan foothills, 65 km south-east of Chzhan'e town, small meadows in mountain valley, 2300 m, July 12; Mon"yuan', Tetung river basin, Ganshiga river valley, 3350–3720 m—Sining, alt. 2900 m, No. 800, Aug. 3, 1930, Hopkinson"— Pai, l.c.).    —*Amdo* ("Ba Valley, Rock"—Rehder and Kobuski, l.c.).
IIIB. Tibet: *Chang Tang* ("33° N. lat., 82°53' E. long., Camp 37, 4600 m, Aug. 30, 1896, Deasy and Pike"—Deasy, l.c.; "Valley of Gugé, 4600 m, Strachey et Winterbottom"—Hemsl. l.c. [1902]; "Camp 211, Tokchen, east of lake Manasarovar, 4654 m, July 24, 1907, Hedin"— Paulsen, l.c.). *Weitzan* (south. bank of Ekspeditsii [Dzharin-Nor] lake, in silty-sandy soil, 4150 m, Aug. 6; [on south. bank of Russkoe lake], 3500–4000 m alt., in silty soil, Aug. 14—1884, Przew.; Yantszytszyan river basin, Donra area, on Dynechu [Khichu] river, hill slopes on humus, 4000 m, July 17, 1900—Lad.; "Zufluss des Noh-Zo, grasiges und versumpftes Tal, 4550 m, Aug. 25, 1906"—Hand.-Mazz. l.c.). *South.* ("Gyantze No. 83, July 30, 1924, Ludlow; Lhasa, 3600 m, No. 8778, July 2, 1942; same loc., 3650 m, Aug. 7, 1943; Reting, No. 8868, July 24, 1942; same loc., 4300 m, July 24, 1944, Ludlow and Scherriff"—Tsoong, l.c. [1954]).
General distribution: West. Sib. (Altay), East. Sib., Nor. Mong. (Fore Hubs, Hent., Hang.), China (North, North-West, South-West), Himalayas (west., east.).

82. **P. siphonantha** Don. Prodr. Fl. Nepal. (1825) 95; Hance in J. Bot. (London) 20 (1882) 292; Franch. in Nouv. Arch. Mus. hist. natur. Paris, sér. 2, 10 (1888) 184; Maxim. in Bull. Ac. Sci. St.-Pétersb. 32 (1888) 534; Prain in Ann. Bot. Gard. Calc. 3 (1890) 113; Kanitz in Szechenyi, Wissenschaft. Ergebn. 2 (1898) 722; Forbes and Hemsley, Index Fl. Sin. 2 (1902) 216; Bonati in Bull. Herb. Boiss. sér. 2, 7 (1907) 541; id. in Bull. Soc. Bot. France, 54 (1907) 183; id. in Notes Bot. Gard. Edinburgh, 5 (1911) 79, 7 (1912) 81, 8 (1913) 37; Limpr. in Beih. Feddes repert. 12 (1922) 484; id. in Feddes repert. 20 (1924) 251, 23 (1927) 338; Marquand in J. Linn. Soc. London (Bot.) 48 (1929) 214; Pai in Contribs Inst. Bot. Nat.Ac. Peiping, 2, 7 (1934) 219; Tsoong in Acta Phytotax. Sin. 3, 3 (1955) 319; id. in Fl. R. P. Sin. 68 (1963) 374.   —*P. delavayi* Franch. ex Maxim. in Bull. Ac. Sci. St.-Pétersb. 32 (1888) 531; Forbes and Hemsley, Index Fl. Sin. 2 (1902) 208; Bonati in Bull. Herb. Boiss. sér. 2, 7 (1907) 541; id. in Notes Bot. Gard. Edinburgh, 5 (1911) 80, 7 (1912) 149, 8 (1913) 37, 15 (1926) 163; Limpr. in Beih. Feddes repert. 12 (1922)) 483; id. in Feddes repert. 20 (1924) 249, 23 (1927) 338; Li in Proc.Ac. Natur. Sci. Philad. 101 (1949) 197.   —**Ic.:** Prain, l.c. tab. 2, A-B; Maxim. l.c. tab. I, fig. 7 (sub nom. *P. delavayi*); Curtis's Bot. Mag. 157 (1934) tab. 9367 (sub nom. *P. delavayi*).

Described from East. Himalayas. Type in London (K).
Alpine meadows, 3000–4600 m.

IIIA. Qinghai: *Nanshan* ("Koko-nor"—Hance, l.c.; "ad ripas lacus Kuku-nor specimen unicum invenile alabastris, Aug. 4, 1879"—Kanitz, l.c.).

General distribution: China (South-West), Himalayas.

## 19. Cymbaria L.
Sp. pl. (1753) 618.

1. Plant whitish-grey due to long silky villous pubescence; leaves (1) 2–5 mm broad. Corolla 3.5–6 cm long, 2.5–3 times longer than calyx; calyx teeth slender long appressed hairs. ...... 1. **C. dahurica** L.
2. Plant green, with short pubescence or subglabrous; leaves 0.5–1 (2) mm broad. Corolla 3–4 cm long, 0.75–2 times longer than calyx; calyx teeth with stiff ciliate pubescence ... 2. **C. mongolica** Maxim.

1. **C. dahurica** L. Sp. pl. (1753) 618; Maxim. in Bull. Ac. Sci. St.-Pétersb. 29 (1881) 64; Forbes and Hemsley, Index Fl. Sin. 2 (1902) 203; Danguy in Bull. Mus. nat. hist. natur. 17 (1911) 554; Pai in Contribs Inst. Bot. Nat. Ac. Peiping, 2, 7 (1934) 225; Kryl. Fl. Zap. Sib. 10 (1939) 2528; Kitag. Lin. Fl. Mansh. (1939) 392; Jernakov in Acta Pedol. Sinica, 2 (1954) 278; Grubov, Konsp. fl. MNR (1955) 247; Golubkova in Fl. SSSR, 22 (1955) 800; Tsoong in Fl. R. P. Sin. 68 (1963) 390; Fl. Kazakhst. 8 (1965) 146.   —**Ic.:** Maxim. lc. tab. IV, figs. 1–10; Fl. R. P. Sin. 68, tab. XCIII, figs. 5–9; Fl. Kazakhst. 8, Plate XVI, fig. 8.

Described from Dauria. Type in London (Linn.).

In forb-feather grass, feather grass and scrub steppes, on rocky and rubble steppe slopes, in rock crevices.

IA. Mongolia: *Cis-Hing., Cent. Khalkha, East. Mong., Val. Lakes, Gobi-Alt., East. Gobi* (with no dates or collection locality—1831, Bunge and 1841, Kirilov; Delger-Hangay mountain range, on rocks and slopes of ravines, Aug. 21, 1926—Lis.; Gurban-Saikhan somon, Sumbur-Ula mountain east of Tabin-Chzhis, rocky slopes and feather grass-pea shrub steppe near foothill, June 23, 1949—Yun.).

General distribution: West. Sib. (Bukhtarma river valley), East. Sib., Nor. Mong. (Hang., Mong.-Daur), China (Dunbei, North).

2. **C. mongolica** Maxim. in Bull. Ac. Sci. St. Pétersb. 29 (1881) 66; Forbes and Hemsley, Index Fl. Sin. 2 (1902) 203; Diels in Filchner, Wissensch. Ergebn. (1908) 263; Pai in Contribs Inst. Bot. Nat. Ac. Peiping, 2, 7 (1934) 226; Kung and Wang in Contribs Inst. Bot. Nat. Ac. Peiping, 2, 8 (1934) 380; Tsoong in Fl. R. P. Sin. 68 (1963) 391. —*C. linearifolia* Hao in Feddes repert. 36 (1934) 224; id. in Contribs Inst. Bot. Nat. Ac. Peiping, 3, 1 (1935) 4; id. in Engler's Bot. Jahrb. 68 (1938) 636. —**Ic.:** Maxim. l.c. tab. IV, fig. 11–20; Fl. R. P. Sin. 68, tab. XCIII, figs. 1–4.

Described from Mongolia (Alashan). Type in Leningrad.

Mountain slopes in semi-deserts and deserts.

IA. Mongolia: *East. Mong.* ("Inner Mongolia"—Tsoong, l.c.). *Alash. Gobi* (Alashan mountain range, in desert near foothills, clayey soil, occasional, July 1873—Przew., typus!; "in clay bank near Choluhsien"—Kung eg Wang, l.c.).

IIIA. Qinghai: *Nanshan* ("Hsining-fu, No. 52, June 21, 1904"—Diels, l.c.; "Lanchow, auf trokenem Platz, 1900 m, No. 714, July 8, 1930, Hao"—Hao, l.c.).

IIIB. Tibet: *Weitzan* (upper Huang He river, July 1880—Przew.).

General distribution: China (North-West).

# Addenda to Russian Original of Vol. 5

## Family Lamiaceae (Labiatae)[1]

### New species for the territory studied and some corrections in synonymy

Dracocephalum argunense Fisch. ex Link, 1822, Enum. Pl. Horti Berol. 2: 118; Gubanov, Konsp. fl. Vnesh. Mong. (1996) 87.
IA. Mongolia: *Cis-Hing.*

Dracocephalum junatovii A. Budantz. in Bot. zhurn. 72, 1 (1987) 92; Gubanov, Konsp. fl. Vnesh. Mong. (1996) 88.
IA. Mongolia: *East. Mong.*

Phlomis tuvinica Schroeter, in Bull. Mosk. obshch. ispyt. prir., otd. biol. 85, 5 (1980) 78.
IA. Mongolia: *Khobd., Mong. Altay, Cen Khalkha.*

Thymus dahuricus Serg. in Sist. zam. Gerb. Tomsk. Univ. 1 (1938) 3; Grubov, Opred. sosud. rast. Mong. (1982) 218; R. Kam. and A. Budantz. in Bull. Mosk. obshch. ispyt. prir., otd. biol. 95, 3 (1990) 93; Gubanov, Konsp. fl. Vnesh. Mong. (1996) 90. —*Th. serpyllum* L. var. *asiaticus* Kitag. in Rep. First Sci. Exped. Manchoukuo, 4, 4 (1936) 92. —*Th. asiaticus* (Kitag.) Kitag. Lin. Fl. Mansh. (1939) 388, non Serg. (1937). —*Th. kitagawianus* Tschern. in Pl. As. Centr. 5 (1970) 87.
IA. Mongolia: *Cis-Hing., East. Mong.*

Thymus komarovii Serg. in Sist. zam. Gerb. Tomsk. Univ. 1 (1938) 5; Grubov, Opred. sosud. rast. Mong. (1982) 218; R. Kam. and A. Budantz. in Bull. Mosk. obshch. ispyt. prir., otd. biol. 95, 3 (1990) 93; Gubanov, Konsp. fl. Vnesh. Mong. (1996) 90.
IA. Mongolia: *East. Mong.*

Thymus minussinensis Serg. in Sist. zam. Gerb. Tomsk. Univ. 6–7 (1936) 5; R. Kam. and A. Budantz. in Bull. Mosk. obshch. ispyt. prir., otd. biol. 95, 3 (1990) 96; Gubanov, Konsp. fl. Vnesh. Mong. (1996) 90.
IA. Mongolia: *Depr. Lakes.*

Thymus baicalensis Serg. 1936, in Sist. zam. Gerb. Tomsk. Univ. 1–2 (1936) 4; R. Kam. and A. Budantz. in Bull. Mosk. obshch. ispyt. prir., otd. biol. 95, 3 (1990) 96; Gubanov, Konsp. fl. Vnesh. Mong. (1996) 90.
*Depr. Lakes.*

Thymus michaelis R. Kam. et A. Budantz. in Bull. Mosk. obshch. ispyt. prir., otd. biol. 95, 3 (1990); Gubanov, Konsp. fl. Vnesh. Mong. (1996) 90. —*Th. mongolicus* Klok. in Bot. mat. (Leningrad), 16 (1954) 311, non Ronnig. 1934. —*Th. gobicus* Tschern. in Pl. As. Centr. (1970) 5: 86, pp; Grubov, Opred. sosud. rast. Mong. (1982) 218 pp.
IA. Mongolia: *Cen. Khalkha, East. Mong.*

Thymus turczaninovii Serg. in Sist. zam. Gerb. Tomsk. Univ. 1–2 (1936) 4; R. Kam. and A. Budants. in Bull. Mosk. obshch. ispyt. prir., otd. biol. 95, 3 (1990) 97; Gubanov, Konsp. fl. Vnesh. Mong. (1996) 90.
IA. Mongolia: *East. Mong.*

Thymus gobi-altaicus (Ulzij.) R. Kam. et A. Budantz. in Bull. Mosk. obshch. ispyt. prir., otd. biol. 95, 3 (1990) 97; Gubanov, Konsp. fl. Vnesh. Mong. (1996) 90. —*Th. gobicus* Tschern. subsp. *gobi-altaicus* Ulzij. in Tr. Inst. Prir. Soed. AN MNR, 1 (1976) 60.
IA. Mongolia: *Gobi Alt.*

---

[1]Prepared by O.V. Tscherneva in 2001

Plate I. 1—*Dracocephalum tanguticum* Maxim.; 2—*Scutellaria viscidula* Bunge; 3—*Phlomis mongolica* Turcz.; 4—*Ph. admirabilis* Tschern.

Plate II. 1—*Lagopsis supina* (Steph.) lk.-Gal.; 2—*Eriophyton wallichii* Benth.; 3—*Nepeta coerulescens* Maxim.; 4—*N. wilsonii* Duthie.

222

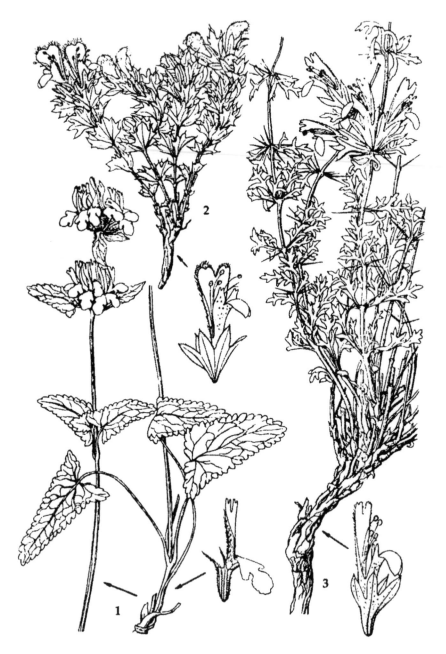

Plate III. 1—*Metastachys sagittata (Rgl.)* Knorr.; 2—*Lagochilus ilicifolius* Bunge; 3—*L. kaschgaricus* Rupr.

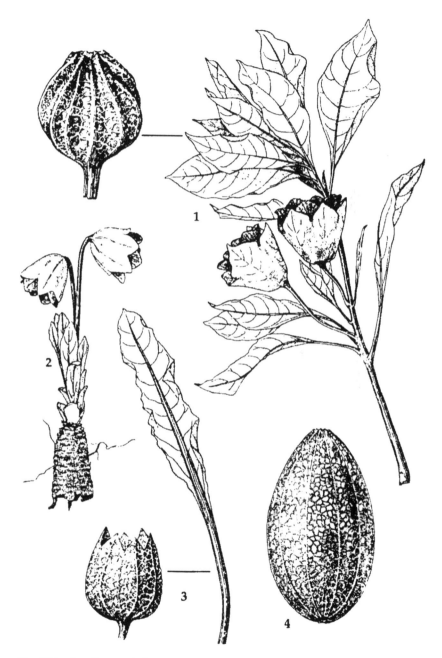

Plate IV. 1—*Scopolia tangutica* Maxim.; 2—*Mandragora caulescens* Clarke; 3—*Przewalskia shebbearei* (C. E. C. Fischer) Grub.; 4—*P. tangutica* Maxim.

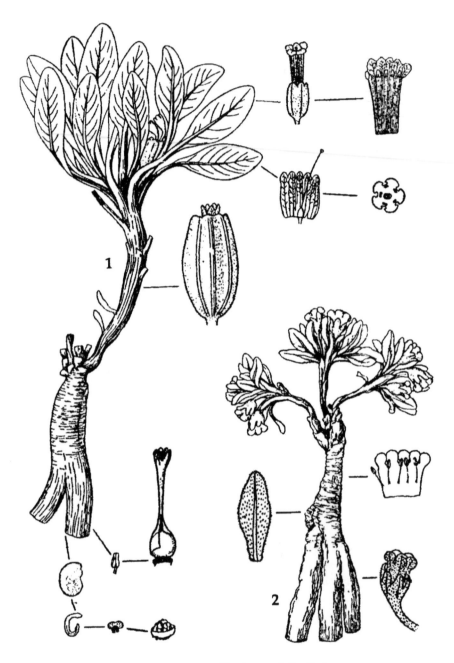

Plate V. 1—*Przewalskia tangutica* Maxim.; 2—*Mandragora tibetica* Grub.

Plate VI. 1—*Scrophularia potaninii* Ivanina; 2—*S. alaschanica* Batalin; 3—*S. przewalskii Batalin.*

Plate VII. 1—*Scrophularia regelii* Ivanina; 2—*S. incisa* Weinm.; 3—*S. kiriloviana* Schischk. var. *flexuosa* Ivanina; 4—*S. pamirica (O. Fedtsch.)* Ivanina.

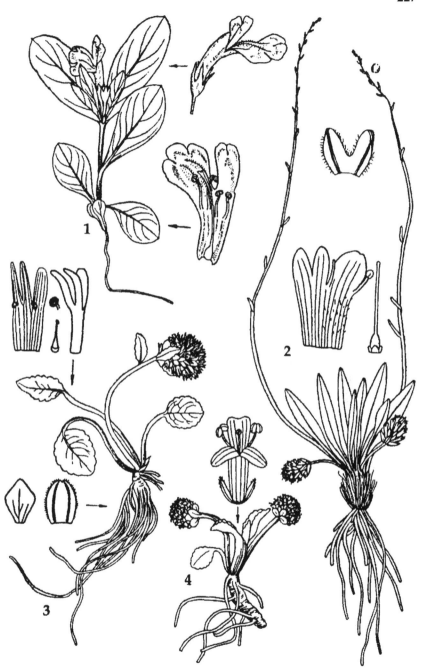

Plate VIII. 1— *Lancea tibetica* Hook. f. et Thoms.; 2—*Lagotis brachystachya* Maxim.; 3—*L. brevituba* Maxim.; 4—*L. ramalana* Batalin.

Plate IX. 1—*Pedicularis cheilanthifolia* Schrenk; 2—*P. ludwigii* Regel.; 3—*P. fetisowii* Regel.; 4—*P. junatovii* Ivanina; 5—*P. alaschanica* Maxim.

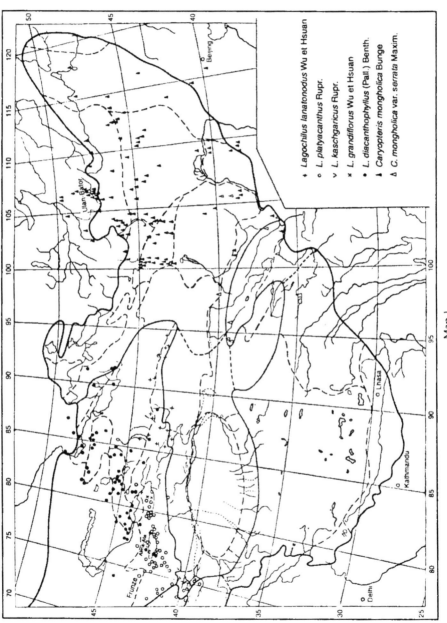

Map 1

+ *Lagochilus lanatonodus* Wu et Hsuan
o *L. platyacanthus* Rupr.
v *L. kaschgaricus* Rupr.
x *L. grandiflorus* Wu et Hsuan
• *L. diacanthophyllus* (Pall.) Benth.
▲ *Caryopteris mongholica* Bunge
Δ *C. mongholica* var. *serrata* Maxim.

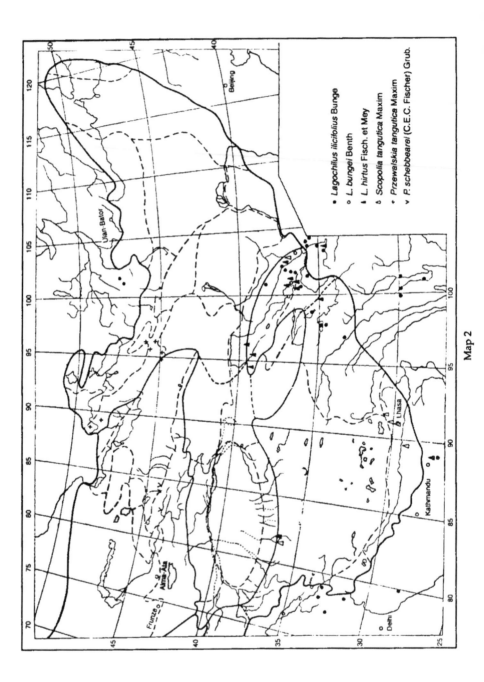

Map 2

- • *Lagochilus ilicifolius* Bunge
- ○ *L. bungei* Benth
- ▲ *L. hirtus* Fisch. et Mey
- ◊ *Scopolia tangutica* Maxim
- + *Przewalskia tangutica* Maxim
- ∨ *P. schebbearei* (C.E.C. Fischer) Grub.

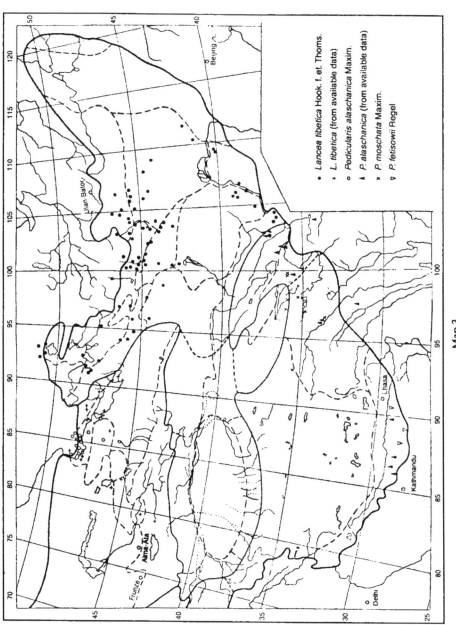

Map 3

# INDEX OF SCIENTIFIC NAMES OF PLANTS

Agastache Clayt. ex Gronov. 10, 26
— *lophanthus* Kuntze 26
— rugosa (Fisch. et Mey.) Kuntze 10, 26
Ajuga L. 8, 12
— lupulina Maxim. 8, 12
— — f. *humilis* Sun. 13
*Alectrolophus* Zinn 168
— *songaricus* Sterneck 168
Amethystea L. 8, 14
— coerulea L. 8, 14
Anaspis (Rech. f.) Juz. subgen. (Scutellaria) 19
*Anisodus* Link et Otto 113
— *tanguticus* (Maxim.) Pascher 113
Antonina Vved. 9, 93
— debilis (Bunge) Vved. 9, 93
*Aphanochilus eriostachyus* Benth. 107
— *fruticosus* (D. Don) Kudo 107
— *polystachyus* Benth. 107
*Apocladus* Li, sect. (Pedicularis) 197

*Ballota lanata* L. 76
— *sagittata* Regel 83
*Bartsia odontites* auct. 167
— *pallida* L. 163
*Belenia* Decne 116
— *praealta* Decne 116
*Betonica laevigata* D. Don 37
*Botryantha* Li, sect. (Pedicularis) 209
*Brachyphyllum* Li, sect. (Pedicularis) 196
*Brunella* Moench 58
— *vulgaris* Moench 59

*Calamintha annua* Schrenk 93
— *debilis* (Bunge) Benth. 93
*Cardiaca quinquelobata* Gilib. 74
Caryopteris Bunge 3, 7
— mongholica Bunge 4, 7
— — var. serrata Maxim. 4
— tangutica Maxim. 3, 4, 7
— — var. *brachyodonta* Hand.-Mazz. 7

Castilleja Mutis ex L. f. 126, 163
— pallida (L.) Spreng. 126, 163
Chaiturus Willd. 12, 71
— marrubiastrum (L.) Spenn. 12, 71
Chamaesphacos Schrenk 11, 86
— ilicifolius Schrenk 11, 86
—*longiflorus* Bornm. et Sint. 86
Cladomania Li, sect. (Pedicularis) 205
Cruciferae 2
Cymbaria L. 1, 126, 217
— dahurica L. 217
— *linearifolia* Hao 218
— mongolica Maxim. 217, 218

Datura L. 109, 124
— stramonium L. 109, 124
*Digitalis glutinosa* Gaertn. 162
Dodartia L. 125, 142
— atrocoerulea N. Pavl. 142
— orientalis L. 1, 125, 142
Dolichomischus Li, sect. (Pedicularis) 213
Dracocephalum L. 10, 43
— *albertii* Regel 50
— *altaiense* auct. 48
— *altaiense* Laxm. 50
— *argunense* auct. 55
— argunense Fisch. ex Link 50, 219
— bipinnatum Rupr. 43, 45
— *breviflorum* Turrill 35
— bullatum Forrest 54, 58
— bungeanum Schischk. et Serg. 44, 45, 46
— *calanthum* C. Y. Wu 58
— *coerulescens* Dunn 35
— discolor Bunge 44, 46, 47
— *erectum* Royle ex Benth. 36
— foetidum Bunge 43, 47, 52
— *fragile* auct. 50
— fruticulosum Steph. 43, 47
— *gobi* Krassan 50
— *grandiflorum* auct. 54
— grandiflorum L. 45, 48, 49, 54

— — var. *purdomii* (Smith W. W.) Kudo 54
— *Hemsleyanum* (Oliv.). Prain ex Marq. 36
— heterophyllum Benth. 43, 49
— — var. *rubicundum* Paulsen 49
— *hookeri* Clarke 57
— *imberbe* auct. 54
— imberbe Bunge 44, 50, 54, 55
— *inderiense* Less. ex Kar. et Kir. 58
— integrifolium Bunge 44, 51
— *junatovii* A. Butantc. 219
— *kaschgariaum* Rupr. 50
— *laniflorum* Rupr. 50
— moldavica L. 43, 51
— *moldavicum* L. β *asiaticum* Hiltebr. 47
— — var. *europaeum* Hiltebr. 51
— *nodulosum* auct. 50
— nodulosum Rupr. 44, 42
— nutans L. 44, 52
— — var. *alpinum* Kar et Kir. 52
— *origanoides* auct. 45
— origanoides Steph. ex Willd. 44, 45, 52
— *pamiricum* Briq. 50
— *paulsenii* Briq. 46, 47
— peregrinum L. 43, 53
— pinnatum L. 46
— — *altaicum* Bunge 53
— — var. *minus* Ledeb. 53
— — var. *pallidiflorum* Kar. et Kir. 53
— *prattii* (Lévl.) Hand.-Mazz. 42
— *pulchellum* Briq. 56
— purdomii Smith W. W. 45, 54, 55
— *rockii* Diels 28
— *royleanum* Benth. 58
— *rupestre* auct. 54
— rupestre Hance 44, 54, 55
— *ruprechtii* Regel 45
— ruyschianum L. 43, 55
— *sibiricum* auct. 42
— *sibiricum* L. 40
— *speciosum* auct. 58
— sp. 46, 47
— stamineum Kar. et Kir. 43, 55
— stellerianum Hiltebr. 49
— tanguticum Maxim. 44, 56, 57
— *truncatum* Sun 54
— *veitchii* (Duthie) Dunn 42
— wallichii Sealy 44, 54, 58
— *wilsonii* (Duthie) Dunn 42
Dysophylla janthina Maxim. 106

Elsholtzia Willd. 1, 9, 105
— *calycocarpa* Diels 106

— ciliata (Thunb.) Hylander 105
— *cristata* Willd. 105
— densa Benth. 105, 106
— *dielsii* Lévl. 107
— eriostachya Benth. 105, 107
— — var. *pusilla* Hook. f. 107
— *exigua* Hand.-Mazz. 107
— fruticosa (D. Don) Rehder 105, 107
— *janthina* (Maxim.) Dunn. 106
— *manshurica* (Kitag.) Kitag. 106
— *patrinii* (Lepech.) Garcke 105
— *polystachya* Benth. 107
— pusilla Benth. 107
— *souliei* Lévl. 107
— strobilifera Benth. 105, 107
— *tristis* Lévl. et Vaniot 107
Eremostachys Bunge 11, 59
— isochila Pazij et Vved. 60
— *laciniata* auct. 60
— moluccelloides Bunge 59, 60
— speciosa Rupr. 59, 60
— *transiliensis* Regel 60
Ericaceae 2
Eriophyton Benth. 11, 68
— wallichii Benth. 11, 68
Euphrasia L. 126, 164
— frigida Pugsl. 166
— hirtella Jord. 164
— maximowiczii Wettst. 164, 165
— *officinalis* auct. 166
— — α pectinata Kryl. 165
— *pectinataeformis* Kryl. et Serg. 165, 166
— regelii Wettst. 164, 165
— *subpetiolaris* Pugsley 165
— syreitschikovii Govor. 164, 165, 166
— — var. pectinataeformis (Kryl. et Serg.)
Ivanina 166
— tatarica Fisch. 164, 166

*Fedtschenkiella staminea* S. Kudr. 56

Galeopsis L. 11, 69
— bifida Boenn. 11, 69
— *tetrahit* auct. 69
Gerardia *glutinosa* Bunge 162
— *parviflora* Bentb. ex wall. 163
Glechoma L. 10, 42
— *complanata* Turrill 28
— *decolorans* Turrill 29
— hederacea L. 10, 42
— *nivalis* Jacquem . ex Benth. 29
— *tibetica* Jacquem. ex Benth. 30

Gratiola L. 125, 144
— officinalis L. 125, 144
Gymnandra Pall. 159
— altaica Willd. 161
— elongata Willd. 161
— integrifolia Willd. 161
— pallasii Cham. et Schlecht. 161

Haplophyllum Li, sect. (Pedicularis) 205
Holophyllum Li, sect. (Pedicularis) 195
Hyoscyamus L. 109, 117
— agrestis Kit. 118
— bohemicus F. W. Schmidt 118
— niger L. 117, 118
— physaloides L. 116
— pusillus L. 117, 118
Hyssopus L. 12, 93
— ambiguus (Trautv.) Iljin 94
— cuspidatus auct. 94
— cuspidatus Boriss. 93, 94
— — var. albiflorus Wu et Li 94
— latilabiatus Wu et Li 94, 95
— lophanthus L. 26
— macranthus Boriss. 93, 94
— ocymifolius Lam. 105
— officinalis auct. 94
— — var. ambigua auct. 94

Inermes Fisch. et Mey., sect. (Lagochilus) 77
Isodon Schard., sect (Plectranthus) 108
Isodon (Schrad.) Kudo 12, 108
— pharicus (Prain) Murata 12, 108

Kokonoria stolonifera Keng et Keng f. 159

Labiatae Juss. 1, 8
Lagochilopsis Knorr. 77
— bungei (Benth.) Knorr. 78
— hirta (Fisch. et Mey.) Knorr. 80
Lagochilus Bunge 12, 77
— affinis Rupr. 82
— altaicus Wu et Hsuan 78, 79
— brachyacanthus Wu et Hsuan 80, 81
— bungei Benth. 77, 78
— — minor Fisch. et Mey. 78
— chingii Wu et Hsuan 79, 80
— diacanthophyllus (Pall.) Benth. 79, 80
— grandiflorus Wu et Hsuan 78, 80
— hirtus Fisch. et Mey. 77, 80, 81
— ilicifolius Bunge 77, 81
— iliensis Wu et Hsuan 82, 83
— kaschgaricus auct. 82

— kaschgaricus Rupr. 78, 81
— keminsis K. Isak. 82, 83
— lanatonodus Wu et Hsuan 78, 82
— leiacanthus Fisch. et Mey. 79, 80
— macrodontus Knorr. 82, 83
— obliquus Wu et Hsuan 79, 80
— platyacanthus auct. 81
— platyacanthus Rupr. 78, 82, 83
— pulcher Knorr. 83
— pungens Schrenk 79
Lagopsis Bunge 9, 23
— eriostachya (Benth.) Ik.-Gal. 23, 24
— flava Kar. et Kir. 23, 24, 25
— incana Bunge 24
— marrubiastrum (Steph.) Ik.-Gal. 23, 24, 25
— supina (Steph.) Ik.-Gal. 23, 25
— viridis Bunge 23
Lagotis Gaertn. 6, 125, 159
— altaica (Willd.) Smirn. 161
— brachystachya Maxim. 159
— brevituba Maxim. 159, 160
— decumbens Rupr. 159, 160
— glauca auct. 161
— — ssp. australis Maxim. 160
— — ssp. borealis var. pallasii Maxim. 161
— — var. pallasii Trautv. 160
— grigorjevii Krassn. 160
— integrifolia (Willd.) Schischk. 159, 160
— pallasii (Cham. et Schlecht.) Rupr. 161
— praecox W. W. Smith 162
— ramalana Batalin 159, 161, 162
Lallemantia Fisch. et Mey. 10, 58
— royleana (Benth.) Benth. 10, 58
Lamiophlomis Kudo 11, 60
— rotata (Benth.) Kudo 11, 60
Lamium L. 11, 70
— album L. 70
— amplexicaule L. 70, 71
— turkestanicum Kuprian. 70
Lancea Hook. f. et Thoms. 1, 125, 143
— tibetica Hook. f. et Thoms. 2, 125, 143
Lasioglossa Li, sect. (Pedicularis) 206
Leonurus L. 12, 71
— bungeanus Schischk. 75
— cardiaca auct. 74
— cardiaca L. 72
— deminutus Krecz. 72, 73
— dschungaricus Regel 69
— glaucescens Bunge 72, 73
— heterophyllus Sweet 75
— lanatus Pers. 76

— *manshuricus* Yabe f. *albiflorus* Nakai et Kitag. 74
— *marrubiastrum* L. 71
— mongolicus Krecz. et Kuprian. 72, 73
— *oreades* Pavl. 74
— panzerioides M. Pop. 72, 74
— quinquelobatus Gilib. 72, 72
— sibiricus L. 72, 74
—   — f. *albiflorus* Wu et Li 74
— *supinus* Steph. 25
— turkestanicus Krecz. et Kuprian. 72, 75
Leptorhabdos Schrenk 125, 163
— *brevidens* Schrenk 162
— *micrantha* Schrenk 162
— parviflora (Benth.) Benth. 125, 163
Limosella L. 125, 144
— aquatica L. 125, 144
Linaria Mill. 125, 129
— acutiloba Fisch. ex Reichb. 129, 131
—   — f. angustifolia Serg. 131
—   — var. pygmaea Ivanina 131
— altaica Fisch. 130, 132
— bungei Kuprian. 134
— buriatica Turcz. 129, 132
— debilis Kuprian. 132
— *dmitrievae* Semiotr. 132
— hepatica Bunge 130, 132, 133
— *macroura* γ *hepatica* (Bunge) Benth. 132
— melampyroides Kuprian. 131
— *odora* auct. 132
—   — α *major* Krylov 132
—   — β *violacea* Ledeb. 134
— pedicellata Kuprian. 130, 133
— *praecox* β *ramosa* Kar. et Kir. 133
— *ramosa* Kuprian. 130, 133
— sessilis Kuprian. 133
— transiliensis Kuprian. 130, 133, 134
— *uralensis* Kotov 132
— *vulgaris* auct. 130
— vulgaris Mill. 131
—   — var. *latifolia* Kryl. 131
Lophanthus Adans. 10, 26
— chinensis (Rafin.) Benth. 26, 28
— krylovii Lipsky 26, 27, 28
— *rugosus* Fisch. et Mey. 26
— schrenkii Levin 26, 28
Lycium L. 3, 108, 109
— barbarum L. 109, 110
—   — var. *chinensis* Ait. 111
— *chinense* auct. 110
— dasystemum Pojark. 110, 111
— flexicaule Pojark. 3, 110, 111

— *halimifolium* Mill. 110
— potaninii Pojark. 3, 110, 111
— *ruthenicum* auct. 110
— ruthenicum Murr. 109, 112
— *talaricum* Pall. 112
— truncatum Wang 109, 113
— *turbinatum* Poir. 110
— *turcomanicum* auct. 110, 111, 113
Lycopus L. 9, 102
— europaeus L. 102
— exaltatus L. f. 102

Macrostachys Li, sect. (Pedicularis) 213
Mandragora L. 3, 108, 122
— *caulescens* Clarke 3, 122
— *shebbearei* C. E. C. Fischer 114
— tibetica Grub. 3, 122, 123
Mannagettaea 2
Marrubium L. 9, 23
— *eriostachyum* Benth. 23
— *flavum* Walp. 24
— *incisum* Benth. 25
— *lanatum* Benth. 24
— vulgare L. 9, 23
Megadenia 2
Mentha L. 9, 103
— arvensis L. 103, 105
—   — ssp. *haplocalyx* auct. 103
— asiatica Boriss. 104
— *austriaca* Jacq. 103, 105
— interrupta Boriss. 105
— *longifolia* auct. 104
— longifolia (L.) Huds. 104
— *patrinii* Lepech. 105
— *sylvestris* auct. 104
Metastachys Knorr. 1, 9, 83
— sagittata (Regel) Knorr. 9, 83
*Molucella diacanthophylla* Pall. 79
— *grandiflora* auct. 78
— *marrubiastrum* Steph. 24
— *mongholica* Turcz. ex Ledeb. 23

Nepeta L. 1, 10, 32
— *annua* Pall. 30
— *botryoides* Sol. 30
— cataria L. 33, 34
— chenopodiifolia Stapf 39
— coerulescens Maxim. 34, 35
— *complanata* Dunn 28
— *cordiifolia* Boiss. 58
— decolorans Hemsl. 29
— densiflora Kar. et Kir. 33, 35

— *discolor* auct. 39
— discolor Royle ex Benth. 33, 36, 37
— erecta (Royle) Benth. 34, 36
— eriostachys Benth. 37
— *fallax* Briq. 40
— *fedtschenkoi* Pojark. 39
— floccosa Benth. 41
— *glechoma* Benth. 42
— hemsleyana Oliv. ex Prain 34, 36
— kokamirica Regel 34, 36
— laevigata (D. Don) Hand.-Mazz. 33, 36, 37
— *lamiopsis* auct. 37
— lamiopsis Benth. ex Hook. f. 34, 37
— *lavandulacea* L. f. 31
— longibracteata Benth. 32, 37
— *lophantha* Fisch. ex Benth. 26
— *macrantha* Fisch. 40
— *maracandia* Bunge 39
— micrantha Bunge 32, 38
— microcephala Pojark. 39
— *multifida* L. 30, 31
— *multifida* L. f. 28
— *nivalis* Benth. 29
— *nuda* auct. 38
— pamirensis Franch. 33, 38
— *pamiro-alaica* Lipsky 35
— pannonica L. 32, 38
— *paulsenii* Briq. 32
— *pharica* Prain 29
— podostachys Benth. 33, 39
— *prattii* Lévl. 42
— *przewalskii* Pojark. 42
— pungens (Bunge) Benth. 32, 39, 40
— *pusilla* Benth. 39
— *reniformis* Briq. 40
— *sabinei* Schmidt 36
— *saposhnikowii* Nick. et Plotn. 35
— sibirica L. 34, 40
— spathulifera Benth. 32, 40
— *spicata* Benth. 37
— sp. 32
— supina Steven 42
— *thomsoni* Benth. 35
— *tibetica* Benth. 30
— transiliensis Pojark. 34, 41
— *turkestanica* Gandog. 38
— ucrainica L. 32, 41
— *vakhanica* auct. 41
— *veitchii* Duthie 42
— wilsonii Duthie 34, 42
— yanthina Franch. 32, 41

Ocimum L. 12, 108
— basilicum L. 12, 108
Odontites Ludw. 127, 167
— *odontites* auct. 167
— *rubra* Pers. 167
— serotina (Lam.) Dum. 127, 167
— *serotina* (Lam.) Reichb. 167
Omphalothrix Maxim. 126, 166
— longipes Maxim. 126, 166
Oresolen Hook. f. 1, 126, 162
— unguiculatus Hemsl. 126, 162
Origanum L. 12, 95
— vulgare L. 12, 95

Panzeria Moench 12, 75
— *alaschanica* Kuprian. 76, 77
— — var. *minor* Wu et Li 76
— *albescens* Kuprian. 76
— *argyracea* Kuprian. 76, 77
— canescens Bunge 75, 76
— *kansuensis* Wu et Li 76, 77
— lanata (L.) Bunge 75, 76, 77
— — var. alaschanica (Kuprian.) Tschern. 77
— — var. argyracea (Kuprian) Tschern. 77
— — var. typica 77
— *parviflora* Wu et Li 76, 77
*Paulseniella pamirensis* Briq. 106
Pedicularis L. 1, 126, 168
— *abrotanifolia* auct. 181, 184
— abrotanifolia M.B. ex Stev. 176, 179
— achilleifolia Steph. ex Willd. 172, 197
— alaschanica Maxim. 179, 180
— albertii Regel 173, 209
— *albida* Pennell 181
— *altaica* auct. 201
— altaica Steph. ex Stev. 172, 197
— *amoena* auct. 190, 192, 193
— amoena Adams 177, 189
— — var. *arguteserrata* (Vved.) Serg. 190
— — var. *elatior* Regel 192
— — var. *violascens* Regel 195
— anas Maxim. 179, 189
— *anthemifolia* auct. 192
— — subsp. *elatior* (Regel) Tsoong 192
— anthemifolia Fisch. ex Colla 177, 190
— *arctica* M.B. ex Stev. 189
— *arguteserrata* Vved. 190
— *armata* auct. 214
— armata Maxim. 174, 214

— *birostris* Bur. et Franch. 215
— *bonatiana* Li 194
— breviflora Regel et Winkl. 170, 197
— *calosantha* Li 194
— *cheilanthifolia* auct. 192
— cheilanthifolia Schrenk 179, 181
— chenocephala Diels 178, 196
— *chinensis* auct. 216
— chinensis Maxim. 175, 214
— «*chorgonica*» 182
— chorgossica Regel et Winkl. 178, 182
— *comosa* L., s. l. 198
— — var. *venusta* Bunge 204
— compacta Steph. ex Willd. 170, 198
— *coppeyi* Bonati 211
— *cranolopha* auct. 214
— cranolopha Maxim. 174, 215
— *crassicaulis* Vaniot ex Bonati 206
— croizatiana Li 174, 215
— *curvituba* auct. 192
— curvituba Maxim. 179, 190
— dasystachys Schrenk 171, 198
— decorissima Diels 174, 215
— *delavayi* Franch. 217
— deltoidea Franch. ex Maxim. 175, 196
— densispica Franch. ex Maxim. 176, 182
— *dielsiana* Limpr. 213
— *dolichorrhiza* auct. 197
— dolichorrhiza Schrenk 172, 199
— elata Willd. 171, 199
— elwesii Hook. f. 170, 213
— *euphrasioides* Steph. ex Willd. 205
— *fedtschenkoi* Bonati 201
— fetisowii Regel 178, 183
— *flava* auct. 201
— flava Pall. 172, 199
— — var. *altaica* Bunge 201
— — var. *conica* Bunge 201
— *floribunda* auct. 192
— *futtereri* Diels 191
— *galeobdolon* Diels 206
— *garnieri* auct. 215
— *garnieri* Bonati 215
— *giraldiana* Diels ex Bonati 192
— globifera Hook. f. 178, 183
— *goniantha* Bur. et Franch. 191
— gracilis Wall. 178, 183
— *hulteniana* Li 165, 189, 190, 192
— ingens Maxim. 169, 206
— *integerrima* Pennell et Li ex Li 195
— integrifolia Hook. f. 178, 195

— interrupta Steph. 182
— *jugentassi* Semiotr. 199
— junatovii Ivanina 178, 184
— *kansuensis* auct. 194
— kansuensis Maxim. 176, 190
— karoi Freyn 169, 205
— *labellata* Jacq. 212
— labradorica Wirs. 169, 205
— *laeta* Stev. ex Claus 198
— lasiophrys Maxim. 169, 207
— lasiostachys Bunge 171, 200
— *leptorhiza* Rupr. 184, 185
— *likiangensis* auct. 193
— *limprichtiana* Fedde 213
— longiflora Rudolph 174, 216
— *ludovicii* Limpr. 213
— ludwigii Regel 175, 184
— lyrata Prain 175, 196
— macrochila Vved. 177, 191
— *macrosiophon* auct. 213
— *magninii* Bonati 182
— mariae Regel 172, 201
— maximoviczii Krassn. 176, 185
— *microphyton* Bur. et Franch. 211
— — var. *purpurea* Bonati 211
— mollis Wall. ex Benth. 176, 185
— moschata Maxim 175, 185
— muscicola Maxim. 173, 213
— muscoides Li 173, 209
— myriophylla Pall. 179, 186
— oederi Vahl 173, 209
— oliveriana Prain 178, 186
— «*olivoiana*» Prain 186
— *palustris* auct. 205
— — subsp. *karoi* (Freyn) Tsoong 205
— pheulpinii Bonati 178, 196
— physocalyx Bunge 171, 201
— pilostachya Maxim. 176, 186
— plicata Maxim. 177, 192
— proboscidea Stev. 170, 201
— przewalskii Maxim. 174, 211
— pseudocurvituba Tsoong 179, 192
— *pseudo-karoi* Bonati 205
— *pubescens* Pai 208
— pubiflora Vved. 171, 202
— pulchella Turcz. 189
— *purpurea* Pennell 181
— *pycnantha* var. semenovii (Regel) Prain 188
— pygmaea Maxim. 175, 187
— resupinata L. 168, 205

238

— retingensis Tsoong 169, 207
— rhinanthoides Schrenk 170, 211
— roborowskii Maxim. 178, 187
— roylei Maxim. 177, 193
— *rubens* auct. 203
— — var. *alatavica* Kar. et Kir. 203
— — var. *altaica* Bunge 203
— rubens Steph. ex Willd. 172, 202
— rudis Maxim. 173, 207
— sceptrum-carolinum L. 173, 208
— scolopax Maxim. 177, 187
— semenovii Regel 176, 187
— semitorta Maxim. 178, 188
— *shawii* Tsoong 193
— *sikiangensis* Li 194
— sima Maxim. 175, 188
— siphonantha Don 174, 217
— *songarica* auct. 202
— spicata Pall. 176, 193
— *stevenii* Bunge 194
— striata Pall. 170, 202
— *stricta* Wall. 183
— *szetschuanica* auct. 191, 194
— szetschuanica Maxim. 176, 193
— *svenhedinii* Pauls. 181
— *tanacetifolia* auct. 198
— *tangutica* Bonati 194
— *tenuicalyx* Tsoong 195
— ternata Maxim. 176, 188
— tibetica Franch. 173, 213
— trichoglossa Hook. f. 169, 208
— tristis L. 169, 208
— *tubiflora* Fisch. 216
— uliginosa Bunge 172, 203
— *venusta* auct. 202
— venusta (Bunge) Bunge 172, 204
— *versicolor* Wahlenb. 210
— *verticillata* auct. 190
— verticillata L. 177, 194
— *verticillata* var. *chinensis* Maxim. 191
— *violascens* auct. 191
— violascens Schrenk 177, 194
— *yargongensis* Bonati 191
Perilla fruticosa D. Don 107
Perovskia Kar. 9, 89
— abrotanoides Kar. 90
— atriplicifolia Benth. 89, 90
Phanerantha Li, sect. (Pedicularis) 213
Phlomis L. 11, 61
— admirabilis Tschern. 62
— *agraria* auct. 67
— agraria Bunge 61, 63

— alpina Pall. 61, 63, 65
— dentosus Franch. 62, 64
— — var. *glabrescens* Danguy 62, 63
— *dszumrutensis* Afan. 65
— kawaguchii Murata 62, 64, 67
— *lamiiflora* Regel 68
— *marrubioides* Regel 69
— mongolica auct. 67
— mongolica Turcz. 61, 64
— *oblongata* Schrenk 69
— — var. *canescens* Regel 69
— oreophila Kar. et Kir. 61, 64, 65
— pratensis Kar. et Kir. 62
— *rotata* Benth. ex Hook. f. 60
— *sagittata* Regel 83
— similis Tschern. 62, 66
— *tuberosa* auct. 64
— tuberosa L. 61, 63
— tuvining schroeter 219
— umbrosa Turcz. 61, 67
— younghusbandii Mukerjee 62, 64, 67
Phlomoides (Moench) Briq., sect. (Phlomis) 64
Phyllophyton Kudo 1, 10, 28
— complanatum (Dunn) Kudo 28
— decolorans (Hemsl). Kudo 28
— nivale (Jacquem.) C.Y. Wu 28
— pharicum (Prain) Kudo 28
— tibeticum (Jacquem.) C.Y. Wu 28, 30
Physalis L. 109, 118
— alkekengi L. 109, 118, 119
— — var. franchetii Hort. 119
— — var. glabripes (Pojark.) Grub. 119
— *glabripes* Pojark. 118, 119
— *practermissa* Pajark 118, 119
— *Physochlaena* Miers 116
Physochlaina G. Don 109, 116
— al biflora Grub. 116
— *dahurica* Miers 116
— *grandiflora* Hook. 116
— *lanosa* Pascher 116
— physaloides (L.) G. Don 116
— praealta (Decne) Miers 116
Platyelasma calycocarpa (Diels) Kitag. 106
— densa (Benth.) Kitag. 106
— *eriostachya* (Benth.) Kitag. 107
— *manshurica* Kitag. 100
Plectranthus L'Herit. 108
— *pharicus* Prain 108
Pogostemon janthinus (Maxim.) Kanitz 106
Prunella L. 11, 58
— *asiatica* Nakai 59

— *japonica* Makino 59
— *officinalis* Güldenst. 59
— *parviflora* Gilib. 59
— vulgaris L. 11, 58
— — var. *japonica* (Makino) Kudo 59
Przewalskia Maxim. 2, 3, 108, 114, 115
— *roborowskii* Przew. ex Batal. 115
— shebbearei (C.E.C. Fischer) Grub. 2, 114, 115
— tangutica Maxim. 2, 114, 115
— — var. roborowskii (Batal.) Grub. 115
*Pseudolophanthus* Levin 28
— *complanatus* Levin 28
— *decolorans* Levin 29
— *nivalis* Levin 29
— *pharicus* Kuprian. 29
— *tibeticus* (Jacquem.) Kuprian. 30
Pungentes Pojark., ser. (Nepeta) 39

Rehmannia Libosch. 125, 162
— *chanetii* (Lévl.) Lévl. 162
— *chinensis* Fisch. et Mey. 162
— glutinosa (Gaertn.) DC. 125, 162
— *glutinosa* (Gaertn.) Fisch. et Mey. 143
Rhinanthus L. 126, 168
— *montanus* auct. 168
— × pseudo-songoricus Vass. 168
— songaricus (Sterneck) B. Fedtsch. 126, 168
— vernalis (Zing.) Schischk. et Serg. 168
Rhododendron 2

Salvia L. 9, 87
— deserta Schang. 87
— *glutinosa* auct. 88
— *hians* auct. 89
— *nemorosa* auct. 87
— nubicola Wall. ex Sweet 87, 88
— prattii Hemsl. 87, 88
— przewalskii Maxim. 87, 89
— roborowskii Maxim. 87, 89
— *silvestris* auct. 87
— wardii Peter-Stibal 87, 89
*Satureja annua* Briq. 93
— *debilis* Briq. 93
Schizocalyx Li, sect. (Pedicularis) 214
Schizonepeta Briq. 10, 30
— annua (Pall.) Schischk. 1, 30
— *botryoides* Briq. 30
— multifida (L.) Briq. 30, 31
Scopolia Jacq. 2, 109, 113, 115

— *praealta* Dun. 116
— sinensis Hemsl. 3
— sp. 115
— tangutica Maxim. 2, 3, 109, 113, 115
Scrophularia L. 125, 134
— alaschanica Batalin 134, 136
— alata Gilib. 135, 136
— altaica Murr. 134, 136
— *aquatica* auct. 136
— canescens Bong. 135, 137
— — var. canescens 137
— — var. glabrata Franch. 137
— cretacea Fisch. 137
— — var. *glabrata* (Franch.) Stiefelh. 137
— *delavayi* auct. 136
— dentata Royle ex Benth. 135, 137, 141
— heucheriiflora Schrenk ex Fisch. et Mey. 135, 137
— *incisa* auct. 140
— — var. *alpina* Kar. et Kir. 139
— — var. *angustifolia* O. Fedtsch. 140
— — var. *pamirica* O. Fedtsch. 140
— — var. *pinnata* Trautv. 139
— incisa Weinm. 136, 138–141
— — var. alpina Kar. et Kir. 139
— *integrifolia* auct. 138
— kiriloviana Schischk. ex Gorschk. 135, 139, 140
— — var. flexuosa Ivanina 140
— koelzii Pennell 141
— *orientalis* auct. 138
— pamirica (B. Fedtsch.) Ivanina 136, 140
— *patriniana* Wyder 138
— *pinnata* Kar. et Kir. 139
— — var. *subpinnata* Fisch. et Mey. 139
— potaninii Ivanina 135, 140
— przewalskii Batalin 134, 141
— regelii Ivanina 135, 141
Scrophulariaceae Juss. 1, 124
Scutellaria L. 8, 14
— alaschanica Tschern. 15, 16
— albertii Juz. 16, 22
— *alpina* auct. 22
— — var. *cordifolia* auct. 20
— — var. *lupulina* Benth. 22
— baicalensis Georgi 15, 17
— *catharinae* Juz. 21
— galericulata L. 16, 17, 18
— — var. *angustifolia* auct. 17
— — var. *angustifolia* Regel 18
— — var. *genuina* Regel 17

— δ *scordifolia* Regel 21
— grandiflora auct. 19
— grandiflora Sims 15, 18, 19
— *karkaralensis* Juz. 19
— kingiana Prain 14, 19
— krylovii Juz. 16, 19
— *lupulina* L. 22
— macrantha Fisch. ex Reichb. 17
— mongolica Sobolevsk. 18
— *oligodonta* Juz. 20
— *orientalis* L. var. *adscendens* Ledeb. 21
— — var. *microphylla* Ledeb. 18
— *oxyphylla* Juz. 22
— paulsenii Briq. 15, 20
— przewalskii Juz. 15, 20
— regeliana Nakai 18
— *rivularis* auct. 16
— rivularis Wall. 16
— scordiifolia Fisch. ex Schrenk 15, 16, 21
— siversii Bunge 16, 21, 22
— *soongorica* Juz. 21
— supina L. 15, 22
— turgaica Juz. 19
— *tuvensis* Juz. 18
— viscidula Bunge 15, 22
*Sideritis ciliata* Thunb. 105
Sigmantha Li, sect. (Pedicularis) 189
Solanaceae Juss. 1, 2, 108
Solanum L. 109, 119
— *alatum* auct. 121
— depilatum Kitag. 120
— *dulcamara* auct. 120
— *nigrum* auct. 107
— nigrum L. 120, 121
— — var. *vulgare* L. 120
— olgae Pojark. 120, 121
— *persicum* auct. 120
— septemlobum Bunge 119, 121
— — var. subintegrifolium Grub. 122
Spicatae (Benth.) Pojark. sect. (Nepeta) 37
Stachyopsis M. Pop. et Vved. 12, 68
— lamiiflora (Rupr.) M. Pop. et Vved. 68
— marrubioides (Regel) Ik.-Gal. 68
— oblongata (Schrenk) M. Pop. et Vved. 68, 69
— — var. *canescens* M. Pop. et Vved. 68
Stachys L. 11, 84
— *affinis* Bunge 86
— *aspera* auct. 85
— *baicalensis* auct. 85, 86
— *baicalensis* Fisch. 85

— *chinensis* Bunge 85
— *japonica* Miq. 85
— *lamiiflora* Rupr. 68
— *modica* Hance 84
— oblongifolia Benth. 84
— palustris L. 84
— riederi Chamisso ex Benth. 84, 85
— setifera Mey. 84, 86
— sieboldii Miq. 84, 86
— silvatica L. 84

Teucrium L. 8, 13
— scordioides Schreb. 8, 13
— *sibiricum* L. 41
— sp. 14
Thymus L. 12, 96
— altaicus Klok. et schost. 97, 99
— *altaicus* Serg. 97
— *asiaticus* Kitag. 99, 219
— *asiaticus* Serg. 105
— baicalensis Serg. 219
— dahuricus Serg. 219
— *debilis* Bunge 93
— disjunctus Klok. 99
— gobi-altaicus (Ulzij.) R. Kam et A. Budantz 219
— — subsp. *gobi altaicus* Ulzij 219
— gobicus Tschern. 96, 97, 99, 219
— kitagawianus Tschern. 96, 99, 219
— komarovii Serg. 219
— marschallianus Willd. 96, 99
— michaelis R. kam. et A. Budantz. 219
— minussinensis Serg. 219
— *mongolicus* Klok. 100, 101
— mongolicus (Ronnig.) Ronnig. 97, 100, 101, 219
— *nerczensis* auct. 97
— petraeus Serg. 96, 101
— rasitatus Klok. 96, 101
— roseus Schipcz. 96, 102
— *serpyllum* auct. 97, 100, 102
— serpyllum L. 96
— — var. *angustifolius* auct. 99, 100
— — var. *asiaticus* Kitagawa 99, 219
— — var. *mongolicus* Ronnig. 100
— — var. *vulgaris* auct. 102
— turczaninovii Serg. 219
Tubiflorae 1

Verbascum L. 124, 127
— blattaria L. 127

— *candelabrum* Kar. et Kir. 128
— *chaixii* Ledeb. 128
— — var. *orientale* Murb. 128
— orientale M.B. 127, 128
— phoeniceum L. 127, 128
— *polystachyum* Kar. et Kir. 128
— *schraderi* G. Mey. 129
— songaricum Schrenk 127, 128
— thapsus L. 127, 129
Verbenaceae Jaume 1, 3
Veronica L. 124, 144
Veronica, subgen. (Veronica) 148
— *acinifolia* var. *glabrata* Trautv. 157
— — var. *karelinii* Trautv. 157
— *anagallis* auct. 148
— anagallis-aquatica L. 145, 148
— angustifolia Fisch. ex Link 153
— arenosa (Serg.) Boriss. 146, 149
— argute-serrata Regel et Schmalh. 148, 156
— beccabunga L. 145, 149
— biloba L. 148, 156
— — var. *dasycarpa* Trautv. 157
— *campylopoda* auct. 156
— campylopoda Boiss. 147, 157
— *cartilaginea* Ledeb. 153
— ciliata Fisch. 145, 150, 151
— dahurica Stev. 147, 151
— densiflora Ledeb. 146, 151
— *ferganica* M. Pop. 158
— *grandis* Fisch. ex Spreng. 151
— incana L. 146, 151
— krylovii Schischk. 145, 152
— laeta Kar. et Kir. 146, 152
— — var. *arenosa* Serg. 149
— linariifolia Pall. ex Link 146, 153
— longifolia L. 147, 153, 156
— — var. *grandis* (Fisch.) Turcz. 151
— *lütkeana* auct. 154
— *macrocarpa* Turcz. ex Steud. 150
— macrostemon Bunge 145, 154
— *maritima* L. 153

— *nudicaulis* auct. 157
— — var. *glabrata* Trautv. 157
— *paniculata* L. 153
— — β. *angustifolia* Benth. 153
— perpusilla Boiss. 147, 157
— persica Poir. 147, 157, 158
— pinnata L. 146, 154
— porphyriana Pavl. 146, 155
— *rubicunda* Ledeb. 153
— rubrifolia Boiss 147, 158
— serpyllifolia L. 145, 155
— sibirica L. 145, 158
— *spicata* auct. 155
— — var. *viscosissima* Kar. et Kir. 155
— spicata L. 155
— *spuria* auct. 153, 155
— — var. *angustifolia* Makino 153
— spuria L. 147, 156
— szechuanica Batalin 151
— *tenella* All. 155
— tenuissima Boriss. 147, 158
— *tetraphylla* auct. 158
— «*tetraphyllos*» 158
— *teucrium* auct. 152
— tournefortii Gmel. 157
— *virginica* auct. 158
— Veronicastrum (Heister) Boriss., subgen. (Veronica) 158
Veronicastrum Heist. ex Fabr. 144
— *sibiricum* (L.) Pennell 158
*Vleckia chinensis* Rafin. 26

Ziziphora L. 12, 90
— bungeana Juz. 90, 91, 92
— *clinopodiodes* auct. 91, 92
— clinopodioides Lam. 90, 91
— — var. *media* Benth. 92
— pamiroalaica Juz. 90, 92
— *pulchella* Pavl. 92
— pungens Bunge 39
— tenuior L. 90, 93
— *tomentosa* Juz. 92

# INDEX OF DISTRIBUTION RANGES

| | Map No. | | Map No. |
|---|---|---|---|
| *Caryopteris mongholica* Bunge | 1 | — *lanatonodus* Wu et Hsuan | 1 |
| — *mongholica* var. | | — *platyacanthus* Rupr | 1 |
| *serrata* Maxim. | 1 | *Lancea tibetica* Hook. f. et Thoms. | 3 |
| *Lagochilus bungei* Benth. | 2 | *Pedicularis alaschanica* Maxim. | 3 |
| — *diacanthophyllus* (Pall.) | | — *fetisowii* Regel | 3 |
| Benth. | 2 | — *moschata* Maxim. | 3 |
| — *grandiflorus* Wu et Hsuan | 1 | *Przewalskia schebbearei* (C.E.C. | |
| — *hirtus* Fisch. et Mey | 2 | Fischer) Grub | 2 |
| — *ilicifolius* Bunge | 2 | — *tangutica* Maxim. | 2 |
| — *kaschgaricus* Rupr. | 1 | *Scopolia tangutica* Maxim. | 2 |

Printed and bound by CPI Group (UK) Ltd, Croydon, CR0 4YY

23/10/2024

01778237-0004